The Hundred Geometrical Questions of Sybrandt Hansz CARDINAEL

Early-Modern Dutch Geometry from the Surveyor's Tradition

by Matthijs Hendricus Sitters and Albrecht Heeffer

Docent Press

Docent Press
Boston, Massachusetts, USA
www.docentpress.com

Docent Press publishes books in the history of mathematics and computing about interesting people and intriguing ideas. The histories are told at many levels of detail and depth that can be explored at leisure by the general reader.

Cover design by Brenda Riddell, Graphic Details.

Produced with TeX.

Contents

Preface **1**

1 Introduction **5**

1.1 About the author, the book, and its context 5

1.1.1 Biography 5

1.1.2 The book and its editions 7

1.1.3 Translations 11

1.2 The context 14

1.2.1 Practical geometry 14

1.2.2 Geometrical algebra 22

2 English translation **27**

2.1 About the translation 27

2.2 Hundred geometrical questions 28

3 Mathematical commentary **165**

3.1 Methods and techniques 165

3.1.1 Point multiplications 165

3.1.2 Geometrical algebra 166

3.1.3 Application of area 168

3.1.4 Two elementary examples of geometrical algebra and parabolic application of area 171

3.1.5 Geometrical identities and the application of area 173

3.1.6 The solution of geometrical systems of two equations with two unknowns 182

3.1.7 Quadratic equations 189

3.1.8 The equivalence of system I and the AH-system 193

3.1.9 Hyperbolic application instead of elliptic application . . 197

3.1.10 Inscribed and circumscribed square 204
3.1.11 Cardinael's methods and techniques in their historical context 206
3.1.12 Some final remarks 208
3.2 Structure and some special themes of the HGQ 209
3.2.1 Classification in groups and subgroups 209
3.2.2 Some special themes from the HGQ and their distribution over the subgroups 213

4 Comments per question **217**
Question 0 217
Question 1 219
Question 2 222
Question 3 225
Question 4 226
Question 5 229
Question 6 232
Question 7 233
Question 8 234
Question 9 235
Question 10 236
Question 11 237
Question 12 239
Question 13 240
Question 14 242
Question 15 248
Question 16 250
Question 17 251
Question 18 252
Question 19 253
Question 20 254
Question 21 256
Question 22 258
Question 23 259
Question 24 261
Question 25 262
Question 26 267
Question 27 268
Question 28 274
Question 29 276

Question 30 279
Question 31 282
Question 32 286
Question 33 291
Question 34 295
Question 35 297
Question 36 300
Question 37 301
Question 38 302
Question 39 304
Question 40 306
Question 41 307
Question 42 310
Question 43 312
Question 44 315
Question 45 319
Question 46 320
Question 47 322
Question 48 323
Question 49 325
Question 50 327
Question 51 330
Question 52 332
Question 53 333
Question 54 335
Question 55 338
Question 56 341
Question 57 343
Question 58 348
Question 59 350
Question 60 351
Question 61 353
Question 62 354
Question 63 357
Question 64 359
Question 65 360
Question 66 362
Question 67 368
Question 68 369
Question 69 373

Question 70 . . . 377
Question 71 . . . 379
Question 72 . . . 380
Question 73 . . . 381
Question 74 . . . 382
Question 75 . . . 384
Question 76 . . . 386
Question 77 . . . 390
Question 78 . . . 392
Question 79 . . . 395
Question 80 . . . 397
Question 81 . . . 400
Question 82 . . . 405
Question 83 . . . 411
Question 84 . . . 416
Question 85 . . . 423
Question 86 . . . 424
Question 87 . . . 427
Question 88 . . . 428
Question 89 . . . 429
Question 90 . . . 431
Question 91 . . . 432
Question 92 . . . 435
Question 93 . . . 439
Question 94 . . . 441
Question 95 . . . 442
Question 96 . . . 446
Question 97 . . . 449
Question 98 . . . 454
Question 99 . . . 455
Question 100 . . . 458
Arithmetical Question . . . 462

A Cross-References **467**
A.1 Propositions from Euclid's *Elements* used by Cardinael . . . 467
Euclid Elements Book I . . . 468
Euclid Elements Book II . . . 469
Euclid Elements Book III . . . 470
Euclid Elements Book IV . . . 470
Euclid Elements Book V . . . 471

Euclid Elements Book VI 471
Euclid Elements Book VII 474
Rules not directly derived from Euclids' *Elements* 474
A.2 Correspondence between Cardinael's and Rudd's problems . . . 475

B Boundary extension **477**
B.1 Method 1. Pasting boundary areas in extension of each other 477
B.2 Method 2. Geometrical algebra, circumferential square or quadrilateral and similarity 477

C Partitioning a triangle **479**

Bibliography **481**

Index **487**

List of Figures

1.1 Frontispiece of the book on surveying by Sems en Dou (copy by one of the authors) . . . 9
1.2 Frontispiece of the book on surveying instruments by Sems en Dou (copy by the authors) . . . 10
1.3 The frontispiece of the German edition (Sächische Landesbibliothek, Dresden) . . . 11
1.4 Sebastian Kurz (University of Oklahoma Libraries History of Science Collections) . . . 12
1.5 Frontispiece of Rudd's adaptation from Cardinael (Early English Books Online) . . . 13
1.6 A legal document describing the partitioning of an irregular piece of land into five parts of equal area (c. $2100 - 2000BCE$, CDLI library) . . . 14
1.7 Friberg shows how the partitioning could have been done consistent with prevailing methods (*Ibid.* CDLI library) . . . 15
1.8 The geometrical interpretation of YBC 6967 by Høyrup . . . 18
1.9 The cut-and-paste solution to YBC 6967 following Høyrup . . 18
1.10 Cardinael's solution to Q51, corresponding to the Old-Babylonian method of YBC 6967 . . . 20

2.1 Question 1 . . . 29
2.2 Question 1, alternative solution . . . 30
2.3 Question 2 . . . 31
2.4 Question 2, alternative solution . . . 32
2.5 Question 3 . . . 33
2.6 Question 4 . . . 34
2.7 Question 4, alternative solution . . . 34
2.8 Question 5 . . . 35
2.9 Question 5, alternative solution . . . 36
2.10 Question 6 . . . 37

2.11 Question 7 38
2.12 Question 8 38
2.13 Question 9 39
2.14 Question 10 40
2.15 Question 11 41
2.16 Question 12 42
2.17 Question 13 43
2.18 Question 14 44
2.19 Question 14, explanation 45
2.20 Question 14, alternative solution 45
2.21 Question 15 46
2.22 Question 16 47
2.23 Question 17 48
2.24 Question 18 49
2.25 Question 19 50
2.26 Question 20 51
2.27 Question 21 52
2.28 Question 21, alternative solution 53
2.29 Question 22 54
2.30 Question 23 55
2.31 Question 24 56
2.32 Question 25 57
2.33 Question 25: solution 58
2.34 Question 25: alternative solution 59
2.35 Question 26 60
2.36 Question 27 61
2.37 Question 27: alternative solution 62
2.38 Question 27: second alternative solution 63
2.39 Question 28 64
2.40 Question 29 65
2.41 Question 30 66
2.42 Question 30: solution by proportions 66
2.43 Question 31 67
2.44 Question 32 69
2.45 Question 32: solution 70
2.46 Question 33 71
2.47 Question 34 72
2.48 Question 35 73
2.49 Question 36 74
2.50 Question 37 75

2.51 Question 38 76
2.52 Question 39 77
2.53 Question 39: alternative solution 78
2.54 Question 40 79
2.55 Question 41 79
2.56 Question 42 80
2.57 Question 43 81
2.58 Question 44 81
2.59 Question 44: solution 82
2.60 Question 45 83
2.61 Question 46 84
2.62 Question 47 85
2.63 Question 48 86
2.64 Question 49 87
2.65 Question 49: alternative solution 88
2.66 Question 50 89
2.67 Question 50: solution 89
2.68 Question 51 90
2.69 Question 52 91
2.70 Question 53 92
2.71 Question 54 94
2.72 Question 54: solution 94
2.73 Question 55 95
2.74 Question 56 96
2.75 Question 57 97
2.76 Question 58 99
2.77 Question 59 100
2.78 Question 60 101
2.79 Question 61 102
2.80 Question 62 103
2.81 Question 63 105
2.82 Question 64 106
2.83 Question 65 107
2.84 Question 66 108
2.85 Question 66: alterative solution 109
2.86 Question 67 110
2.87 Question 68 111
2.88 Question 68: alternative solution 112
2.89 Question 68: alternative solution 113
2.90 Question 69 114

2.91 Question 69: demonstration 115
2.92 Question 70 117
2.93 Question 71 118
2.94 Question 72 119
2.95 Question 73 120
2.96 Question 74 121
2.97 Question 75 122
2.98 Question 76 123
2.99 Question 77 124
2.100 Question 78 126
2.101 Question 79 127
2.102 Question 80 128
2.103 Question 81 130
2.104 Question 82 131
2.105 Question 83 132
2.106 Question 84 134
2.107 Question 85 136
2.108 Question 86 137
2.109 Question 87 138
2.110 Question 88 139
2.111 Question 89 140
2.112 Question 90 142
2.113 Question 91 143
2.114 Question 91: transforming a quadrangle to a triangle 144
2.115 Question 92 145
2.116 Question 92: solution 146
2.117 Question 93 147
2.118 Question 94 148
2.119 Question 95 149
2.120 Question 96: the given catheti 150
2.121 Question 96: the construction 151
2.122 Question 96: the final triangle 152
2.123 Question 97: the given sides 153
2.124 Question 97: the construction of the diagonals 153
2.125 Question 97: the construction of the squares 154
2.126 Question 97: the quadrangle inscribed in a circle 155
2.127 Question 98 156
2.128 Question 99 157
2.129 Question 99: solution 158
2.130 Question 100 159

2.131 Question 100 160
2.132 Question 101 161

3.1 Construction of the mean proportional 167
3.2 Application of area 169
3.3 Parabolic application 169
3.4 The Pythagorean theorem. 171
3.5 Solving linear equations geometrically 172
3.6 Elem.II-4 175
3.7 Elem.II-7 175
3.8 Elem.II 5-6 176
3.9 Gnomon and rectangle 177
3.10 The hyperbolic identity 178
3.11 Constructing the golden section 180
3.12 The system I(DA, DB, c, b) 183
3.13 Hyperbolic application of area 185
3.14 Quadratic equations, system III 187
3.15 Quadratic equations, system IV 188
3.16 Quadratic equations, type I 190
3.17 Quadratic equations, type II 192
3.18 The AH-system 194
3.19 The case where $BA \geq BC$ 198
3.20 The case where $AB \leq BC$ 199
3.21 Circumscribed and inscribed square 205
3.22 Stewart's theorem 213

4.1 Question 0: subgroup 0, the Pythagorean theorem 217
4.2 Question 1: subgroup 1, altitude in acute triangle 219
4.3 Question 1, alternative solution 221
4.4 Question 2: subgroup 1, altitude in obtuse triangle 222
4.5 Question 2, alternative solution (letters B and C should be interchanged to correspond with the previous figure) 224
4.6 Question 3: subgroup 2, triangle with two lines known parallel to an altitude and how these lines cut the base 225
4.7 Question 4: subgroup 1, altitude of a right-angled triangle unknown but the segments of the base divided by this altitude known 226
4.8 Question 4: With K such that $DA = DK$ and $CI = AD$ . . . 227
4.9 Question 5, subgroup 1: right-angled triangle with base and altitude known 229

4.10 Question 5, $AHID$, $CMNK$ and $IOFL$ are squares, E is the midpoint of AC and of MD 230
4.11 Question 6: subgroup 2, right-angled triangle with perpendicular and one segment of the base known 232
4.12 Question 7: subgroup 1, altitude AD in right-angled triangle and the difference of the segments DC and DB of the base known 233
4.13 Question 8: subgroup 2, right-angled triangle with the sides and the direction of a transversal known 234
4.14 Question 9: subgroup 2, tower with known height of which the length of the ladder to the window equals the distance from the window to the top 235
4.15 Question 10: subgroup 2, triangle with sides known and one side is extended with a known segment. Point I is chosen so that EI is parallel to FC. 236
4.16 Question 11: Transforming a triangle without changing its area by removal of two vertices along their line of connection with the third vertex 237
4.17 Question 12: subgroup 2, a tower at an unknown distance and of an unknown height 239
4.18 Subgroup 1, a bisector bisecting the opposite side 240
4.19 Subgroup 2, right-angled triangle with hypotenuse and the sum of the sides of the right angle know 242
4.20 Question 14, explanation 242
4.21 Question 14, alternative solution 244
4.22 Subgroup 2, right-angled triangle with sum of the sides and the area known 248
4.23 Subgroup 2, right-angled isosceles triangle with the sum of the sides known 250
4.24 Subgroup 2, measuring the height of a tower by means of two sticks 251
4.25 Subgroup 2, calculating the sides of a right-angled triangle given the area of two parts 252
4.26 Subgroup 1, isosceles triangles 253
4.27 Subgroup 1, the median of a triangle 254
4.28 Subgroup 1, a line from the top of a triangle divides the base in segments in proportion of 1 to 2 256
4.29 Question 21, alternative solution 257

4.30 Subgroup 2, calculation of the distance of a point inside a triangle to the third vertex with the distances to both other vertices known . . . 258
4.31 Subgroup 2, determining the third side of a triangle when the distance of a point within the triangle to the vertices and two sides are known . . . 259
4.32 Measuring a distance blocked by water . . . 260
4.33 Subgroup 2, calculating the distance of two points on two different sides of a given triangle . . . 261
4.34 Subgroup 8, a right-angled triangle with a right angle side known and the base divided by the perpendicular according to the Golden Section . . . 262
4.35 Question 25: solution . . . 263
4.36 Question 25: alternative solution . . . 265
4.37 Subgroup 2, the length of the two intersecting perpendiculars on the extension of two of the sides of a triangle . . . 267
4.38 Subgroup 2, two ladders at the same point leaning against opposite walls . . . 268
4.39 Question 27: alternative solution . . . 270
4.40 Question 27: second alternative solution . . . 272
4.41 Subgroup 2, ladders of equal length from the same point against opposite walls . . . 274
4.42 Subgroup 2, a right-angled triangle with sum of the right angle sides, the area and the hypotenuse known . . . 276
4.43 Subgroup 12, calculation of a triangle with the perpendiculars given . . . 279
4.44 Question 30: solution by proportions . . . 280
4.45 Subgroup 1, a triangle with base, perpendicular and sum of the other sides known . . . 282
4.46 Subgroup 1, a triangle with its base, perpendicular and the ratio of the other sides known . . . 286
4.47 Question 32: solution . . . 288
4.48 Cardinael's geometrical solution completed . . . 289
4.49 Subgroup 1, triangle with transversal , segments of the base and the proportion of the slanted sides known . . . 291
4.50 Subgroup 1, a triangle with its slanted sides and the ratio of the transversal to the segments of the base known . . . 295
4.51 Subgroup 8, isosceles triangle with two angles of 72 degrees and the equal sides known . . . 297
4.52 Subgroup 3, square battalion of musketeers encircled by lancers 300

4.53 Subgroup 3, a squared battalion of 120 lancers covering and unknown number of musketeers 301
4.54 Subgroup 3, the area given of the rectangle on BE on the diagonal BD and of EF on the line AF perpendicular on BD 302
4.55 Subgroup 3, the area of a 'pentagon' calculated by the diagonal 304
4.56 Subgroup 4, a rectangle with its area and the ratio between length and breadth known 306
4.57 Subgroup 4, rectangle with the ratio of the sides and the ratio between area and length known 307
4.58 Solution to question 41 308
4.59 Subgroup 4, a rectangle with known length, breadth unknown, divided into a left part and a smaller right part 310
4.60 Subgroup 4, a rectangle with known length, breadth unknown, divided in left part and a larger right part 312
4.61 Completion of Cardinael's diagram of figure 4.60 313
4.62 Subgroup 4, complement within a square of two connecting squares on the diagonal of that square 315
4.63 Question 44: solution 316
4.64 Solution to problem 44 317
4.65 Subgroup 5, similar triangles contained in one rectangle . . . 319
4.66 Subgroup 5, a rectangle with the same ratio as a truncated segment 320
4.67 Subgroup 5, a right-angled garden in which is contained a similar flower-bed 322
4.68 Subgroup 4, a farmland with a ditch of prescribed area 323
4.69 Subgroup 3, an unknown number of lancers cover a known number of musketeers 325
4.70 Question 49: alternative solution 326
4.71 Subgroup 3, a known number of musketeers covered by an unknown number of lancers 327
4.72 Question 50: solution 328
4.73 Subgroup 3: a given number of lancers covered by four rows of a known number of musketeers, the ratio of the sides of the rectangle unknown 330
4.74 subgroup 3, a rectangle divided by a ditch 332
4.75 Subgroup 3, a circumscribed rectangle of a quadrangle, with their diagonals perpendicular to each other 333
4.76 Subgroup 6, two circles with their difference in area and difference in diameters given 335
4.77 Question 54: solution 336

4.78 Subgroup 6, two circles with the difference of the diameters known and the width of the 'lunar' difference known 338
4.79 Q55 figure 2 340
4.80 Subgroup 6, two chords known, perpendicular on the diameter of a circle and at a given distance of each other 341
4.81 Subgroup 6, a semi-circle with an inscribed triangle with as top a point on the diameter and as base a chord parallel with that diameter 343
4.82 Subgroup 6, an isosceles triangle with two inscribed circles . . 348
4.83 Subgroup 6, an isosceles triangle with an inscribed circle . . . 350
4.84 Subgroup 6, a circle with an inscribed square and the size of the remaining segments of the circle known 351
4.85 Subgroup 6, a circle with an inscribed octagon with known sides353
4.86 Subgroup 6, a circle with an inscribed square itself containing an inscribed isosceles triangle 354
4.87 Subgroup 6, a semicircle with known radius with four chords given and an unknown remainder 357
4.88 Subgroup 6, a quadrant of a circle 359
4.89 Subgroup 6, a triangle with an inscribed and circumscribed circle 360
4.90 Subgroup 7, a triangle with two known sides and a given circumscribed circle 362
4.91 Problem 66: alterative solution 363
4.92 Question 66: elaborating on Cardinael's picture 364
4.93 Subgroup 13, three marksmen with the same distance from a parrot 368
4.94 Subgroup 7, a circle with sides of an inscribed quadrilateral known 369
4.95 Problem 68: alternative solution 371
4.96 Subgroup 8, a pentagon and a decagon in a circle with a given diameter 373
4.97 Problem 69: demonstration 374
4.98 Subgroup 9, a quadrangle with a pair of opposite sides of equal length given, $\sqrt{106}$, the diagonals given with lengths 12 and 10 and the other pair of opposite sides of equal length 377
4.99 Subgroup 9, a quadrangle with three sides and the two diagonals known 379
4.100 Subgroup 9, a quadrangle with its four sides and one diagonal known 380

4.101 Subgroup 9, a quadrangle with its four sides known and a right angle . . . 381
4.102 Subgroup 9, a quadrangle with its four sides known and a right angle . . . 382
4.103 A quadrangle with a right angle, with area and three sides given 384
4.104 Subgroup 13, a tetrahedron with four congruent planes and three pairs of edges with three marksmen on known distances from a parrot and at the same distance from each other . . . 386
4.105 Question 76: solution . . . 388
4.106 Subgroup 11, partitioning a triangle in three parts . . . 390
4.107 Subgroup 11, partitioning of a right-angled trapezium into two equal parts from the center of the right angled side . . . 392
4.108 Subgroup 11, partitioning a right-angled trapezium from the centre of the slanted side into two parts . . . 395
4.109 Subgroup 10, partitioning a trapezium such that the greatest trapezium has the same area as the greatest triangle and for that same reason the smallest trapezium has the same area as the smallest triangle . . . 397
4.110 Subgroup 11, partitioning a rectangle into two equal parts by means of a constant width area along the two succeeding sides 400
4.111 Subgroup 11, partitioning a right-angled trapezium into two parts with a constant boundary area along the sides of the acute angle . . . 405
4.112 Subgroup 11, partitioning a right-angled trapezium into two parts by a constant boundary area along the sides of the right angle . . . 411
4.113 Subgroup 11, partitioning any given convex quadrilateral into two parts by constant boundary area along two succeeding sides 416
4.114 Subgroup 11, partitioning a quadrilateral into two equal parts with a line parallel to a side . . . 423
4.115 Subgroup 11, partitioning a quadrilateral into two equal parts with a line perpendicular to one side of the quadrilateral . . . 424
4.116 Subgroup 10, transforming a trapezium into a trapezium with common angle and equal area . . . 427
4.117 Subgroup 10, transforming a trapezium into a triangle with the same base vertex . . . 428
4.118 Subgroup 10, a square of maximal area inscribed inside a triangle 429
4.119 Subgroup 11, partitioning a quadrilateral at a point on one side into two parts according to a given ratio . . . 431

4.120 Subgroup 11, partitioning a hexagon into two parts with the ratio of one to two by means of a boundary area of a prescribed size . . . 432
4.121 Question 91: transforming a quadrangle into a triangle 433
4.122 Subgroup 11, partitioning a triangle by a line segment through a given point inside the triangle . . . 435
4.123 Question 92: solution . . . 436
4.124 Question 92: the construction of a parallelogram . . . 437
4.125 Subgroup 11, partitioning a quadrilateral at a point inside the quadrilateral into two equal parts . . . 439
4.126 Subgroup 11, replace an angled ditch dividing a land into two parts by a rectilinear one so that both parts keep their size . 441
4.127 Subgroup 11, partitioning any convex quadrangle into two parts by means of a constant border area along two succeeding sides . . . 442
4.128 Subgroup 12, constructing a triangle with given catheti 446
4.129 Question 96: the construction of three line segments 447
4.130 Question 96: solution . . . 448
4.131 Subgroup 12, construction quadrilateral inscribed in a circle . 449
4.132 Question 97: constructing the diagonals . . . 450
4.133 Question 97: the construction of the squares . . . 451
4.134 Question 97: the quadrangle inscribed in a circle . . . 452
4.135 Subgroup 13, the volume of a tetrahedron with two congruent lateral faces and three pairs of edges of 13, 14 and 15 454
4.136 Subgroup 13, the volume of any inclining tetrahedron, with the perpendicular from the top outside of the base . . . 455
4.137 Question 99: solution . . . 457
4.138 Subgroup 13, determining the height of a tower and its distance to two points . . . 458
4.139 Question 100: solution . . . 459
4.140 Solving an arithmetical question in a geometrical way 462

Preface

This book brought together two scholars who independently had a long-time interest in the "Hondert Geometrische questien met hare solutien" (*Hundred geometrical question and their solutions*, henceforth *HGQ*) by Sybrandt Hansz. Cardinael. The first, Matthijs H. Sitters from Holland, started an in-depth study of the book after the discovery in 1999 of a manuscript by the surveyor Nanninghsz. at the State Archives of North-Holland (Algemeen Rijksarchief in Noord-Holland) which contained a written copy of the *HGQ*. Several years studying Cardinael, the historical context, his books and manuscripts, lead to a PhD dissertation [Sitters 2007] and a book publication in Dutch [Sitters 2008]. The second author, Albrecht Heeffer from Belgium, learned about Cardinael during his PhD study on the emergence of symbolic algebra [Heeffer 2006], where Cardinael presented a curious objection to the use of algebra. It was decided that Cardinael's work deserved a closer look later. Through contacts with Jan van Maanen in 2012, Albrecht occasionally spent time on the book, using some of the questions in his course, History of Mathematics for master students in the Mathematics program at Ghent University. He also translated selected problems into English for discussing some of Cardinael's questions with English-speaking scholars. The translation gradually expanded and at the beginning of 2014 both authors came into contact and explored the plan to publish a critical English edition of the *HGQ* with mathematical commentary.

The Dutch edition [Sitters 2008] was the base for the new book, but from the beginning the plan was not to do a mere translation but to follow the successful format of the Docent press editions: presenting a critical edition of a historical work with an introduction, describing the historical context and a comprehensive mathematical commentary to understand the particular techniques and methods employed by the original author for solving the questions. Though the book represents the the fruitful collaboration of both authors, chapters 3, 4 and 5 mostly rely on the work of Sitters while Heeffer was mostly responsible for the other chapters.

Chapter 1 presents a short introduction on the life and works of Sybrandt Hansz. Cardinael based on the research by Matthijs Sitters and others. Then the *HGQ* is discussed with its editions and translations, introducing some new findings on Thomas Rudd, the English translator. Further, the book is situated within the specific context of practical mathematics and surveyor's geometry of the early seventeenth century. While many of the problem solving methods of Cardinael can be compared with Euclid's *Elements*, as listed in Appendix 5.1, some techniques for the division of plane figures bear a remarkable resemblance with cut-and-paste surveyor geometry dating back four thousand years. Cardinael's affinity with Old-Babylonian methods is surprising and invites further study on the continuous dissemination of mathematical knowledge by lay surveyors. Chapter 1 is concluded by situating this work's historiographical approach within the ongoing discussion on the nature of geometrical algebra.

Chapter 2 provides a full edition of the *HGQ* in modern English, closely following the source text and preserving as much of the original style as possible. The original drawings by Cardinael have been included by manual enhancing and cleaning up the images from a digital reproduction.

Chapter 3 presents a systematic overview of the methods and techniques employed by Cardinael, with many improvements on the earlier [Sitters 2008]. The terminology and methodology closely follows the lesser-known and undervalued edition of Euclid's *Elements* [Dijksterhuis 1929-30]. The last part of this chapter reveals the interdependencies between Cardinael's questions and the common themes by structuring them in five groups and 14 subgroups.

Chapter 4 gives for each of the questions a very comprehensive mathematical commentary with additional figures to elucidate some of the more intricacies of Cardinael's problem solving skills. For each of the questions references are added to the techniques employed and to similar solutions to other questions.

Chapter 5 groups several appendices such as lists of the use of propositions that correspond to Euclid's *Elements*, tables with cross-references between Cardinael and Rudd and some more technical comments.

We hope that this edition will serve as a valuable source for historians of mathematics, studying the history of geometry, surveyor geometry and the evolution of geometrical algebra. As has been demonstrated by future teachers of mathematics in Albrecht's course, Cardinael's questions provide ample useful material for classroom teaching. His geometrical solutions show an alternative to algebraic problem solving and may even elucidate algebraic solutions by making them more tangible.

The authors wish to thank Jan Van Maanen for the support and Charlotte Sitters-Zwolsman, for the considerable effort and patience to do the base conversion to LaTeX of the mathematical commentaries. The contributions by Albrecht Heeffer have been funded by the research project "An epistemological study of material models from practical mathematics in early-modern natural philosophical debates" of the Research Foundation Flanders (FWO Vlaanderen) and supervised by Maarten Van Dyck and Danny Segers.

Chapter 1

Introduction

1.1 About the author, the book, and its context

1.1.1 Biography

Sybrandt Hansz. Cardinael was born in Harlingen in 1578 as the son of Hans Sybrandt and Ymcke Emesdr, hence the name Hanszoon (son of Hans) and Emesdochter (daughter of Emes).[1] Harlingen is a small fishing and shipping city in the shadow of the university city Franeker in the province of Friesland in the northern Netherlands. The parents belonged to the Mennonite church of the Netherlands (in Dutch 'doopsgzinden' or 'baptism-minded'), named after Menno Simons (1496 – 1561), a Roman Catholic priest from the Province of Friesland who converted to Anabaptism around 1536. Several mathematicians and engineers of that time were Mennonites, such as Willem Jansz. Blau (1571 – 1638), Cornelis Drebbel (1572 – 1633), Jacob Metius (1575 – 1650), Abraham de Graaf (1635 – 1713) and Gerard Kinckhuysen (1625 – 1706).

After father Hans died (probably in 1600), Sybrandt Hansz. moved to Amsterdam in 1605. He married Levijntje Panten, the daughter of a reckoning master Lieven Panten in Amsterdam in 1607. The Panten family moved from Bruges in Flanders to Harlingen because of the religious persecutions at the end of the sixteenth century. Levijntje gave birth to six children between 1610 and 1625. Sybrandt took over the reckoning school of his father-in-law, probably soon after arriving in Amsterdam, since Lieven Panten died in 1603. He taught geometry, arithmetic, surveying, astronomy, navigation and book-

[1] The genealogy of the family Cardinael and several related family members has been studied by De Jong [De Jong 1996].

keeping. The first proof of his mathematical activities in Amsterdam stems from a recreational problem he publicly announced in 1606 about the shadows of three sticks of unequal length.[2] In the year 1606 he also published a rabat book, with tables of discounts useful for merchants. From 1608 he was active as a surveyor, as several maps are extant from his hand. In 1613 he was commissioned for the mapping of the third enlargement of Amsterdam ([Jansen 1960], 51).

Not much is known about Sybrandt Hansz. until his first book book on geometry, *Hondert Geometrische questien met hare solutien* (*HGQ*), was published in 1612. After the establishment of *Costers' Nederduytsche Academie* in 1617 he became one of the teachers. This Academy taught in Dutch for a growing class of merchants and artisans in Amsterdam, but did not last long because of political and religious turmoil. In 1620 he published a book on arithmetic (*Arithmetica ofte Reecken-konst*), the same year that also a new edition of the *HGQ* appeared. Apart from his surveying work, he might also have been active as a wine gauger at that time since his name appears in an official document.[3] In 1635 he published a book in Dutch on cosmology rejecting Copernican astronomy. His main argument against Copernicanism, as expounded in the 2nd and 3rd chapter, stems from Aristotelian notions on natural and forced motion. From 1639 onwards several editions appear of his earlier work on arithmetic edited for their use in schools. He died in 1647 and was buried on November 26 in the family tomb which he had constructed the year before.

While Cardinael currently is not recognized in the history of mathematics at the level of his Dutch contemporaries, such as Willebrord Snellius, Frans van Schooten or Christiaan Huygens, he certainly was well regarded in his time. As a witness of his recognition we can mention several lesser known historical facts (for further references, see [Sitters 2008], $59-62$). To resolve a polemic on the theory of navigation, he was invited by the Admiralty of Amsterdam to seat in a commission together with Plancius, Blaeu, Stevin, Snellius and Dou during the years $1612-1614$, which clearly places him at the center of the scientific elite. Also, the curious story, recorded by Dirck Rembrandtsz. van Nierop, that René Descartes took the trouble to visit Cardinael, trying to convince him to accept the Copernican cosmological model. While Descartes' effort was to no avail, it shows that he was a person to be accounted for. Jan Stampioen de Jonge (1610-1653), who was commissioned by Constantijn Huygens to tutor his son, advised the young Christiaan Huygens to study Cardinael's *HGQ* for

[2] Published as appendix 4 in a book by Kinckhuysen [Kinckhuysen 1661, 42-5, figure on p.36].

[3] Reported by [Dillen 1929] and discussed by [Sitters 2008], 51.

an understanding of geometry. And further, as discussed below, he was also internationally recognized through the translations of his work in German and English. Rudd's adaptation of Cardinael's work was considered as one of the most influential geometry book of the seventeenth century in a study published in *Isis* [Karpinski and Kokomoor 1928].

Finally, his fame in the Dutch Republic can be illustrated through a poem dedicated to Cardinael by the most prominent Dutch poet and playwright, Joost van den Vondel. With some irony, Vondel reveals his admiration for Cardinael as a mathematician:

De Vriesche Euklides hangt alleen
Van cijferletters hecht aen een
Bewaert toch Sybrand met u allen,
Bewaert dien Rekenschat getrou;
Viel Cardinael van t plat,
hij sou / Aen cijferletters stucken vallen.

(our translation):

The Frisian Euclid hangs closely
together by numerals solely
Thou all together cherish Sybrand
Faithfully conserving his reckoning treasure
Would Cardinael fall down from the balcony
He would shatter apart in digits

1.1.2 The book and its editions

None of the editions of *Hondert Geometrische questien* (Hundred geometrical questions) list a printing date, which makes it difficult to identify and distinguish the different print runs. However, there appear to be at least two editions, one printed by Willem Janszoon, 'op het water in de vergulde sonnewyser' at Amsterdam and a second one by Jan Janszoon, also in Amsterdam 'op het water in de Pas-Caert'. Both printers are of the Blaeu family. Several historians have suggested dating based on secondary sources. The pioneer of Dutch history of mathematics, Bierens de Haan, included biographical data on Cardinael in his *Bouwstoffen* at the end of the nineteenth century ([Bierens de Haan 1878-87, 13-72], and an incomplete bibliography in his *Bibliographie Néerlandais historique scientifique* ([Bierens de Haan 1960, 52]. Jan van Maanen, in his PhD, [Van Maanen 1987] as well as in a later article dedicated to Cardinael [Van Maanen 2003], dates the first edition in or before 1612. In an article of 2003, Matthijs Sitters ([Sitters 2003]) follows the date of 1612 for the first edition but in his PhD on Cardinael, he consistently uses 1614 as the publication date of the book, which appeared bound together with the second edition of the *Practijck des landmetens* published in 1614. A bibliographical study by Klaas Hoogendoorn [Hoogendoorn 2006-7] provides us with the most reliable data on the publication history.

The book of Cardinael most often appeared together with the several editions of the *Practijck des landmetens* by Johan Sems and Jan Pieterszoon Dou, as is mentioned on its frontispiece: "Expanded by Hundred geometrical questions, by Sybrandt Hansz. Reckoning master at Amsterdam". Sems and Dou published their book on surveying first without Cardinael's work in 1600 which was titled *Practijck des Landmetens: Leerende alle rechte ende cromzydige Landen, Bosschen, Boomgaerden, ende ander velden meten, soo wel met behulp des quadrants, als sonder het selve* (*The practice of surveying*) and printed at Leyden. At least two editions of the Willem Janszoon prints of Cardinael's book can be identified: one with and one without a dedication to Prince Mauritz and Duke Willem Lodewijk. The one without the dedication appears to be the later one.

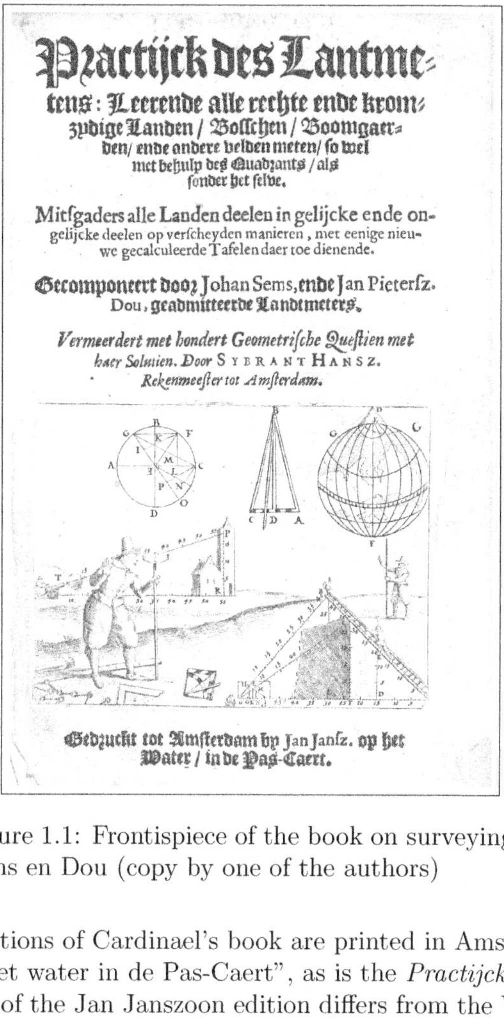

Practijck des Lantme-
tens: Leerende alle rechte ende krom-
zydige Landen / Boſchen / Boomgaer-
den / ende andere velden meten / ſo wel
met behulp des Quadrants / als
ſonder het ſelve.

Mitſgaders alle Landen deelen in gelijcke ende on-
gelijcke deelen op verſcheyden manieren, met eenige nieu-
we gecalculeerde Tafelen daer toe dienende.

Gecomponeert door Johan Sems, ende Jan Pieterſz.
Dou, geadmitteerde Landtmeters.

*Vermeerdert met hondert Geometriſche Queſtien met
haer Solutien. Door* SYBRANT HANSZ.
Rekenmeeſter tot Amſterdam.

Gedruckt tot Amſterdam by Jan Janſz. op het
Water / in de Pas-Caert.

Figure 1.1: Frontispiece of the book on surveying by Sems en Dou (copy by one of the authors)

The later editions of Cardinael's book are printed in Amsterdam with Jan Janszoon, "op het water in de Pas-Caert", as is the *Practijck des landmetens.* The typesetting of the Jan Janszoon edition differs from the Willem Janszoon editions but the woodcuts and copper plates appear to be identical. Also of the Jan Janszoon print of the *HGQ*, two different editions can be identified. The later one counts 118 numbered pages while the first has 127 pages (16 octavo quires of 8 pages each, numbered A1 to H8). The later edition is mostly found bound together with the *Practijck des landmetens* but there exist also separate copies. Ghent University library owns three copies, one by Willem Jansz. and two by Jan Jansz. (118 p.). Only one of the latter is bound together with the *Practijck des Landmetens.*[4]

[4]Most of the reproductions in our text are taken from a copy of this latter edition owned

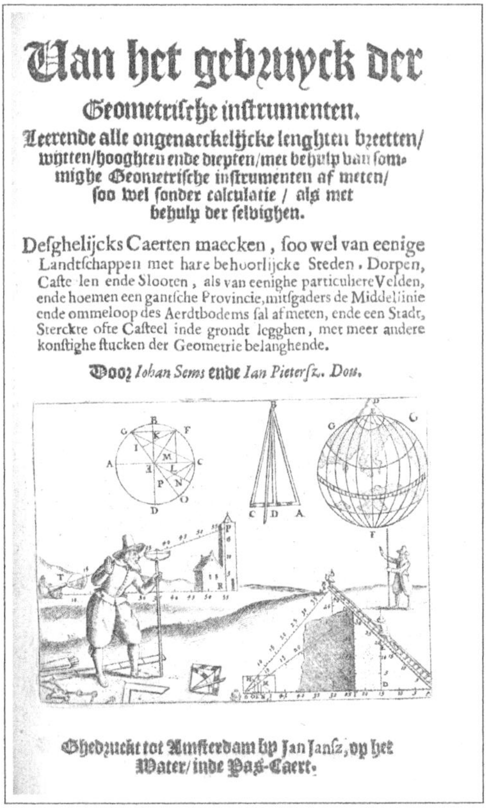
Van het gebruyck der
Geometrische instrumenten.
Leerende alle ongenaeckelijcke lenghten breetten/ wijtten/hooghten ende diepten/met behulp van sommighe Geometrische instrumenten af meten/ soo wel sonder calculatie / als met behulp der selvighen.

Desghelijcks Caerten maecken, soo wel van eenige Landtschappen met hare behoorlijcke Steden, Dorpen, Caste len ende Slooten, als van eenighe particuliere Velden, ende hoemen een gantsche Provincie, mitsgaders de Middellinie ende ommeloop des Aerdtbodems sal afmeten, ende een Stadt, Sterckte ofte Casteel inde grondt legghen, met meer andere konstighe stucken der Geometrie belanghende.

Door *Iohan Sems* ende *Ian Pietersz. Dou.*

Ghedruckt tot Amsterdam by Jan Jansz, op het Water/ inde Pas-Caert.

Figure 1.2: Frontispiece of the book on surveying instruments by Sems en Dou (copy by the authors)

Hoogendoorn [Hoogendoorn 2006-7] determined an upper bound for the first edition through a reference to his book in the dedication to *Het gulden zegel der grote zeevaart* (The golden seal of great navigation) by Jan Hendrik Jarichs van der Leys, dated 1614. The publication date of 1612, proposed by Bierens de Haan and Van Maanen, is therefore most likely correct. The second edition by Willem Janszoon is dated around 1620 because from 1623 Willem started adding Blaeu to his name. However, because the German edition by Willem Janszoon appeared in 1617, we expect the Dutch version to have been published before or in the same year. The two editions by Jan Janszoon are dated by Hoogendoorn much later than previously assumed, based on book

by one of the authors, Albrecht Heeffer.

catalogues of that era: around 1635 and 1645. However, this seems somewhat incoherent with the publication by Jan Janszoon of the German editions of *Practijck des Landmetens* in 1617.

1.1.3 Translations

Soon after the publication of the first Dutch edition of the *HGQ*, the book was translated into German by Sebastian Kurz (1576-1659, in Latin known as Curtius), a reckoning master from Nurnberg, who earlier published several works on merchant arithmetic, geometry and algebra [Curtius 1604], [Curtius 1610]. Kurz was a friend of Faulhaber and they exchanged hundreds of letters. He also translated Dou's book on mathematical instruments [Curtius 1615] and on surveying [Curtius 1615]. These last two were published in Amsterdam and may have been commissioned by the printer.

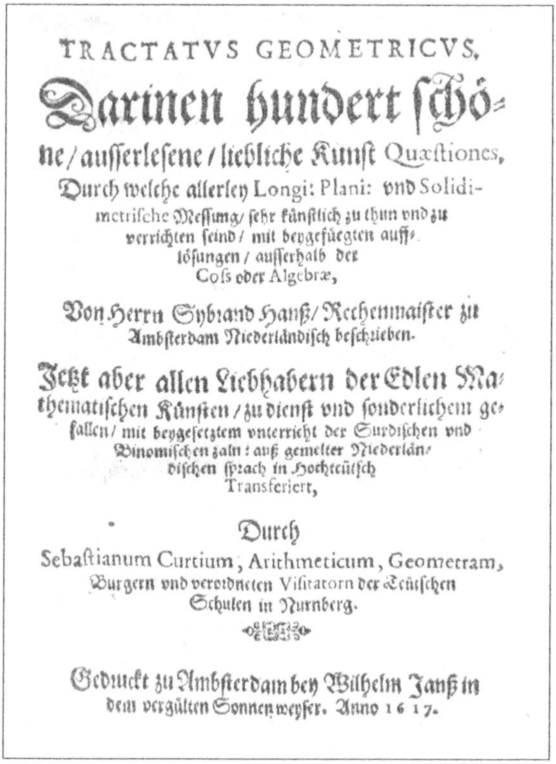
TRACTATVS GEOMETRICVS,
Darinnen hundert schö-
ne/ausserlesene/liebliche Kunst Quæstiones,
Durch welche allerley Longi: Plani: vnd Solidi-
metrische Messung/ sehr künstlich zu thun vnd zu
verrichten seind/ mit beygefüegten auff-
lösungen/ ausserhalb der
Coss oder Algebræ,

Von Herrn Sybrand Hanß/ Rechenmaister zu
Ambsterdam Niderländisch beschrieben.

Jetzt aber allen Liebhabern der Edlen Ma-
thematischen Künsten/ zu dienst vnd sonderlichem ge-
fallen/ mit beygesetztem vnterricht der Surdischen vnd
Binomischen zaln: auß gemelter Niderlän-
dischen sprach in Hochteütsch
Transferiert,

Durch
Sebastianum Curtium, Arithmeticum, Geometram,
Burgern vnd verordneten Visitatorn der Teütschen
Schulen in Nurnberg.

Gedruckt zu Ambsterdam bey Wilhelm Janß in
dem vergülten Sonnenweyser. Anno 1617.

Figure 1.3: The frontispiece of the German edition (Sächische Landesbibliothek, Dresden)

It appears only one edition of the German translation was published. The book was printed in Amsterdam by Willem Janszoon and has its dedication dated at 1 January 1617.

Figure 1.4: Sebastian Kurz (University of Oklahoma Libraries History of Science Collections)

The German translation adds an introduction on operations with surds at the beginning, referring to the works of Christof Rudolff and Michael Stifel. At the end of the book, twelve geometrical and arithmetical problems are added acknowledging Johann Heer and Paulo Schmidt, possibly students of Kurz. Otherwise, the translation is very faithful to the original, using the same woodcuts and the exact same pagination.

Another translation of the *HGQ* was in the English language by Thomas Rudd (1583?-1656). Rudd was an English military engineer who learned the Dutch language through his service in the Low Countries from around 1612 to 1627.[5] In 1627, King Charles I appointed him as his first chief military engineer in charge of the fortifications in Wales. In 1659 he wrote a supplement to a handbook for military engineers, first published in 1650, Elton's *Complete Body of the Art Military.* According to Frances Yates ([Yates 1972], 258), Thomas Rudd was an enlightened Rosicrucian and possibly Dr. Rudd, the author of

[5]Historical documents on the city of Breda mention Rudd as an official surveyor and cartographer in 1618 ([Aa 1845], 14). A PhD dissertation by [Schaefer 2001] lists Rudd as official surveyor at the province of Brabant as early as 1612.

occult works such as *Treatise on Angel Magic.* In 1651, he added an edition of John Dee's well known *Mathematical preface* to his book on the first six books of Euclid's *Elements.*

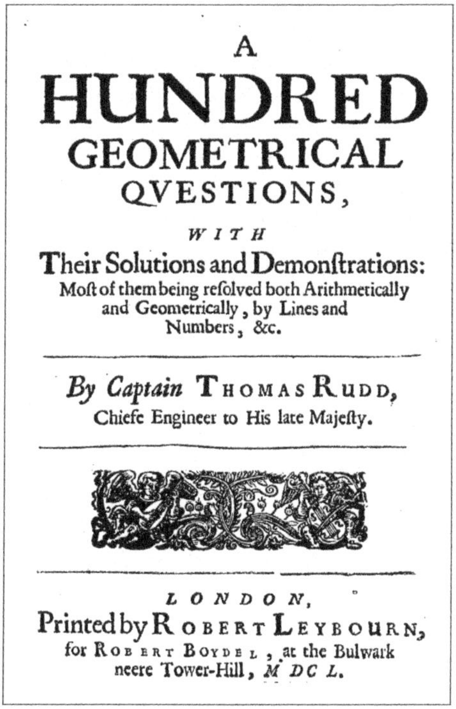

A
HUNDRED
GEOMETRICAL
QVESTIONS,
WITH
Their Solutions and Demonſtrations:
Moſt of them being reſolved both Arithmetically and Geometrically, by Lines and Numbers, &c.

By Captain THOMAS RUDD,
Chiefe Engineer to His late Majeſty.

LONDON,
Printed by ROBERT LEYBOURN,
for ROBERT BOYDEL, at the Bulwark neere Tower-Hill, *M DC L.*

Figure 1.5: Frontispiece of Rudd's adaptation from Cardinael (Early English Books Online)

Rudd's adaptation of the *HGQ* was published in 1650 by Robert Leybourn in London. The book commences with an introduction of over 50 pages explaining techniques typical for the surveyor. For the second part he heavily borrowed from Cardinael, translating 79 problems and adding some others by his own.[6] The translation is close to the original but adapted to the English units of measures. Rudd also occasionally changes the context. Cardinael's question 48 concerns delving a ditch in a farmland which becomes a fish pond in Rudd's question 87. However, Rudd did not take much effort to conceal his source. Most of the drawings and the values used for the questions are identical to those of Cardinael.

[6]See Appendix 5.2 for a cross table.

1.2 The context

1.2.1 Practical geometry

Almost all of the questions posed by Cardinael are set into a practical context and many concern surveying in a direct or indirect way. Direct references to surveying pertain to the measurements of heights with the help of instruments (**Q9, 12, 17, 100**). Indirect references to surveying are the questions in which certain sides of a regular or irregular piece of land have been *measured*, and in which the length of unknown sides have to be established (as in **Q3**).[7] The whole of subgroup 10, transformation of geometrical shapes with conservation of area, are classic problems from the surveyors tradition. Also subgroup 11, partitioning a convex polygon into equal parts or parts with prescribed ratios, belongs to the oldest practices of land administration. As an illustration, figure 1.6 shows a Babylonian tablet from the Ur-III period (c. $2100 - 2000 BCE$).

Figure 1.6: A legal document describing the partitioning of an irregular piece of land into five parts of equal area (c. $2100 - 2000 BCE$, CDLI library)

[7]See section 3.2.1 for a classification of similar problems in groups and subgroups.

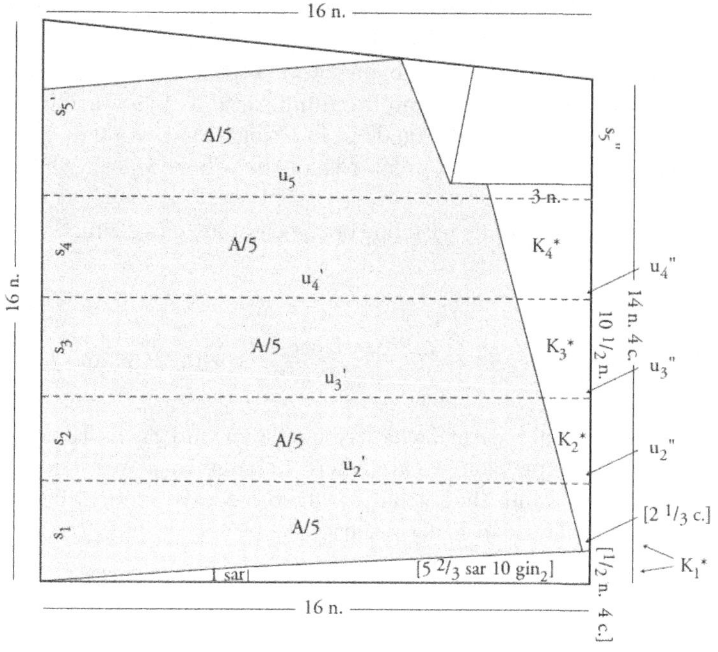

Figure 1.7: Friberg shows how the partitioning could have been done consistent with prevailing methods (*Ibid.* CDLI library)

Although this precedes Cardinael by more than three millennia, it concerns problems which undoubtedly were also part of early-modern practices. The tablet is a juridical document describing how an irregular piece of land needs to be partitioned into five parallel strips of equal area. It is discussed in detail by Jöran Friberg [Friberg 2002]. Given the area of the outer quadrangle, the area of a piece of land can be determined by subtracting the calculated area of the neighbouring triangles and trapezoids. Hence, with the area known and length of 16 n given (see Figure 1.7) the problem is reduced to determining the widths of the five pieces, S_1 to S_5. Though the extant document does not specify how the land can be divided into pieces of equal area, Friberg convincingly argues that doing so requires knowledge of solving quadratic problems. Since the document lists some of the values S_1 to S_5, we can assume that someone was able to do the calculation. Friberg shows in detail how the calculation can be performed by geometrical algebra. The techniques needed for solving the division problem are shown to be consistent with the lay surveyors methods from around 1800 BCE.

The discovery that Old-Babylonian surveyors could solve quadratic problems goes back to the late 1930's in Göttingen. It was Otto Neugebauer who first described that quadratic problems were solved on tablet YBC 6967 of around 1800 BCE. The problem can be summarized as follows. Given that the product of two unknown values equals 1 (in sexagesimal, or 60 in decimal) and that one unknown (*igibūm*) is 7 greater than the other (*igūm*), what are these values?

This problem corresponds with one of the eighteen cases which we will treat in section 3.1:

$$\textbf{system II}(DA, DB, c, b) \left\{ \begin{array}{l} DA \cdot DB = c \\ DA - DB = b \end{array} \right. \quad \text{with } c = 60 \text{ and } b = 7 \qquad (1.1)$$

In his book together with Sachs ([Neugebauer and Sachs 1945], 129-130), he writes that "The problem treated here belongs to a well known class of quadratic equations" and gives a purely algebraic/arithmetical interpretation of the tablet, as is shown in his translation:[8]

1. [The igi. b]*ūm* exceeded the *igūm* by 7
2. What are [the *igūm* and] the *igibūm*?

3-5. As for you – halve 7, by which the *igibūm* exceeded the *igūm* and (the result is) 3;30

6-8. Multiply together 3;30 with 3;30 and (the result is) 12;15

9. add [1,0, the product and (the results is)] 1,12;15
10. What is [the square root of 1],12;15? (Answer:) 8;30
11. Lay down [8;30 and] 8;30, its equal, and then

1-2. Subtract 3;30, the *takilūm*, from the one,

3. add (it) to the other
4. one is 12, the other is 5
5. 12 is the *igibūm*, 5 is the *igūm*

[8] Numbers are shown in base 60 system. Where we normally use the decimal point, in the sexagesimal system it is customary to use a semicolon. Higher-level multiples and lower-level fractions are separated by a column. The sexagesimal number 3;30 thus corresponds to 3 units and $\frac{30}{60}$ fractions of a unit, or in decimal $3\frac{1}{2}$. The square brackets in the text fill in text missing by damage on the tablets. The front text on the tablet is numbered $1-11$, the lines on the back from 1 to 4.

Neugebauer believes that the operations in text correspond with the solution to the system of two equations, amounting to the calulation

$$\left.\begin{matrix} x \\ y \end{matrix}\right\} = \sqrt{\left(\frac{7}{2}\right)^2 + 1;0} \pm \frac{7}{2} \tag{1.2}$$

When comparing Neugebauer's translation with this formula one can indeed see the relation. The difference between the two unknowns being 7, is divided in half (line 3 – 5), which is 3;30 (or $3\frac{1}{2}$ decimal). This is squared, which becomes 12;15 (line 6 – 7) or $(3\frac{1}{2})^2 = 12\frac{1}{4}$ decimal. To this is added 1;0 which is the product of the two unknowns. In line 10 the root is taken from this sum, being 8;30 (or $8\frac{1}{2}$ decimal). The two unknowns then are 8;30 plus or minus 3;30, being 12 and 5.

From the end of the 1970's, Jens Høyrup published several articles on Old-Babylonian mathematics, advancing the thesis that instead their algebraic methods depend on lay surveyors practices. Through a thorough semantic analysis of the Akkadian text, he determined that certain verbs, interpreted by Neugebauer as arithmetical operations, instead refer to concrete operations. Where Neugebauer reads 'to subtract', Høyrup understands 'to tear off', 'to halve' becomes 'to break in two' and 'adding together' becomes 'pasting together'. In this context, the 'lay down' now makes sense as a concrete action. Instead of solving quadratic equations, Old-Babylonian surveyors thus used a cut-and-paste geometry to solve problems, very much as Cardinael is doing.

The two unknown values *igūm* and *igibūm* can be considered as the sides of a rectangle with area 1 (or 60 decimal). In a first step the area exceeding the square with side *igūm* is cut in half (see figure 1.8).

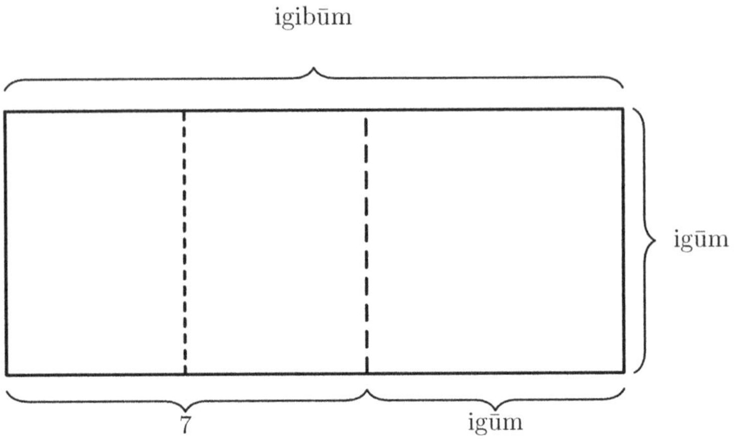

Figure 1.8: The geometrical interpretation of YBC 6967 by Høyrup

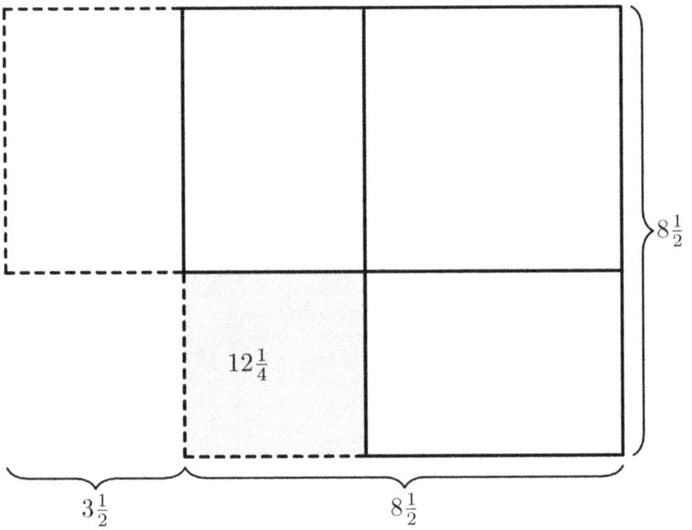

Figure 1.9: The cut-and-paste solution to YBC 6967 following Høyrup

In a following step, half of the rectangle which was cut is pasted onto the bottom of the original rectangle (see figure 1.9), which gives meaning to the term 'laying down' which puzzled Neugebauer. The area of the original rectangle remains unchanged (the white area in figure 1.9). The open space then is a square of side $3\frac{1}{2}$. So the shaded square in figure 1.9 has an area of $12\frac{1}{4}$. The area of the completed square is therefore known: $72\frac{1}{4}$. From this, the side can be derived, which is $8\frac{1}{2}$. It now becomes evident that the two unknown values are $8\frac{1}{2} \pm 3\frac{1}{2}$ or 12 and 5.

This new view makes Neugebauer's symbolic interpretation, as a problem belonging to "a class of quadratic equations", inconsistent with Høyrup's lexical analysis. When substituting $y = x + 7$ in $xy = 60$, one arrives at the quadratic equation $x^2 + 7x = 60$. The roots for the quadratic equation $ax^2 + bx + c = 0$ are

$$\left.\begin{matrix} x_1 \\ x_2 \end{matrix}\right\} = -\frac{b}{2a} \pm \sqrt{\frac{b^2 - 4ac}{4a^2}}$$

with $a = 1$, $b = 7$ and $c = -60$, this amounts to

$$x_1 = \frac{-7}{2} + \sqrt{\frac{(-7)^2 + 240}{4}} = 5$$

and

$$x_2 = \frac{-7}{2} - \sqrt{\frac{(-7)^2 + 240}{4}} = -12$$

The negative solution only arises in the context of symbolic algebra and is avoided in the Old-Babylonian cut-and-paste geometry as well as in Cardinael's techniques for solving problems. By applying case 12, system-II as in (1.1) Cardinael would arrive at the positive solution $DA = 12$ and then determine that $DB = DA - 7 = 5$.

Curiously, this geometrical solution to the problem of tablet YBC 6967 is identical to case 12 of the application of area discussed below in §3.1.5, as can be seen by comparing figure 1.9 with figure 3.13. Cardinael applied this method in **Q51** as shown in figure 1.10.

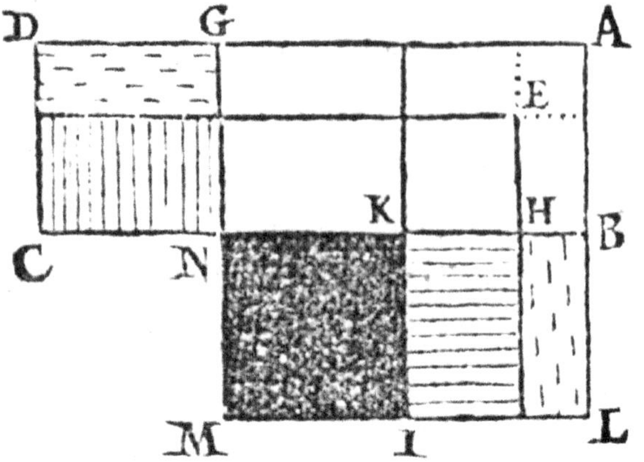

Figure 1.10: Cardinael's solution to Q51, corresponding to the Old-Babylonian method of YBC 6967

Though the problem of **Q51** is more complex than tablet YBC 6967, Cardinael's solution is based on the same technique: cutting off an area $DGNC$ from the given rectangle, pasting it at the bottom, completing the square $GALM$, determining its side AG which allows to calculate the solution. We give here the corresponding explanation by Cardinael to show how his reasoning is remarkably similar to the practices known to Old-Babylonian surveyors (see figure 1.10):

> So, cut off of the quadrangle $ABCD$ as much as $GDCN$ and paste it to BK as $BKIL$. So, then $GALM$ is a square, making AG or AL 40, as BAD is equal to LAG, the square $AGML$ then is 1600.

Equally noteworthy is that the Old-Babylonian problem also appears with the same values as problem **Q14** otherwise, with a rectangle of area 60 and sides 5 and 12, though the solution method is slightly different.

But, as there are almost no diagrams on these tablets, the geometrical interpretation of Old-Babylonian eluded historians for many years. The close correspondence between the techniques employed by Cardinael and those of Old-Babylonians algebra are quite remarkable. Despite the three millennia separating Old-Babylonian mathematics from early-modern surveyors geometry, there seems to be a continuity in the surveyor's tradition.

So, is it possible that traces of Old-Babylonian surveyor techniques survived up to the seventeenth century? Recent scholarship in the history of mathematics gives support to such a remarkable relationship. Firstly, there is book II of the *Elements*. In a recent publication, Friberg [Friberg 2007] goes quite far in claiming a close relationship. He meticulously shows how propositions II.2-3, II.4,7, II.5-6, II.8, II.9-10, II.11,14 and II.12-13 correspond to surveyor techniques found on Old-Babylonian tablets. In his view, these propositions by Euclid are a reformulation of the Old-Babylonian geometrical models but in abstract terms, without the use of numbers. Cardinael's methods and techniques are obviously related to Euclid's *Elements*, as we show in §3.1.5, but in spirit they may be even closer to surveyor formulations of Old-Babylonian mathematics, including the use of concrete numbers and the reference to tangible operations. Take a typical example such as **Q35**, where Cardinael uses the terms 'to halve' (*halveer*), 'to cut off' (*neemt weg*) and 'to paste' (*addeer*).

Apart from an influence through the scholarly tradition, passed on through authoritative texts such as from Euclid, there may have been an important but neglected influence by the oral tradition, passed down from master surveyors to their apprentices. The enigma of the origin of Arabic algebra at the beginning of the ninth century can be partially explained from the Old-Babylonian heritage of surveyors practices. The myth that Arabic algebra descended from the *Arithmetica* by Diophantus, created by 15th century humanists, is currently unsustainable [Heeffer 2009]. That Arabic algebra originated from India, as did the Arabic numerals, has already been refuted more than a century ago by Léon Rodet [Rodet 1878]. With Ancient Greece and India crossed off from the list, few options remained, until the Dutch historian Hubertus Busard noted a close correspondence between a Medieval Latin text and Old-Bablyonian methods [Busard 1968]. The text he edited is the *Liber mensurationem*, attributed to Abū Bakr, of which the Arabic original has been lost. The first part is on *misāḥa* (surveying), and bears a close resemblance to Old-Babylonian problems. Jens Høyrup further developed the theory that Old-Babylonian surveyor techniques survived the decline of the Sumerian empire and were passed on through the oral tradition of lay culture. Its influence is also notable in the tradition of Medieval works on practical geometry, of which Fibonacci's *Practica Geometriae* is the main proponent ([Hughes 2008], further discussed in Appendix §5.4).

From Fibonacci's *Practica Geometriae* to Cardinael, the line of influence becomes more direct. Practical geometry expanded rapidly during the Renaissance. While during the Middle-Ages, the discipline depended mostly on surveying, it soon expanded to several other areas: fortification, perspective, ballistics, gauging, navigation, cartography, hydrostatics. From the fifteenth century onwards, in Italy the traditional empirical engineers such as Taccola (Mariano di Jacopo, 1382– c. 1453) were transformed into mathematically literate military engineers such as Buonaittu Lorini and Vittorio Zonca (1568 – 1603). In a well-informed study by Mario Biagioli [Biagioli 1989] it is argued that these engineers not only migrated upwards in social status but they also significantly advanced the role and recognition of practical mathematics in Renaissance society. Also in France, military engineers firmly established themselves in service of the Dukes by the end of the sixteenth century, while publishing works on fortification, surveying, mathematical instruments and often also their own editions of Euclid's *Elements*. Some representative figures are Joseph Boillot, Claude Flamand, Jean Errard and Salomon de Caus. In the Low Countries the engineering tradition first flourished in Flanders through Simon Stevin (1548 – 1620) and Phillipe van Lansberge (1561 – 1632). With the fall of Antwerp in 1585 many engineers and practitioners of practical geometry fled to the North, including Stevin and Van Lansberge, and Holland became the new center for engineers. Prince Maurits asked Stevin to found a new school of engineering at Leyden to teach sons of artisans and surveyors in the Dutch language, rather than in Latin. The school was established in 1600 with Ludolf van Ceulen and Simon Fransz van Merwen as the first teachers, to be followed by father and son Van Schooten. Stevin's book [Stevin 1605-8] was used for teaching and provides a good idea of the curriculum: geometry, surveying, trigonometry, arithmetic, perspective, cosmology and book-keeping. Also the *Practijk des lantmetens* of Dou and Sems was used as a textbook throughout the seventeenth century. Although Stevin devoted a significant part of his l'*Arithmetique* of 1585 to algebra, the Rule of Coss as it was called, was not commonly taught to engineers or surveyors. It thus comes as no surprise that Cardinael took the pragmatic choice of solving his hundred geometrical problems without the use of the Rule of Coss. It was pedagogically and commercially sound given his audience of surveyors and engineers and it fitted his personal preference for ancient Greek ruler and compass constructions. While Cardinael composed his book during the Twelve Years' Truce (1609 – 1619) of the Eighty Years' War, he even accounted for the military context by including problems on military formation as in subgroup 3 (**Q36-39**, **Q49-53**).

1.2.2 Geometrical algebra

Since we will use the term 'geometrical algebra' for many of the techniques employed by Cardinael and modern symbolism in our mathematical commentaries discussing his solutions and techniques, this section is intended as a disclaimer to the reader. Let us start with a notable quote by Michael S. Mahoney in his book on Fermat ([Mahoney 1973], XII-XIII):

> Any historian of mathematics conscious of the perils and pitfalls of Whig history quickly discovers that the translation of past mathematics into modern symbolism and terminology represents the greatest danger of all. The symbols and terms of modern mathematics are the bearers of its concepts and methods. Their application to historical material always involves the risk of imposing on that material, a content it does not in fact possess.

We are aware that our use of modern symbolism in the mathematical commentaries poses the risk of obtruding an algebraic content into the geometrical techniques employed by Cardinael. However, we believe it is beneficial to the reader to convince himself of the validity of Cardinael's geometrical reasoning by means of modern symbolism. On the one hand, we provide a literal translation of the Dutch text, on the other hand a modern rendering of the mathematical reasoning, within the context of the systematic treatment of techniques in section §3.1. When we use AB^2 in our commentary, this corresponds to Cardinael's somewhat misleading phrasing "the square AB", where AB is actually the side of a square. Cut-and-paste operations on, for example two squares, will be represented by operators as in $AB^2 - CD^2$ for "subtracting the square of CD from the square of AB" or $AB^2 + CD^2$ for "adding the squares of AB and CD together". In addition, for some cases we also provide an alternative algebraic solution which will be clearly denoted as such by the use of the symbol x for an unknown line segment, as AB.

As for our use of the term "geometrical algebra" it may be appropriate to give the reader some background on an important controversy in the history of mathematics which started some forty years ago.

In 1975, Sabetai Unguru started the polemic in a long article in the prestigious journal *Archive for History of Exact Sciences.* Unguru is a Romanian classicist and historian of mathematics who migrated to Israel, best known for his edition of the *Conics* by Apollonius. Starting with a critical assessment of several claims by van der Waerden in his *Science Awakening*, he considers geometrical algebra "a monstrous, hybrid creature, a contradiction in terms, a logical impossibility" ([Unguru 1975], 77) and argues that there is no algebra in Euclid's *Elements.*

During the following years, three respected mathematicians with a longtime interest in the history of mathematics responded furiously to Unguru's article. The first [van der Waerden 1976] responded by demonstrating that mathematicians can easily see and interpret an algebraic basis in Euclid's *Elements.* Obviously, this is a fact Unguru would not deny. His objection was that such interpretation is historically not correct. By defining algebra as "the art of handling algebraic expressions like $(a+b)^2$ and solving equations like $x^2 + ax = b$", van der Waerden did not serve his case very well. What mathematicians *see* as equations are often not symbolic equations at all, as illustrated by the example of Old-Babylonian algebra in the previous section.

Hans Freudenthal also responded in the same journal, claiming that "Elements V is algebra and nothing else" [Freudenthal 1977]. The third mathematician to respond was André Weil, in a short note [Weil 1978] asking to "let us stop the disease before it gets fatal". Unguru did not succeed in getting a response to his critics published in the same journal, but it was accepted by *Isis* [Unguru 1979]. By the end of the seventies, it became widely accepted that an algebraic interpretation of Euclid, which started during the Arabic period and was pursued by Renaissance mathematicians, is historically unsound. Historians of mathematics should especially be careful when using symbolic representations of geometrical operations from the *Elements*. Let us discuss one subtle example.

Euclid's *Elements*, book IX, prop. 21 states that "if any multitude whatsoever of even numbers is added together, then the whole is even". The proof is very simple and does not depend on any other propositions, only on the definition of number. A number is even when it has a half part. Given numbers AB, BC, CD and DE (shown by Euclid as line segments) being even, the whole AE (the concatenation of line segments) must also be even since it has a half part, being the sum of the half parts of the original numbers. A symbolic rendering of the reasoning could be: suppose an arbitrary long finite sum of even numbers $2a + 2b + 2c + \ldots + 2n$. This can be rewritten as $2(a + b + c + \ldots + n)$ and is obviously even. However convincing the mathematical reasoning here is, the proof differs from Euclid's original. In a second try, we could involve the definition of an even number in our symbolic representation and write an even number a as $c+c$, thus exemplifying that it has a half part c. Suppose we have two even numbers $a = c+c$ and $b = d+d$, whose sum is $a+b = (c+c)+(d+d)$. This we can rewrite as $a + b = (c + d) + (c + d)$ showing that the sum consists of two equal parts, which makes it even. Again, this proof is different from Euclid's original. In the two examples we have either used the commutativity or the associativity of addition, while Euclid's proof depends only on the definition of even numbers. Thus by representing the original proof in a symbolic way we have made it a different proof, sound but depending on valid truths other than the original one.

After this historiographical contextualisation we can now address the question of what geometrical algebra could entail. The criticism of Unguru was directed towards the unjustified interpretation of geometrical concepts, operations and proofs in ancient Greek mathematics as algebra and more specifically symbolic algebra. However, if we understand algebra as an analytic problem solving method for arithmetical or geometrical problems which uses some abstract representation of unknown values to reason, then geometrical algebra makes sense. Symbolic algebra depends on a symbolic model for representations [Heeffer 2010]. Symbolism is a script, which delegates the cognitive load of problem solving to an external environment, thus facilitating our reasoning. An alternative model is the geometrical model. Difficult arithmetical problems can be solved algebraically by using a cut-and-paste geometry as in Old-Babylonian algebra and as Cardinael is doing. As traced in a recent publication by Jens Høyrup [Høyrup 2016], the concept of geometrical algebra was introduced by Hans Georg Zeuthen, not Paul Tannery as often assumed. Høyrup points out that Zeuthen's original conception of geometrical algebra was that of "a tool for visualization of these as well as other operations in form of the geometrical representation and handling of general magnitudes and the operations to be undertaken with them". This original conceptualization of geometrical algebra is close to what we have described as a geometrical model for algebra, parallel to the symbolic model. So, Unguru's criticism should be narrowed down to symbolic translations of ancient Greek mathematics. There was a geometrical algebra practiced by Old-Babylonian scribes and it apparently survived in the surveyors tradition to the time of Cardinael.

Chapter 2

English translation

2.1 About the translation

The translation and pagination is from the second edition of *Hondert Geometrische Questien met hare solutien*, by Sybrandt Hanszoon van Harlingen, printed by Jan Janszoon, 'op het water in de pascaert' (probably in 1620). We have used a copy from the Bayerische Staatsbibliothek, shelf no. Math. p. 201 p., digitized by the Münchener Digitalisieringszentrum. All figures used in this edition are based on this digitized version. They have been enhanced for contrast and manually cleaned of irregularities.

We have opted for a very literal translation to preserve as much of the original nuances in terminology. However, to improve on the readability, punctuation and grammar have been adapted. Printing errors or mistakes have been corrected and are shown in footnotes. Some expressions have been silently modified for correctness. For example, Cardinael uses "the square of AB" interchangeably with "the square AB" when referring to the area of a square with side AB. Also, words have been added between square brackets when they are needed to interpret the enunciation of a question and some part of a solution. Somewhat oddly, Cardinael uses the term "contents" (in Dutch: inhoud) for the area of a plane figure as well as the volume of a solid. We have used the modern terms to avoid confusion.

2.2 Hundred geometrical questions

Hundred
Geometrical questions
with their solutions

by

Sybrandt Hanszoon van Harlinghen,
Reckoning master at Amsterdam.

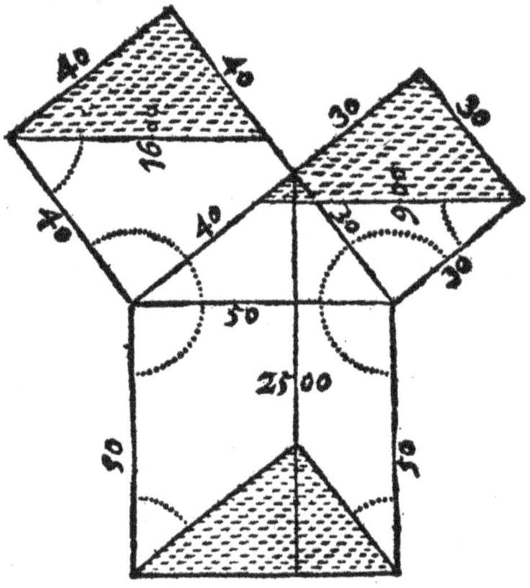

At Amsterdam,
Printed by Willem Janszoon on the water
at the Golden Sun Dial

Foreword

Thou art loving reader, here you have hundred geometrical questions, with instructions on how to solve them, some by lines, some by numbers (but none with cossic numbers), of much utility for all those who aim to practice the art of geometry, because in the same are exposed and treated the most prominent parts of this art, to perform various measures most artfully, to discover unknown and hidden lines with ease and dexterity. If we find that this pleases you, we hope (God willing) to expose them henceforth more and better. In the mean time, fare well.

Hanszoon Sybrandt

Question 1

[5] There is a triangle ABC of which AB makes 13, AC 15 and BC 14. The question is, what will be the length of the perpendicular AD?

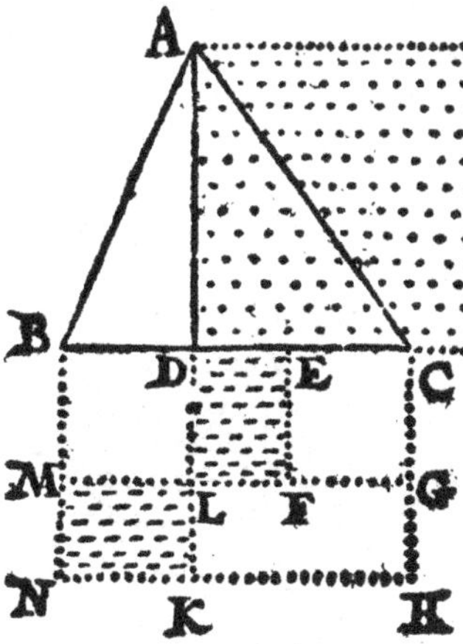

Figure 2.1: Question 1

To do this, notice that when the square of AB is subtracted from the square of AC, there will remain as much as when one subtracts the square of BD from the square of DC, as by prop. 47 from the I of Euclid, the squares of AD and DB are together as much as the square of AB. Also, the squares of AD and DC together are as much as the square of AC. Now, the square of AB is 169 and the square of AC 225. Subtracting this 169 from 225 there remains

56 for the try-square or gnomon $DEFGHKD$, which try-square is of equal size as $MGHN$, because $DEFL$ equals $LKNM$. Therefore $MGHN$ makes 56 and the length NH or BC becomes known, being 14. Therefore, divide 56 by 14 and one arrives at 4 for MN, being equal to DE. Now, subtract DE 4 from BC 14, there remains 10 for BD and EC together. As BD and EC are equal, halve 10 and one arrives at BD 5. And because of the aforementioned proposition, AD becomes known, subtracting the square of BD being 25, from the square of AB, being 169, there remains 144 for the square of AD. The $\sqrt{\mathfrak{z}}$ of this is 12 for AD, as one required.

Otherwise

Add the square of AB 169 to the square of BC 196, becomes together 365 for the three squares of AD, BD and BC because by prop. 47 of Euclid I, the squares of BD and of AD together are equal to the square of AB. Now, subtract from 365 the square of AC, being 225, there remains 140 for the parts E and F, as the squares of AD and of DC together are equal to the square of AC, by prop. 47 of Euclid I. Therefore, take the half of 140 and one arrives at 70 for E or F. This divided by 14, the length of F, one arrives at 5 for the breadth GH, as much as is BD.

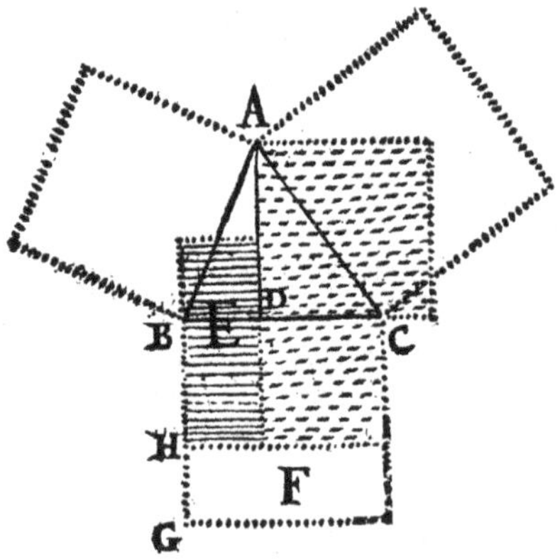

Figure 2.2: Question 1, alternative solution

Question 2

From the triangle, shown above, the three sides are known, AB being 15, AC 13, and 4 for the base CB. The question is, how much is the perpendicular falling down the triangle, being AD?

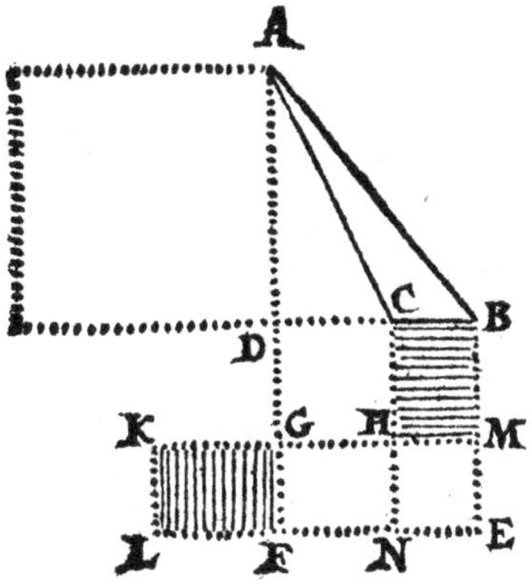

Figure 2.3: Question 2

The square of AD and the square of DB together equal the square of AB by proposition 47 of Euclid I. Also, the square of AD and the square of DC together equal the square of AC. If the square of AC is subtracted from the square of AB there remains as much [7] as one would subtract the square of DC from the square of DB. Now, the square of AB being 225 and that of AC is 169. Subtracting this 169 from 225, there remains 56, as much as the square of AB is greater than the one of AC, or as the square of DB is more than the square of DC, which is the gnomon $CBEFGHC$. The area of this gnomon being 56, as much as is also $KMEL$ and because $CBMH$ equals $LKGF$, $KMEL$, also makes 56. And also LK is known because LK equals CB, being 4. So that 56 divided by 4, gives 14 for LE. From this NE 4 is equal to CB or FN, which leaves 10 for LN. Half of this is 5 for LF or FN and this is as much as DC. Now that DC is known, so is AD known by earlier prop. from Euclid. Subtracting the square of DC being 25 from the square of AC 169, there remains 144 for the square AD. The $\sqrt{\mathfrak{z}}$ from this is 12 for AD, as one required.

Otherwise

Look at the following figure.

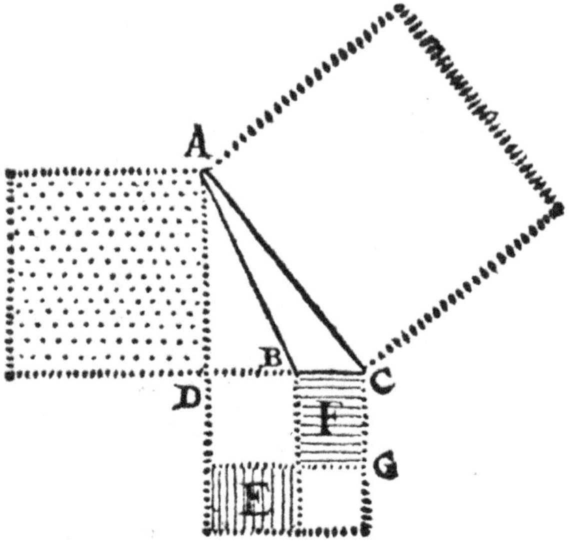

Figure 2.4: Question 2, alternative solution

Add the square of AB 169 to the square of BC 16, which is together 185, as much as the squares of AD, DB and BC are together, because of prop. 47 of Euclid I, the square of AB is as much as the squares of AD and DB together. Now, subtract 185 from the square of AC 225, there remains 40 for the parts E and F. Because of aforementioned prop. of Euclid, as much as the square AC is greater than the squares of AD, DB and BC, so is the square of AC more than the squares of AD and DC together. [8] The half of 40 is 20 for the part E or F. Now divide the part of 20 by the breadth BC 4 and the length CG will be 5, also as much as DB.

Question 3

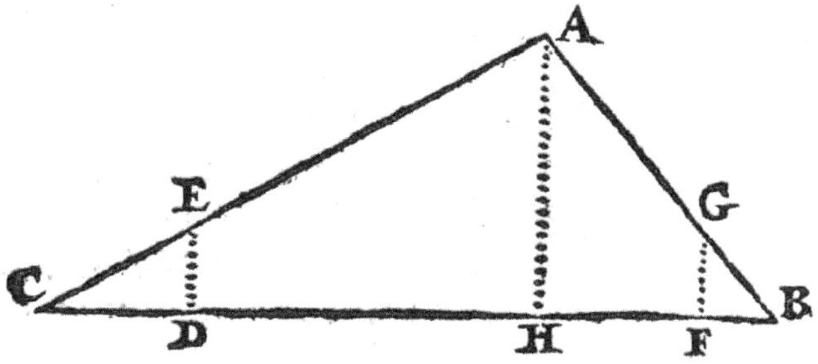

Figure 2.5: Question 3

There is a triangular piece of land ABC for which I measured from B to H and, arriving at H, I found that the cathetus of the angle A to the point H measures 50 rods.[1] I also found that the measure from B to F equals 5 rods, that the perpendicular FG will be 12 rods, equal if measuring from C to D as 12 rods. So is the perpendicular [9] DE 9 rods. The question is how much will the sides AB, BC and AC be?

Say, BF 5 and given FG 12, what shall be BH 50? We shall arrive at 120 for AH. Hence, look for HC and say DE 9 and given DC 12, what will give me AH? 120, as 160 for HC and also BH is 50, so is BC 210. Now as BH, AH and HC are known, both AB and AC can be known by the 47th prop. of Euclid I. Adding the square of BH, being 2500, to the square of AH, being 14400, together this is 16900. The $\sqrt{\mathfrak{z}}$ is 130 for AB. Also adding the square of AH, being 14400 to the square of HC being 25600, together is 40000, the $\sqrt{\mathfrak{z}}$ is 200 for AC.

[1] A rod, or in Dutch, 'roede' is a unit of length as well as a unit of area. Depending on the region in the Netherlands its value is between 7 and 21 feet. In Amsterdam, where Cardinael worked, a rod was 4.53 m. When used as a measure of area a rod corresponds with the area of a square with 1 rod as the side. Cardinael uses rods for units of lengths and area interchangeably.

Question 4

In the triangle ABC, shown below, the angle B is a right angle, of which the perpendicular BD is unknown, but the parts of the base, on which the perpendicular descends are known, being AD 18 and DC 32. The question is, what is the perpendicular BD?

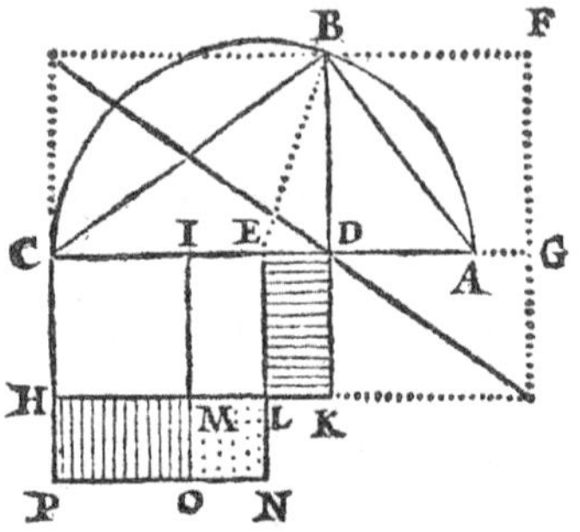

Figure 2.6: Question 4

Because AD is 18 and DC 32, so is the whole line AC 50, and its half 25 for AE, as much is also BE, because ABC is a right angle. Now, subtract AD 18 from AE 25, there remains 7 for DE. Now as BE is also known, being 25, so BD becomes known by the 47th prop. of Euclid I. Subtracting the square of DE, being 49, from the square of BE, being 625, so there remains for the square of BD 576. The $\sqrt{ʒ}$ is 24 for BD, the perpendicular as one required. [10]

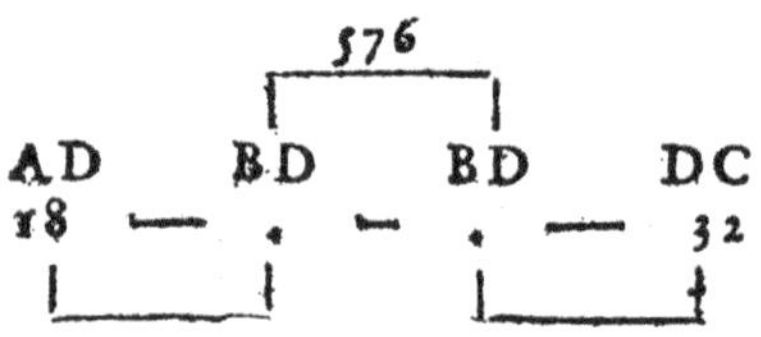

Figure 2.7: Question 4, alternative solution

Otherwise [1]

In the right triangle AD is also in proportion to DB as DB is to DC. Then one can construct a right-angled quadrangle from AD and DC, as the square can be constructed from BD, or the multiplication of AD by DC equals the multiplication of BD by BD. Now AD 18 multiplied by DC 32 becomes 576. The $\sqrt{\mathfrak{z}}$ is 24 for BD, as before.

To prove that the square $BDGF$ is equal to the quadrangle $DKHC$, make $EIML$ equal to $DELK$, so that IC is equal to CH, or equal to AD. Cut off $DELK$ and add it to MH, as MH is equal to CH or KD, to become $MHPO$. To complete this we need the square $LMON$, which is equal to the square of DE, to arrive at the square $ECPN$. When we now remove $LMON$ from the square $ECPN$, there remains the quadrangle $DKHC$. Also, as we subtract from the square that can be made from EB, the square of DE, equal to $LMON$, so there remains, by the 47th prop. of Euclid I. the square $BDGF$. And as EC is also equal to EB, so are their squares equal. And if from both sides the square of DE is subtracted, the remainders will stay equal, knowing to be the square $BDGF$ equal to the quadrangle $DKHC$, which we needed to prove. [11]

Otherwise[2]

Or pose 18 and 32 for the smallest proportion, one arrives at $1\frac{7}{9}$. Now because AD is to DB as DB is to DC, therefore take $\sqrt{\mathfrak{z}}$ of $1\frac{7}{9}$ which is $1\frac{1}{3}$. This, multiplied by 18, will be 24 for BD.

Question 5

There is a triangle ABC with a right angle in B, with AC as base making 50 and the cathetus BD making 24. Asked is the size of each of the parts AD and DC of the base in particular?

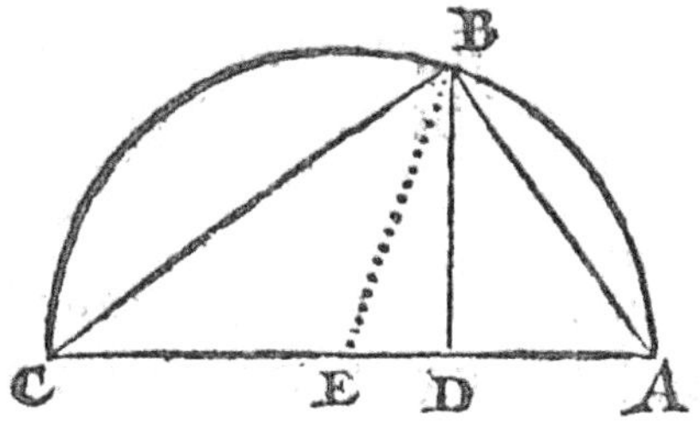

Figure 2.8: Question 5

To solve this, one takes the half of AC being 50, arriving at 25 for AE, also as much as EB is, because ABC is a right angle. As now both BD and BE

are known, so can DE be known by the 47th prop. of Euclid I. Subtracting the square of BD, being 576 from the square of BE being 625, there remains 49 for the square of DE. So is DE 7. This subtracted from AE 25, there remains 18 for AD, or added to EC 25, makes 32 for DC, and also AD and DC are known as required.

Otherwise

Look at the following figure.

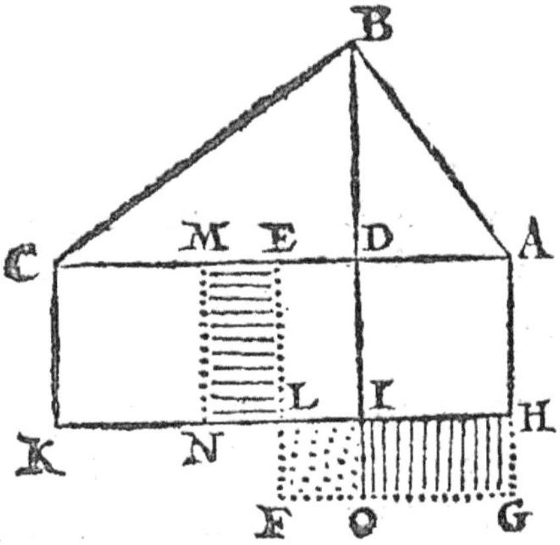

Figure 2.9: Question 5, alternative solution

The square of BD in the aforementioned triangle equals the multiplication of AD by DC. Therefore, the right-angled quadrangle which can be constructed from AD and AC equals the square of BD and is 576. So the half of AC becomes AE 25. This squared becomes 625 for the square $AEFG$. Cutting off from this $DIKC$ being 576, so there remains the [12] square $ILFO$ 49 as $HIOG$ has been made equal to $LEMN$. So is gnomon $AGOILEA$ also equal to $DIKC$ 576. Now, if the square $ILFO$ makes 49, so is also OF 7 as much as DE. This [subtracted] from AE leaves 18 for AD. So DC is 32 as before.

Question 6

There are two cities H and F, and on their straight line standing at point A, I have measured the rectangular side to B as 500 rods and I found that B makes a right angle with H and F. Henceforth, I measured the rectangular side to C as 200 rods, so that C makes a right angle with H and B. I require to know how many rods there are from H to F.

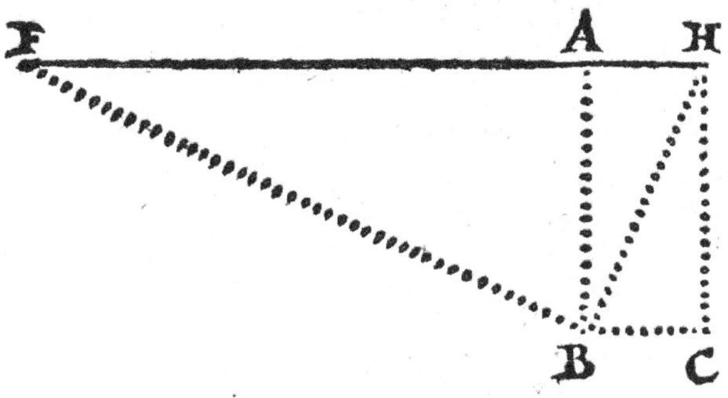

Figure 2.10: Question 6

Say: AH 200, and given that AB 500 (because AH equals BC) what will be AF 500? One will arrive at 1250 for AF because the triangle HBF is rectangular in B and AB falls perpendicular on HF. So line HA is to AB as AB is to AF. Now, add to the found AF as much as to AH, knowing to be 200. So we come to 1450 rods for the distance from H to F.

Question 7

[13] I have a right triangle ABC with the right angle in A, of which the sides are unknown to me, where the cathetus AD is known, and it makes 24. I also know that part DC of the base is 14 longer than BD. I am asking how much the sides of such triangle will be.

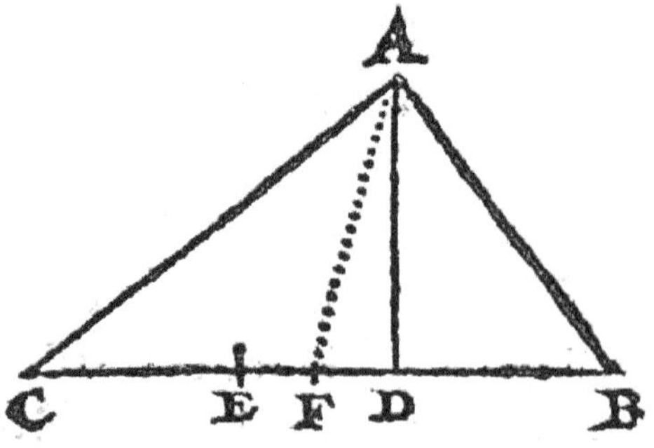

Figure 2.11: Question 7

Let EC be equal to BD, so DE is 14, its half 7 for DF and F is the midpoint between B and C. This squared and added to the square of AD makes 625 for the square of AF. The $\sqrt{\mathfrak{Z}}$ is 25 for the line AF. So much is also BF or FC. Now finally, subtract DF being 7 from BF 25 and there remains 18 for BD. So is DC 32 and by the 47th prop. of Euclid I, AB and AC become known, arriving at AB 30 and AC 40.

Question 8

A farmer ordered a gate being made as shown in the figure, of which for example AB measures 9 feet, BC 10 feet and CD 6 feet. The question is the size of EA, EC, EB and ED, and as the wooden supports AC and BD cross each other, also how much are EF and GD?

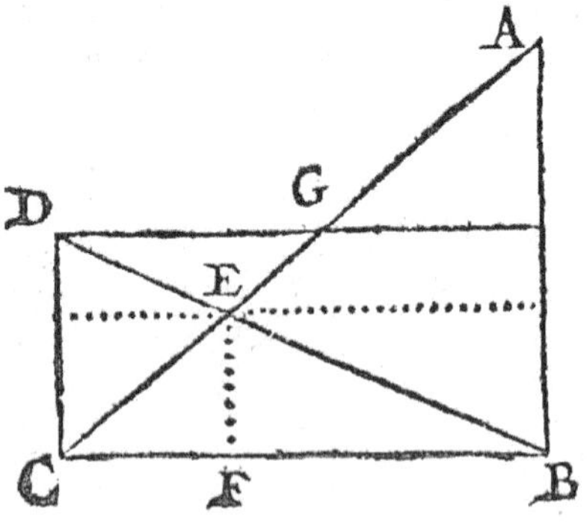

Figure 2.12: Question 8

Add 9 and 6 together, which makes 15. Say: 15 gives me 10, what shall give me 6? This makes 4 for FC. Or, 15 gives me 10, what shall give me 9? Arriving at 6 for BF. So the line BC is divided into two parts so that BF stands to AB as [14] FC to CD. Furthermore say: BC 10 gives me BA 9, what shall give me FC 4? Arriving at $3\frac{3}{5}$ for EF and because of the 47th prop. of Euclid I, EC and EB are also known, making EC $\sqrt{28\frac{24}{25}}$, EB $\sqrt{48\frac{24}{25}}$ and thus ED is $\sqrt{21\frac{19}{25}}$ and EA $\sqrt{65\frac{4}{25}}$. As EC stands to EB as EG to ED, and EB to ED, so [stands] BC to DG, so that DG can also be known. Say: EB $\sqrt{48\frac{24}{25}}$ gives me ED $\sqrt{21\frac{19}{25}}$, what shall give me 10 BC? Arriving at $\sqrt{44\frac{4}{9}}$ for GD.

Question 9

There is a tower whose height is known and it makes 75 feet, on which is mounted a ladder, so that it stands 15 feet away from the tower, with the purpose to climb into a window. The question is, if the distance of the window to the top of the tower equals the length of the ladder, how many feet is it from the bottom to the window and how many feet is the ladder?

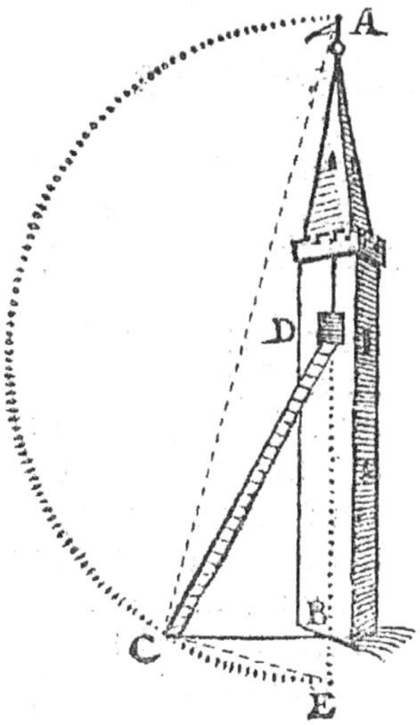

Figure 2.13: Question 9

Multiply BC, being the distance that the ladder stands from the tower, by itself, making 225. This divided by AB 75, the height of the tower, makes 3 from BE. The reason for this result is that AB stands to BC as BC stands to BE, which is the case because DA equals the length of the ladder DC. Now that BE is known, so add BE 3 to AB 75 which together is 78 for AE. So is its half 39 for AD or the length of the ladder DC. And if one subtracts BE 3 from [15] DE 39, there remains 36 for DE, the height of the window.

Question 10

From the triangle ABC shown below, the sides make AB 13, BC 14, AC 15, in which BC is extended to D, and BD makes 10. Also, from D is extended a line through E into F, so that BE makes 5. The question then is, what will make AF and FC each by itself?

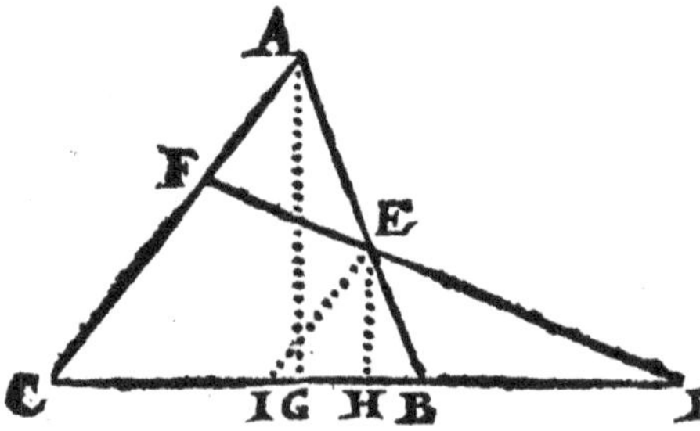

Figure 2.14: Question 10

Say, given AB 13 and EB 5, what shall give me BC 14? One shall arrive at $5\frac{5}{13}$ for BI. Now find EI as follows: say AB 13 gives me EB 5, what shall give me AC 15 and one arrives for EI at $5\frac{10}{13}$. So is triangle EBI of the same proportion as ABC and is also the triangle DEI of the same proportion as the triangle DFC. Therefore say now: DI $15\frac{10}{13}$ gives me EI $5\frac{10}{13}$, what will give me DC 24? And one shall arrive at 9 for FC, so is AF 6. One can then also easily find DE and EF.

Question 11

Look at the following figure.

There is a triangle ABC, AB making 26, BC 30, and AC 28, of which the line AB is extended to E, and if one would draw a line from E to D, knowing to be ED, so will the area of the triangle BED be equal to the area of the triangle ABC. The question is, if BD makes 18 how much will make BE and ED?

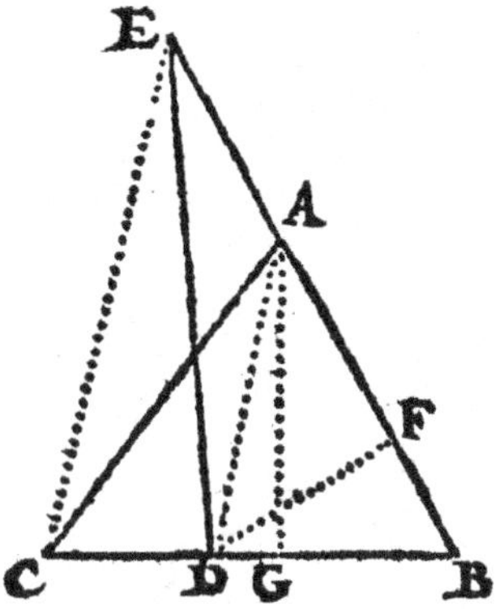

Figure 2.15: Question 11

Here, notice that if the triangles ABC and EBC are of the same size, then line AD must also be parallel to line EC, and because triangles AED [16] and ACD are equal and stand on the same base AD, so must triangles ABD and EBC be of equal proportion. Therefore now say: BD 18 gives me BA 26, what shall give BC 30? Arriving at $43\frac{1}{3}$ for BE. Hence, now proceed as [in] our first question, how much will be BG and one shall arrive for BG $13\frac{1}{5}$. Having found this, we thus say: AB 26 gives me BG $13\frac{1}{5}$, what shall give me BD 18? And we shall arrive at $9\frac{9}{65}$ for BF. This subtracted from BE $43\frac{1}{3}$, there remains $34\frac{38}{195}$ for FE. The square of BF subtracted from the square of BD, there remains for the square of FD $240\frac{2064}{4225}$. Added to this the square of FE, together one arrives at $1409\frac{1183}{1522}$ for the square of DE. So DE is $\sqrt{1409\frac{1183}{1522}}$.

Question 12

Look at the following figure.

Having positioned a mirror at a specific location (as in this example in C), to measure the height of the tower AB. I then retreated backwards on the straight line CB, 8 feet from C to D, so that I saw the top A of the tower above the stick DE. And without moving the stick I retreated backwards another 9 feet onto F, so that I could see the top A of the tower above the stick ED. The question then is, if the stick ED measures 6 feet, how many feet is the height of the tower?

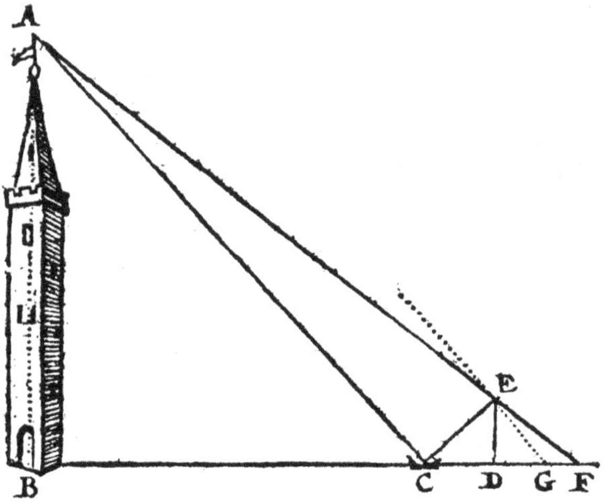

Figure 2.16: Question 12

To find the height of this tower, I subtract CD 8 from DF 9 and there remains 1 for GF. Now say: GF 1 gives me ED 6 feet, what shall give me CF 17? And one shall arrive at 102 feet for the height of the tower [17] AB. The reason for this result is that the triangle CDE is of the same proportion as ABC and the triangle EDF [is] of the same proportions as ABF. Therefore, if we now make EDG equal to CDE, so is EGF of the [same] proportion as ACF because EG is parallel to AC, so that GF stands to ED as CF stands to AB, the height of the tower.

Question 13

There is a triangle ABC in which AB makes 13, BC 14 and AC 15, and in which the line AD is drawn such that the angle BAC is divided into two equal parts. The question is the length of such line.

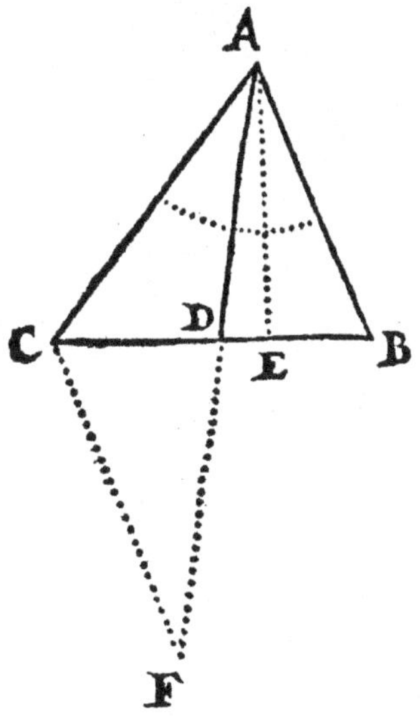

Figure 2.17: Question 13

Draw CF parallel to AB and extend AD unto F. Then CF must be equal to AC since the triangle [18] BDA is of the same proportion as the triangle CDF, and the angle DFC is equal to to DAC. Therefore, the line CF is equal to AC and FC stands to CD as AB to BD. So add FC 15 to AB 13, [which] together is 28. This gives me 14 for BC, what shall give me AB 13? One shall arrive at $6\frac{1}{2}$ for BD. Or, what shall give me AC 15? One shall arrive at DC $7\frac{1}{2}$. Now look for BE and EA as in our first question and one shall arrive at BE 5 and AE 12. Subtract BE 5 from DB $6\frac{1}{2}$, there remains $1\frac{1}{2}$ for ED. Adding this square to the square of AE, one shall arrive at $146\frac{1}{4}$ for the square of AD and so AD is $\sqrt{146\frac{1}{4}}$.

Question 14

I have a triangle ABC, with B being a right angle, of which is known to me the sides AB and BC, which together make 17 and AC alone makes 13. I desire to find how much AB and BC make each in particular.

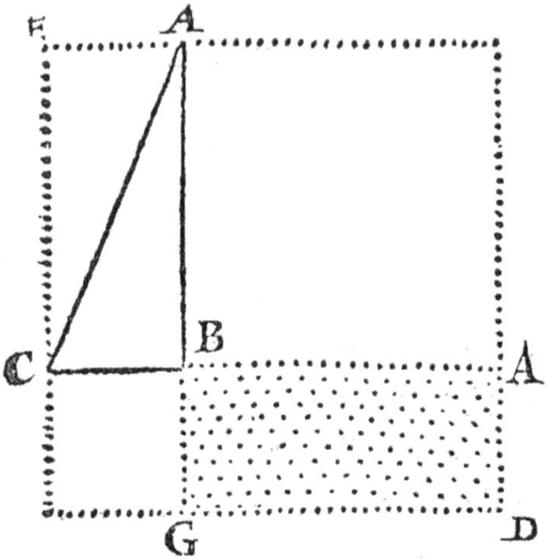

Figure 2.18: Question 14

Make a square of AB and BC (together being 17), the area of such a square then is 289. Now, also multiply 13 AC as a square, arriving at 169 for the area of both the squares of AB and BC, because by 47th prop. [19] of Euclid I, the squares of AB and BC together equal the square of AC alone. Subtract this 169 from 289, there remains 120 for so much the square of the line ABC is greater than the sum of the square of AB and the square of BC together. Half of 120 is 60 for the part $ABGD$ or for $ABCE$, which parts are double the size of triangle ABC and are the mean proportional between the square of AB and the square of BC. Now seek how much CB and BA are by themselves, following our fifth question, and you shall find 25 for the square of CB and for [the square of] BA 144. So the line AB is 12 and BC 5.

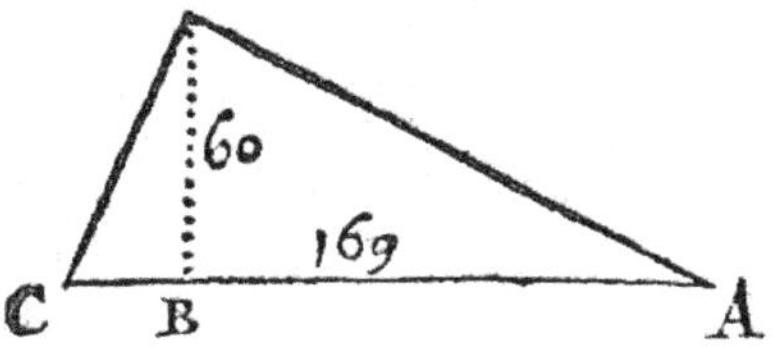

Figure 2.19: Question 14, explanation

[Otherwise]

Or else, one takes the aforementioned part $ABCE$, being 60, and so is BC 17 and EF its half $8\frac{1}{2}$. The square of EF then is $72\frac{1}{4}$ of which 64 subtracted, there remains $12\frac{1}{4}$ for the small square G. The $\sqrt{\text{℥}}$ is $3\frac{1}{2}$ for the sides, thus for HF, which we subtract from EF $8\frac{1}{2}$, there remains 5 for AH, or for EA or for BC. But adding the $3\frac{1}{2}$ to $8\frac{1}{2}$, one arrives at 12 for AB.

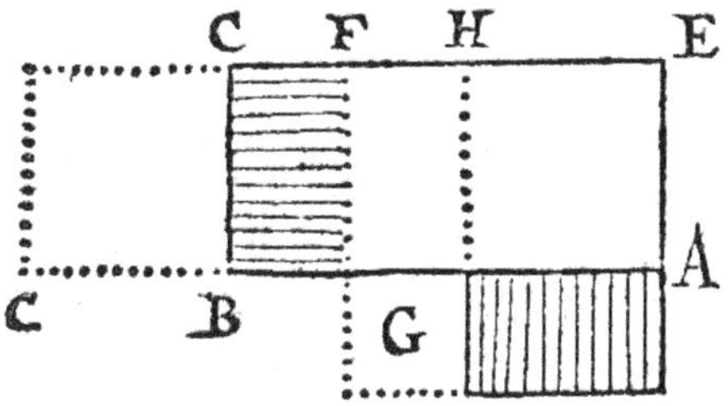

Figure 2.20: Question 14, alternative solution

Question 15

Look at the following figure.

There is a given triangle ABC right-angled in A, whose sides together are 120 and its area is 600. The question is what are the sides each apart?

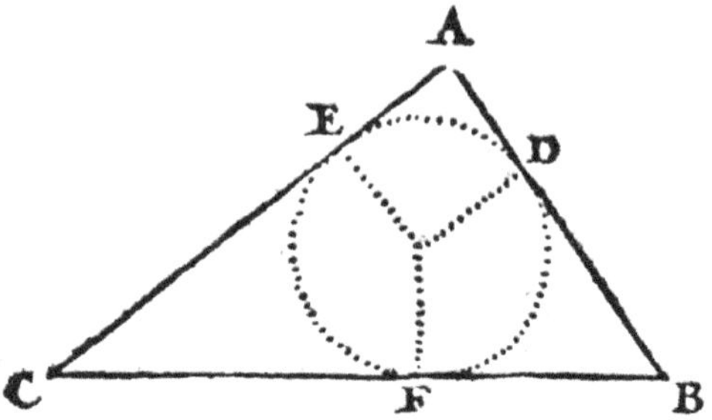

Figure 2.21: Question 15

Double 600, arriving at 1200 and this divided by 120 [20] makes 10 for AD or AE. Hence, DAE is 20. This, subtracted from 120, there remains 100 for $DBCE$. Its half is 50 for BC as much as is also BD and EC together. Then subtract this 50 from 120, there remains 70 for AB and AC together. And to find AB and AC each in particular, proceed as we have learned from in fourteenth example: square half of 70, being 35, makes 1225, subtract from this 1200, there remains 25. The $\sqrt{\mathfrak{z}}$ from this is 5, this subtracted from 35, there remains 30 for AB. And added to this 35, this makes 40 for the side AC, and as such all three sides are known, as one required to know.

Question 16

In the triangle ABC here shown, B is a right angle, of which the sum of all three sides is known, being $4+\sqrt{8}$. The question is, if the sides AB and BC are equal, how much then is each side in particular?

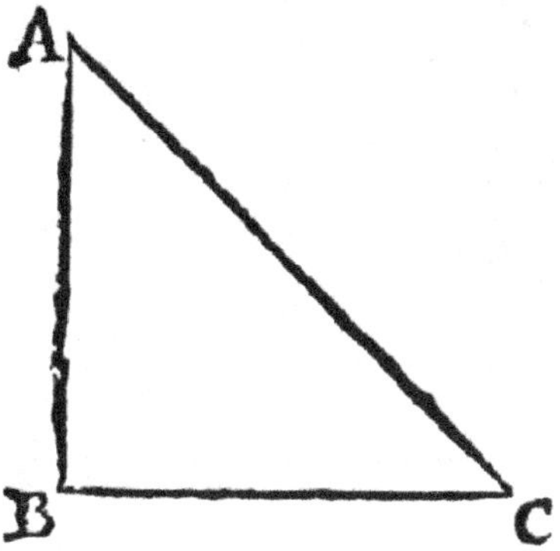

Figure 2.22: Question 16

Pose that AB and BC each are 1, then AC is $\sqrt{2}$. The sum of these three sides then is $2+\sqrt{2}$. Now, $2+\sqrt{2}$ stands to $4+\sqrt{8}$, as 1 stands to the side AB or BC. Thus say: $2+\sqrt{2}$ gives me $4+\sqrt{8}$, what shall give me 1? And one shall arrive at 2 for AB or BC, so then AC is $\sqrt{8}$.

Question 17

I placed a staff perpendicular in the ground at a certain place (as in this example) at point C [21] (here drawn as CD), 20 feet above the earth, and I found that the sight above the staff to the top of the tower AB, comes to E, with EC being 25 feet. Then, I have found that when the staff was four feet shorter as from C to G, that the sight above such staff to the top of the tower would come to F. The question is: if the staff is placed from F to C over a distance of 17 feet, how many feet is the tower AB high?

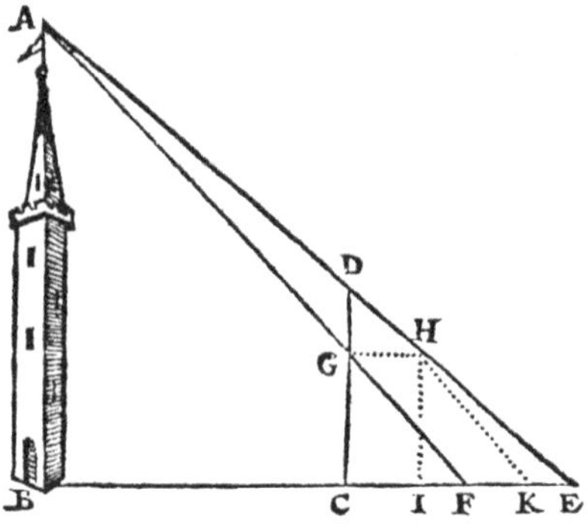

Figure 2.23: Question 17

To find this, thus say: CD 20 gives me CE 25, what shall give me GD 4? One shall arrive at 5 for GH or FK. This subtracted from FE, there remains 3 for KE. Now say: KE 3 gives me 16 for HI or GC, what shall give me FE 8? And one shall arrive at $43\frac{2}{3}$ feet for the height of this tower AB, as one required. The distance from B to E, or from A to E and from A to F can also be known, so one desires. [22]

Question 18

There is a triangle ABC right angled in B, in which there is drawn a line DE parallel to AB, so that BD makes 4 feet and that the area of $ABDE$ makes 30. The area of the remainder of the cut-off triangle EDC makes 24. I ask for the three sides of such triangle.

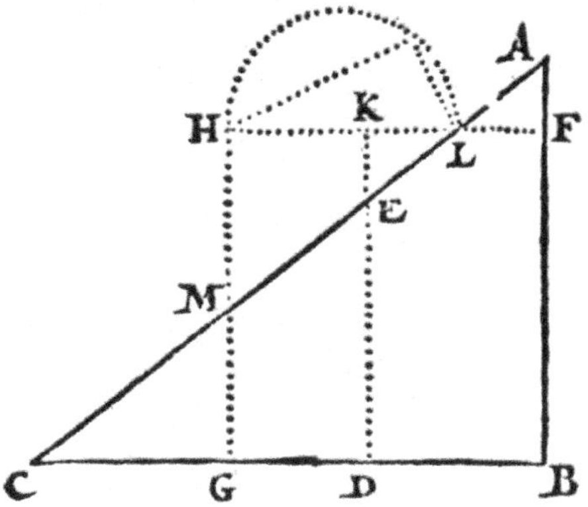

Figure 2.24: Question 18

Bring this part $ABDE$ into a quadrangle whom's breath is BD, the length then will be BF or DK, so that BF or DK makes $7\frac{1}{2}$, which becomes known as the area of $ABDE$ is known (being the same as $BDKF$) and the breadth BD. Therefore, divide 30 by 4 and one arrives at $7\frac{1}{2}$ for BF or DK. Also, make $KDGH$ equal to the triangle EDC, being 24 which is done as such: divide 24 for KD by $7\frac{1}{2}$, one arrives for DG or KH at $3\frac{1}{5}$. So the quadrangle $BFHG$ then is equal to the triangle ABC and the triangles AFL and GCM together are equal to the triangle LHM by itself. And as these triangles AFL, LHM and MGC have equal angles, they therefore are of the same proportion. Now halve BD 4, one arrives at 2 for FL or LK, then KH makes $3\frac{1}{5}$, making together $5\frac{1}{5}$ for LH. Subtract the square of FL 2 from the square of LH $5\frac{1}{5}$, there remains the square of GC, the $\sqrt{\mathfrak{z}}$ then is $4\frac{4}{5}$ for the line GC. Add this to DG $3\frac{1}{5}$ and to BD 4 and this makes together 12 for BC, its half being 6. Divide 54, the area of triangle ABC by this, so AB makes 9 and AC then is also known as 15 feet by the 47th proposition of the first book of Euclid.

Otherwise

Add 30 and 24, making 54, this divided by 24 makes $2\frac{1}{4}$, so many times is the triangle ABC larger than the triangle EDC. From this $2\frac{1}{4}$ take the $\sqrt{\mathfrak{z}}$, making $1\frac{1}{2}$ and so many times is line BC larger than DC. Now let DC be 1, so is BC $1\frac{1}{2}$ and subtract 1 from $1\frac{1}{2}$ there remains $\frac{1}{2}$ for what BD would be when DC equals 1. Therefore say thus: $\frac{1}{2}$ stands to 4 as 1 stands to DC, so one arrives at 8 for DC. So BC is 12 as one required.

Question 19

There is a triangle ABC of which AC equals CB, in which there is drawn the line CD with length $\sqrt{106}$ so that DA makes 10 and BD makes 12. The question is the size of perpendicular CE?

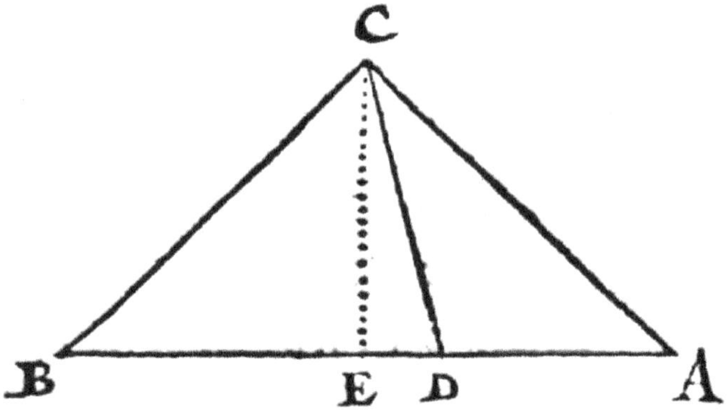

Figure 2.25: Question 19

Add 10 and 12 so that the whole line AB is 22. Its half is 11 for AE or EB. From this [subtract] 10, there remains 1 for DE. Subtracting the square of 1 from the square DC, there remains 105 for the square of EC. The $\sqrt{\mathfrak{z}}$ from this is $\sqrt{105}$ for the line EC. So are the sides AC and BC also known by the 47th propos. of the first of Euclid.

Question 20

Look at the following page.

In a triangle ABC, in which AB makes 10 and AC makes 12, a line AD is drawn on the middle [24] of line BC at D. The question is, if AD makes $\sqrt{106}$, how much will then make the parts BD or DC from line BC?

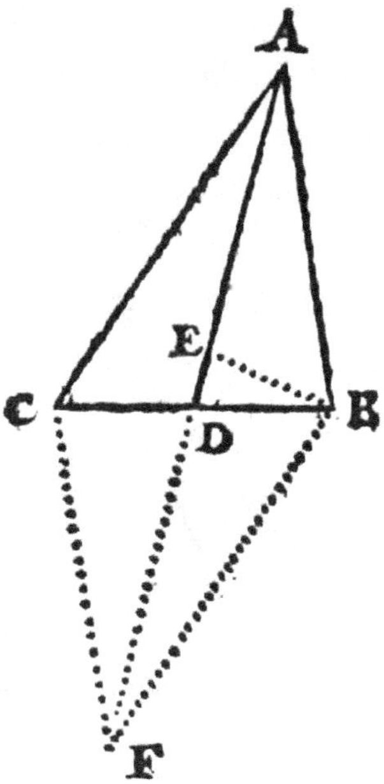

Figure 2.26: Question 20

Make the line AD as long as ADF, so is AF $\sqrt{424}$. BF now becomes equal to AC 12 because BD and DF are equal to DC and DA [respectively], with equal angles. We thus have a triangle AFB of which the three sides are known, being AF $\sqrt{424}$, AB 10 and BF 12, so that can also be known, by our first example, AE, EF and EB, that EA becomes $\sqrt{85\frac{15}{106}}$, EB $\sqrt{14\frac{91}{106}}$. Subtract EA $\sqrt{85\frac{15}{106}}$ from AD $\sqrt{106}$, there remains $\sqrt{1\frac{15}{106}}$ for ED. The square of ED added to the square of EB $\sqrt{14\frac{91}{106}}$ arrives at 16 for the square of BD. The $\sqrt{ʓ}$ of this is 4, for the line BD or for DC, so that the whole of the base BC is 8.

Question 21

Of this given triangle two sides are known, making AB 13, BC 15, and in angle A a line is drawn to the point D on base BC, the length being $\sqrt{162\frac{7}{9}}$, so that CD is $\frac{1}{3}$ of the whole of the base. One desires to know the length of the whole of the base BC.

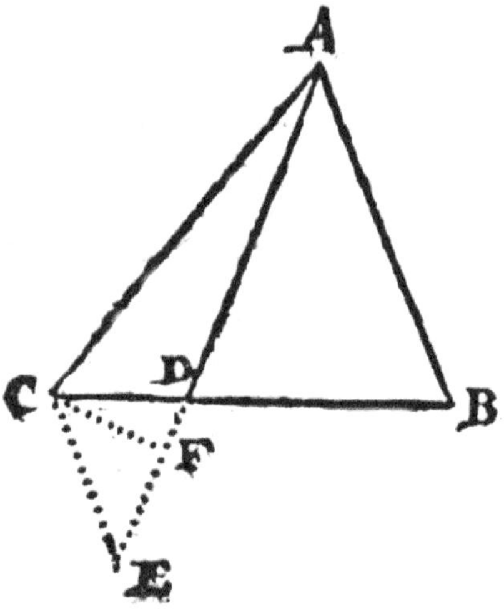

Figure 2.27: Question 21

Extend AD as much as when a line CE is drawn from C parallel to AB, that the extension of AD cuts this parallel line [25] in E by which the triangle DCE becomes of the same proportion as the triangle DBA. So stands BD to DC as 2 to 1. Also is AB to CE what AD is to DE. Therefore say: 2 gives me 1, what will give me 13 AB, arriving at CE $6\frac{1}{2}$. Likewise, what shall give me $\sqrt{162\frac{7}{9}}$ AD, arriving at $\sqrt{40\frac{25}{36}}$ for DE. Add $\sqrt{40\frac{25}{36}}$ to $\sqrt{162\frac{7}{9}}$ arriving together at $\sqrt{366\frac{1}{4}}$ for AE. Now, look for FE as in our first example, arriving at $\sqrt{22\frac{5769}{5860}}$, so is also known CF, making CF $\sqrt{19\frac{389}{1465}}$. Next, subtract EF from ED, there remains for FD $\sqrt{2\frac{6754}{13185}}$. This square of FD added to the square of FC arrives at the square DC $21\frac{7}{9}$, the $\sqrt{\mathfrak{z}}$ of which is $4\frac{2}{3}$ for the line CD. So is BD $9\frac{1}{3}$ and the whole of BC is 14.

Otherwise

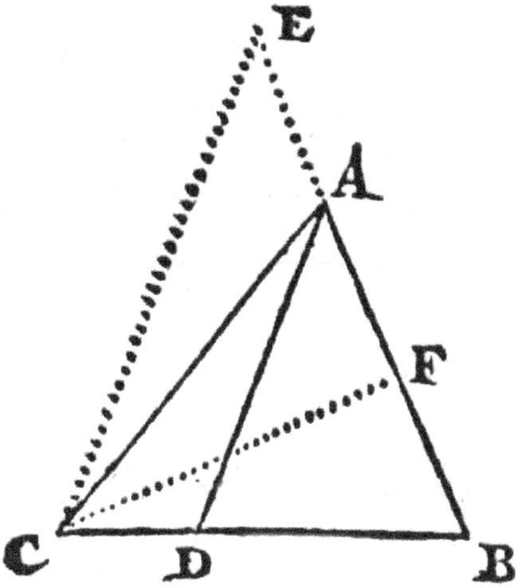

Figure 2.28: Question 21, alternative solution

Extend BA to E as in the figure [2.28], as long as the line CE drawn parallel to DA cutting the line BA at point E. So is the triangle BEC of the same proportion as BAD. Then make AB 13, so shall BE make $19\frac{1}{2}$, or AE $6\frac{1}{2}$, because DC is to BD as 1 is to 2. So is also AE $6\frac{1}{2}$ to AB 13. While three sides of the triangle EAC are known, so can be found AF and FC by our second example, and we shall arrive for AF $7\frac{8}{13}$. So is FB $5\frac{5}{13}$. The square of FB is $28\frac{168}{169}$, this added to the square of FC being $167\frac{1}{169}$ arriving together at 196 for the square of BC. The $\sqrt{\mathfrak{z}}$ of this is 14 for the line BC, as one required. [26]

Question 22

There is a triangular piece of land ABC of which are known all 3 sides, AB makes 26, AC 40, BC 42, and D being a pit so that from the pit to B is $\sqrt{512}$ and from the pit to C is $\sqrt{932}$ rods. Now, one wants to delve a ditch from the pit to the corner A. The question is: how many rods will the ditch be, being AD?

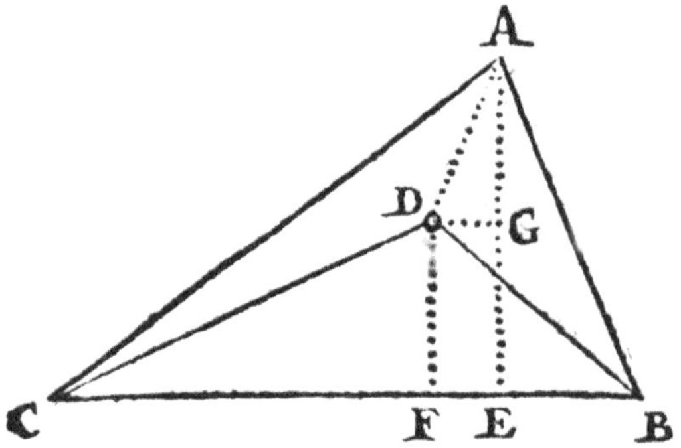

Figure 2.29: Question 22

Looking for BE as in our first example for how much the cathetus AE will be, one shall arrive at BE 10, and for AE 24. Then look for BF 16 and for DF 16. Now subtract BE from BF, remaining 6 for EF, as much as is GD. Also, subtract FD from EA, remaining 8 for GA. The square of this added together with the square of GD arrives at 100 for the square of AD. So the line AD is 10, the length of the ditch.

Question 23

There is a triangle of ABC of which AD makes 14, BC 15, which is divided into three other triangles by the lines DA, DB and DC. The question is how much the side AC is if AD makes $\sqrt{101\frac{1}{4}}$, DB $\sqrt{45\frac{1}{4}}$ and DC $8\frac{1}{2}$?

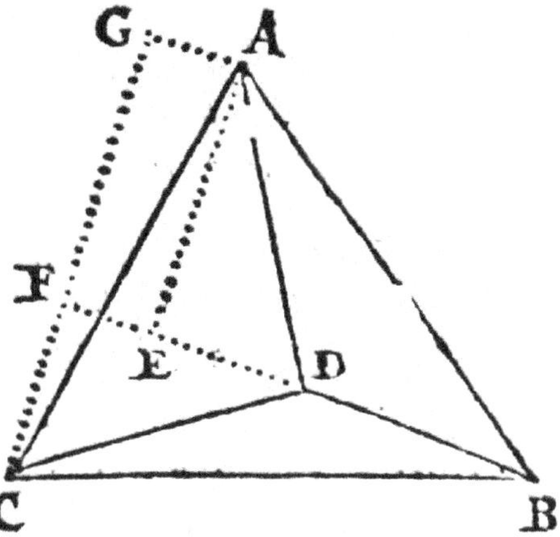

Figure 2.30: Question 23

First look for the lines BE and AE by our second example, one shall arrive at BE [27] $108\frac{52}{181}$ and for AE $\sqrt{87\frac{129}{181}}$. Also, by the same means, look for BF and CF, arriving at BF $\sqrt{216\frac{108}{181}}$ and for CF $\sqrt{8\frac{73}{181}}$. Add FG, being equal to AE, to FC arriving together at $\sqrt{150\frac{75}{181}}$ for the whole of the line GC. Now, subtract BE $\sqrt{108\frac{52}{181}}$ from BF $\sqrt{216\frac{108}{181}}$, remaining $\sqrt{18\frac{106}{181}}$ for EF or for AG. At last, add the square of AG to the square of GC, [which] together is 169 for the square of AC. The $\sqrt{\mathfrak{Z}}$ of this is 13 for the line AC as one required to know.

Question 24

Someone owns a triangular piece of land, as for example ABC, of which the sides are known: AB makes 13, BC 14 and AC 15 rods. Now, one desires to known how many rods the length of ditch EF is (if B to C measures 4 rods, knowing that one digs a straight ditch from E to F, so that AF makes 5 rods)?

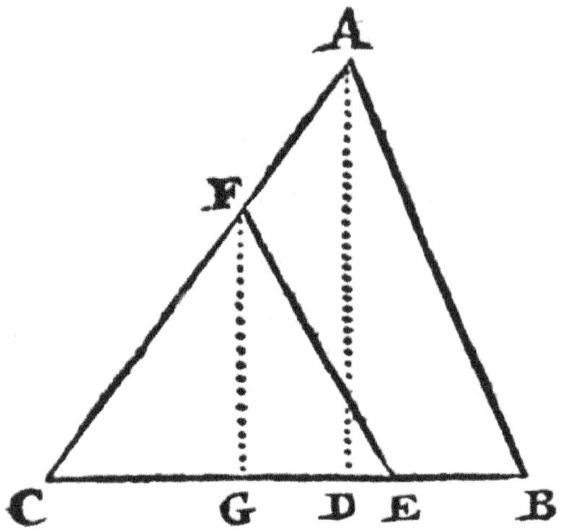

Figure 2.31: Question 24

First, look for the altitudes AD and DC as in our first example. You shall find for DC 9 and for AD 12. Then say as thus: AC 15 gives me AD 12, what shall give me CF 10? One shall arrive at 8 for GF. Also say: CA 15 gives me DC 9, what shall give me FC 10? Arriving at 6 for GC, so GC and BE together is 10. This subtracted from BC 14, there remains 4 for EG. Now add the square of EG to the square of GF which makes together for the square of EF 80, so is line EF or the length of the ditch $\sqrt{80}$ as one required. [28]

Question 25

In the triangle ABC shown, right-angled in C, the side BC is known, making $10 - \sqrt{20}$ and the perpendicular from C falls on base AB at point D. Also: that BD stands to AD as AD to the whole line AB. The question is how much are the sides AB and AC, as well as the parts of the base AD and DB?

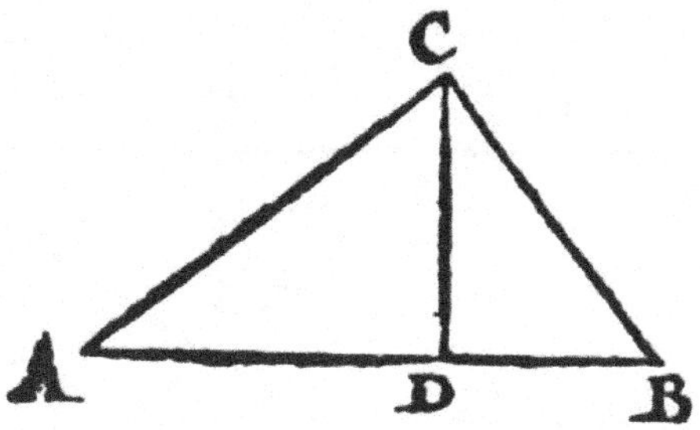

Figure 2.32: Question 25

If making a square of line AB, as in the second figure $ABCF$ and cutting off a part of this square, as for example $DMOC$ which is as big as the square that can be made from AM or OB, being $OBKL$, such line AD or BC will thus be cut into two equal parts, as OC and OB, such that OC stands to OB as OB stands to the whole of the line BC, as being proven by the eleventh proposition of the second book of Euclid, and by the 30th of the 6 book, because OL and OB are equal and NL stands to LO as OB stands to BA.

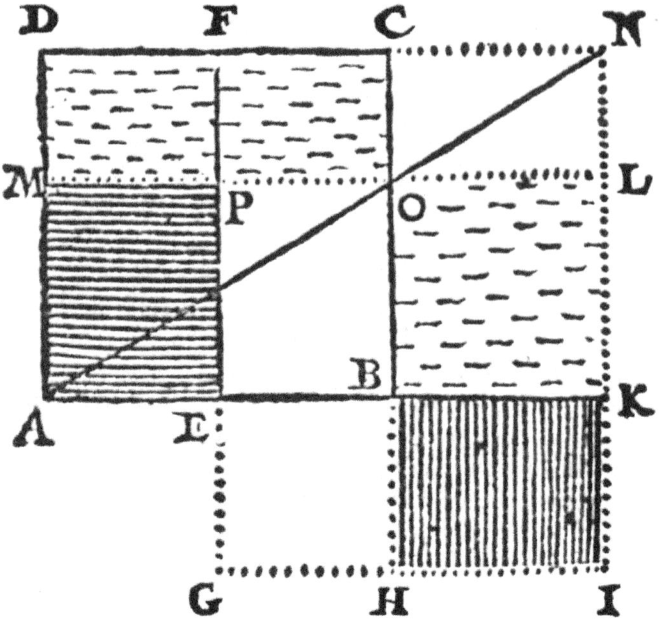

Figure 2.33: Question 25: solution

Let use now pose 2 for AB, the side of the square. So is EB 1. The square of EB, being $EBHG$ then makes 1 and the large square $ABCD$ then makes 4. The two parts $EBOP$ and $HBKI$ together with the square $BOLK$ then also makes 4 together. To this [29] add the square $EBHG$, being 1, which is together 5 for the square $GPLI$. So is line GI $\sqrt{5}$ as much as is also EK. From this EB 1 subtracted, there remains for BK, or for BO $\sqrt{5}-1$. This subtracted from the whole line AB or BC so there remains for OC $3-\sqrt{5}$. And now also, CO $3-\sqrt{5}$ stands to BO $\sqrt{5}-1$ as BO $\sqrt{5}-1$ stands to the whole line CB 2. Thus say: $\sqrt{5}-1$ gives me BO $10-\sqrt{20}$, what shall give me 2? And one shall arrive at $\sqrt{80}$ for AB, the base of the triangle.

Now subtract the square of CB from the square of AB, there shall remain $\sqrt{8000}-40$ for the square of BC. Also, because BD stands to BC as BC stands to AB, AD is therefore equal to BC and thus makes $10-\sqrt{20}$ and DB will then be $\sqrt{180}-10$.

Otherwise

As CB is known to me, making $10 - \sqrt{20}$, so is also AD $10 - \sqrt{20}$. As BD stands to BC so does BC stand to the whole line AB. While DB stands to AD as (according to the enunciation) AD stands to the whole line AB.

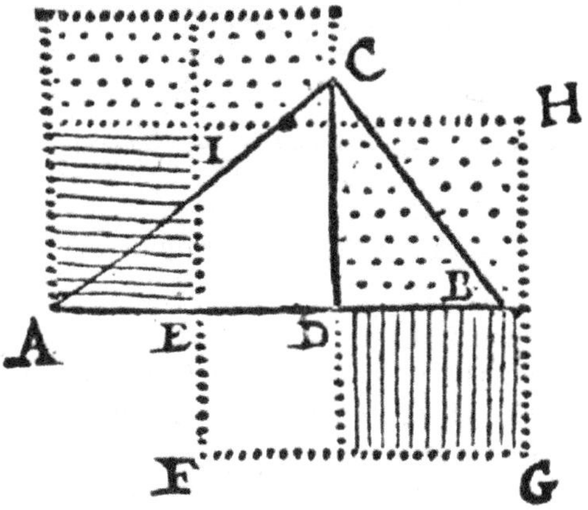

Figure 2.34: Question 25: alternative solution

From this follows that AD must be equal to BC. Let us pose that the half of AD, being $10 - \sqrt{20}$, arriving at $5 - \sqrt{5}$ for ED, this multiplied as a square, arriving at $30 - \sqrt{500}$ for the square of ED. To this add the area of square AD, being $120 - \sqrt{8000}$, making together $150 - \sqrt{12500}$ for the area of the square of EB, which is $FGHI$. The $\sqrt{\mathfrak{z}}$ of of this then is this number $\sqrt{1255} - 5$ for the side of such square, or for the line EB. From this subtract [30] at last ED $5 - \sqrt{5}$ so there remains DB $\sqrt{180} - 10$, or add EB $\sqrt{125} - 5$ to ED which is equal to EA $5 - \sqrt{5}$, arriving together at $\sqrt{80}$ for AB as above.

Question 26

There is a triangle ABC of which the sides are known, making AB 10, BC $\sqrt{640}$ and AC 30, and if one extends line AB 5 to F, and CA 11 to E and one draws a line FG from F and EG from G so that AEG and AFG both have right angles and that they cross each other at point G. Question: how much will make FG and EG in particular?

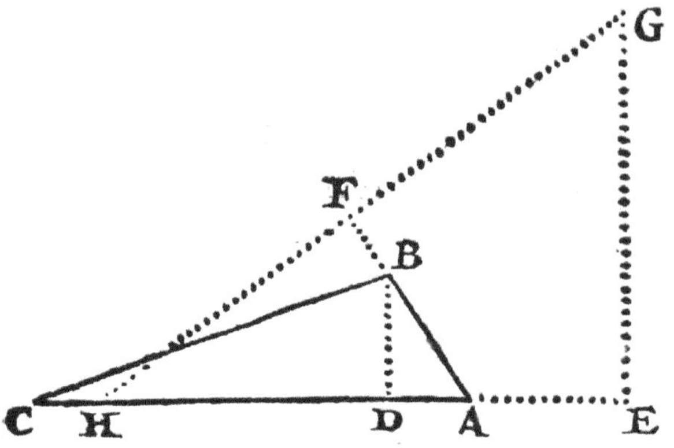

Figure 2.35: Question 26

First look for AD and the perpendicular BD as in our first example. Arriving at AD 6 and for the perpendicular BD 8. Now extend GF to H, so that the triangle AHF is of the same proportions as the triangle GEH, as also the small triangle ADB. Therefore say: AD 6 gives me AB 10, what shall give me AF 15? One shall arrive at 25 for AH. To this add AE 11, which is together 36 for EH. Also say: AD 6 gives me DB 8, what shall give me AF 15? Arriving at FH 20. And to find EG say: BD 8 gives me AD 6, what shall give me EH 36? Arriving at 27 for EG. Also: BD 8 gives AB 10, what shall give EH 36? Arriving at 45 for GH. From this subtracted FH 20, there remains 25 for GF and thus are found both lines FG 25 and EG 27 as one required. [31]

Question 27

There are two walls AB and CD at a distance of 20 feet from each other, which I will represent in the next figure by BC. At AB stands a ladder touching [the wall] perpendicularly above the earth at 16 feet. The feet of the ladder is drawn at E in which point another ladder is placed, this one leaning against the other wall DC at a height of 6 feet above the earth so that AED makes a right angle. The question is: at how many feet are the ladders placed in point E from both sides, being: how much are EB and EC? Also, how many feet are the ladders AE and ED of length?

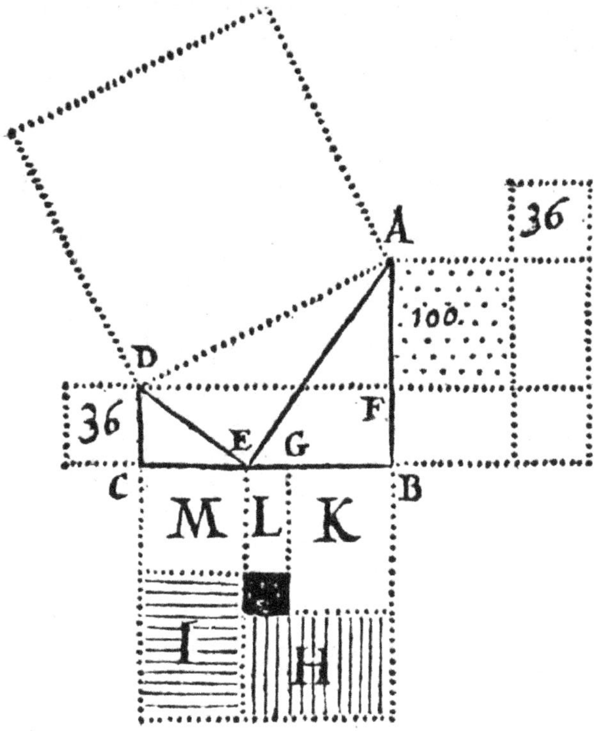

Figure 2.36: Question 27

The squares of AB and BE together with the squares of EC and CD have all 4 together the same area as the square of AD alone, by the 47th propos. of 1 of Euclid. Also the squares of AF and BC are together as much as that of AD alone. From this then follows that these two last squares of AF, BC are as much together as the first 4 squares from AB, BE, EC and CD. If one subtracts the square of AF from the square of AB so must be the remainder together with the three [32] squares of BE, EC and CD be as much as the square of BC and aforementioned remainder with the small square of CD then be as much as the two longer parts H and I. Also, in the square of BC should be comprised four such longer parts, two of which are represented by the letters H and I and the others by K and LM. As such can the area of these four equal parts be known. One doubles 16, arrives at 32, this multiplied by BF, being 6 as is CD, arriving at 192 for the two parts H and I. This doubled becomes 384 for the parts H, I, K, and LM. Now, the square of BC is 400, from this subtracting 384 there remains 16 for the square of GE. The $\sqrt{ȝ}$ of this is 4 for GE. This subtracted from BC 20, there remains 16, its half is 8 for BG or EC. So is EB 12. The length of the ladders are known by the 47th propos. of 1 of Euclid. Arriving at a length of AE 20 and ED 10.

Otherwise [1]

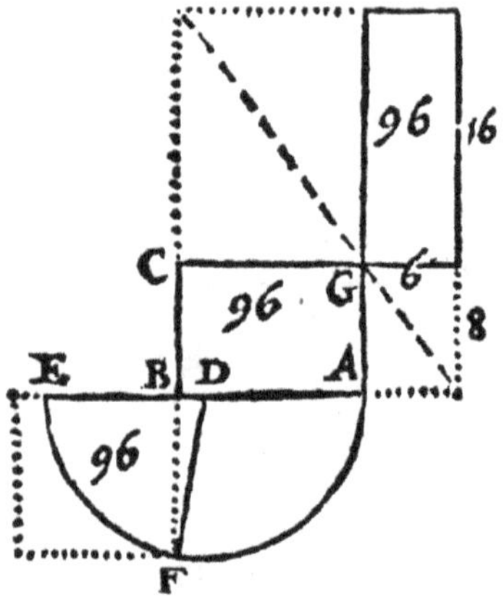

Figure 2.37: Question 27: alternative solution

The line AB 16 stands to BE as EC stands to CD 6. The multiplication of AB 16 by CD 6 is therefore equal to the multiplication of BE by EC, each multiplication arriving at 96. Now, while BE and EC together make 20, so much as the whole of BC which, in the figure below, we represent by AB and BC, so that AB and BC together make 20. Therefore, make BE equal to BC, so that here AE is 20, its half then is 10 for AD, as much as is also DF or DE. [33] This multiplied as a square arrives at 100 for the square of DF. From this [subtract] the square of BF (being 96, equal to the quadrangle $ABCG$ (BE previously being made equal to BC) so is the square of BF equal to $ABCG$ by the 17th proposition of 6 of Euclid), so there remains 4 for the square of DB. So line DB is 2. This added to AD 10, one arrives at 12 for AB, or subtract 10 from DE, there remains 8 for BE or BC, and so we have found AB 12 and BC 8, which are BE and EC in the previous figure, which we had to find.

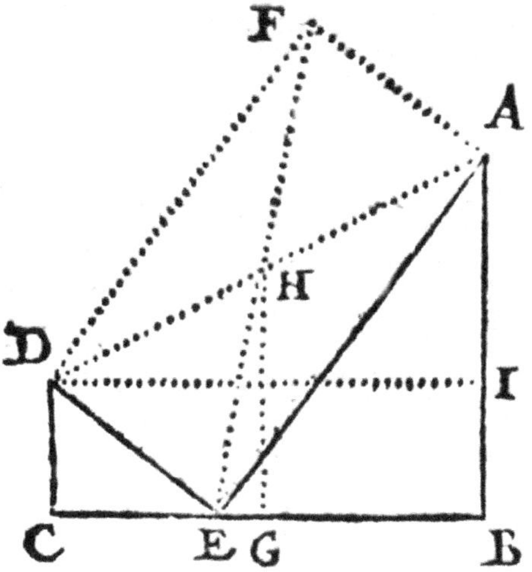

Figure 2.38: Question 27: second alternative solution

Otherwise [2]

Let (as in the previous figure) one wall be AB and the other DC, EA the one ladder and ED the other ladder, BC the distance and AED a rectangle. This makes AD or FE the diagonal of the quadrangle $AEDF$ with H as the center point. Now, subtract CD 6 from AB 16, there remains AI 10 and ID is 20, as much as the distance between the wall BC, of which AD is known by the 47th proposition of book 1 of Euclid. It makes $\sqrt{500}$, the half of AH or HD then is $\sqrt{125}$, as much as is also HE. Now, add DC 6 to AB 16, arriving at 22, its half is 11 for HG. As H is the center of AD and G the center of BC, and also HG and HE are both known, so is GE also known. GE then makes 2, this subtracted from GC 10, there remains 8 for EC. Or added to BG, arriving at 12 for BE.

Question 28

A stonemason has placed a ladder at a certain place to the wall of a room, as in our example at point C, touching the wall AB at a height of 20 feet. But if he placed the same ladder against the opposite wall ED staying with its feet at the same point C, it will touch the wall at a height of 15 above the earth.

The question is, if these walls are separated on a straight line BCD by 35 feet, how many feet is the length of the ladder?

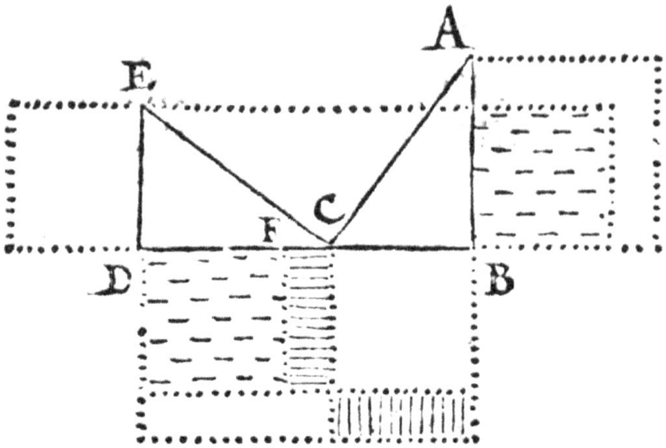

Figure 2.39: Question 28

The squares of AB and BC are together equal to the square of AC. Also, the squares of CD and DE are together equal to the square of CE. And as CE is equal to CA, the squares of AB and BC are together equal to the squares of CD and DF. If one subtracts the square ED from [the square of] AB, there remains a gnomon of 175 which is equal to the square of BC subtracted from [the square of] CD. So also there remains a gnomon of 175, as CD is larger than BC. Because, while [the squares of] ED and CD are equal to the squares of AB and BC, and because what AB is larger than ED, the square of CD must again be larger than [the square of] BC. So, now divide the given 175 by 35 BD, arriving at 5 for CF. This subtracted from 35 there remains 30 for BC and FD, which are equal. Thus halved, one arrives at 15 for BC and so is CD 20. [35]

Question 29

There is a triangle ABC, right-angled in B, of which the sides AB and BC make as much as one adding them together with the area, the sum is 154 and AC makes $\sqrt{605}$. The question is, how much each of the sides AB and BC make in particular?

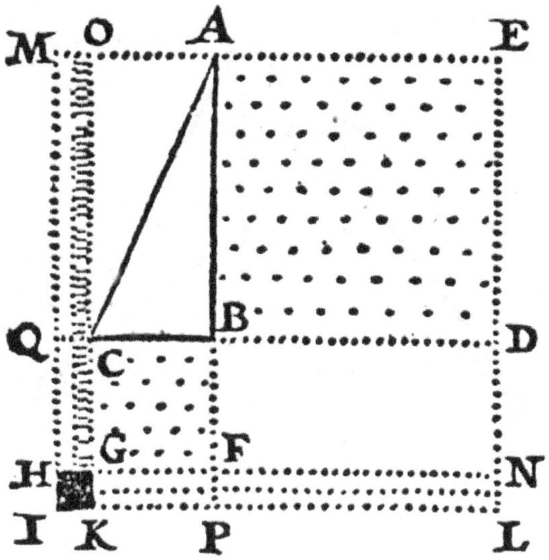

Figure 2.40: Question 29

Double 154, arriving at 308 for $ABCO$ and $OCGHQMO$ together. So much is also $BFND$ and $NFGKPLN$ because OP is 1 and OM 2. Now, the squares of AB and BC are together 605 by the 47th proposition of 1 of Euclid, because the square of AC is 605. Now, while line AC is known, being $\sqrt{605}$, so double 308, arriving at 616 and add this to 605. And to this add also the area of the small square $GHIK$, being 4, the sum is 1225 for the area of the large square $EMIL$. The $\sqrt{\mathfrak{z}}$ of this is 35 for ME, or for one of the other sides of this square. From this subtracted MO 2, there remains 33 for OE or DC and so much are the sides AB and BC of the proposed triangle together. As these together are known, now look how much AB and BC are in particular from our fourteenth example, and one shall find 22 for AB and 11 for BC.

Question 30

Look at the following figure.

Of this triangle ABC are known the three perpendicular lines, AD makes 60, BF 56 and CE $64\frac{8}{13}$. The question [36] is how much the sides of the triangle then make?

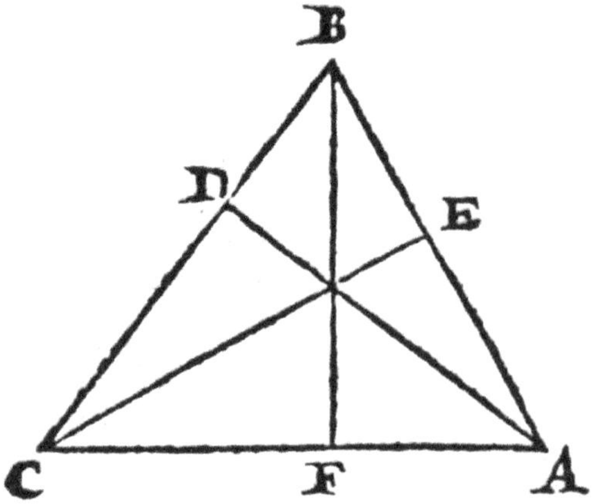

Figure 2.41: Question 30

As AD stands to EC also AB is to BC. Also, as AD stands to BF, also AC is to BC. Therefore say: BC $64\frac{8}{13}$ gives me BC 56, what shall give me AB in the same proportion? And one shall arrive at 52 for AB. The proportion of the three sides of this triangle are thus known to me: if AB makes 52, so is BC 56 and AC 60.

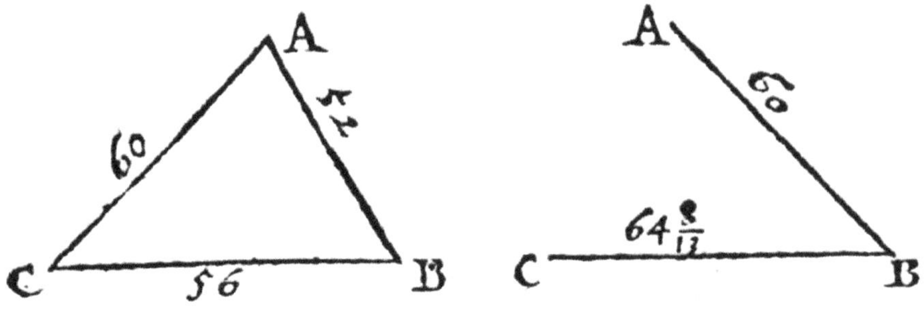

Figure 2.42: Question 30: solution by proportions

Now look for the size of the perpendicular AD in the same proportion, as in our first example. You will arrive at AD 48. Now say: AD 48 gives me AB 52, AC 60, BC 56, what shall give me AD 60? From what is given as known in our triangle ABC and we desire to know its sides, and we shall arrive at 65 for the side AB, for BC 70 and for AC 75.

Otherwise

Divide the three numbers of the known perpendicular lines 60, 56, $64\frac{8}{13}$ by any number as you like, let my choice be 1, so I arrive at $\frac{1}{60}$, $\frac{1}{56}$, $\frac{13}{640}$. So, you also have the proportion of the sides and BC is $\frac{1}{60}$, as AC is $\frac{1}{56}$ and AB is $\frac{13}{640}$. These numbers in their smallest whole then form the proportion, arriving at AB 13, BC 14 against AC 15, for the proportion of the sides [37] in whole numbers. Now, look for the perpendicular of such triangle with sides 13/14/15 as in our first example. So, from A in the same proportion to BC you will arrive at 12 for AD. Say now as we have learned before: 12 gives me the sides 13/14/15 what shall the sides be as that same perpendicular becomes AD 60? And you will arrive at the sides AB 52, BC 56, AC 75 as above.

Question 31

There is a triangle ABC of which the sides AB and AC together make 28. But the base BC makes 14 and the perpendicular AD makes 12. The question is, how much make the sides AB and BC each in particular?

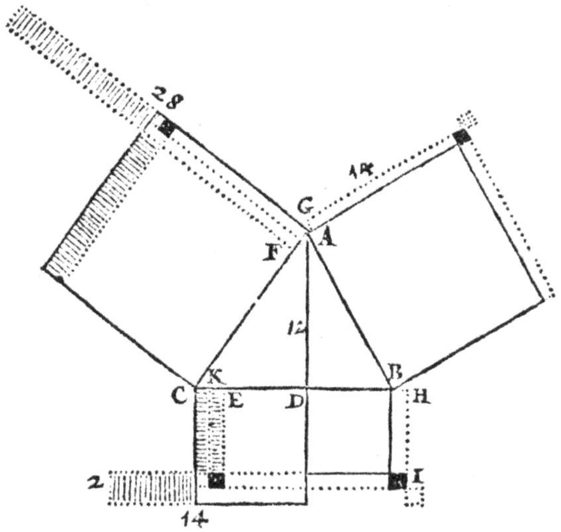

Figure 2.43: Question 31

As can be noticed from our first example, that the square of AC is as much larger than the square of AB as the square of DC is larger than [38] the square of BD. Furthermore, that the sides BA and AC together, being 28, stand in proportion to the base BC 14 as EC stands to AF, being 2 to 1. If one enlarges the square of AB by the half of AF, that is AG and the square of BD enlarges as much as the half of EC, which is HB, then such square of BG will be larger than the squares of HD and DA together as much as the small square of HB will be larger than that of AG. But since HB is 1 against EC 2 and AG $\frac{1}{2}$, the square of HB is 1, [the square] of AG is $\frac{1}{4}$ and the square of HB $\frac{3}{4}$ larger than the square of AG, and this in the same proportion as EC 2 is to AF 1. Also, BG makes 14 as the right half of 28 and its square 196, and AD 12 and its square 144, HD 7 because BC makes 14 and HB is the half of EC. Therefore HD 7 or DK 7. Also, that HK 14, is equal to BC, the square of HD is 49. To this add 144 from AD, together being 193. This subtracted from 196, the square of BG there remains 3, for what the square of BG is larger than that of HD and DA together. From this follows that the aforementioned $\frac{3}{4}$ for the square of HB is as much as 3 such whole square of HB to 4. The $\sqrt{\mathfrak{z}}$ from this is 2 for HB, so BD is 5 and DC 9. And by the 47th proposition of 1 of Euclid, as AD is known and it makes 12, then also AB is known and makes 13. And hence AC makes 15, by which each in particular has been found as one required.

Question 32

From the following triangle ABC, the perpendicular is known, AG makes 12 and the base BC makes 26. Question: how much that AB and AC shall make if AC stands to AB as 3 stands to 2?

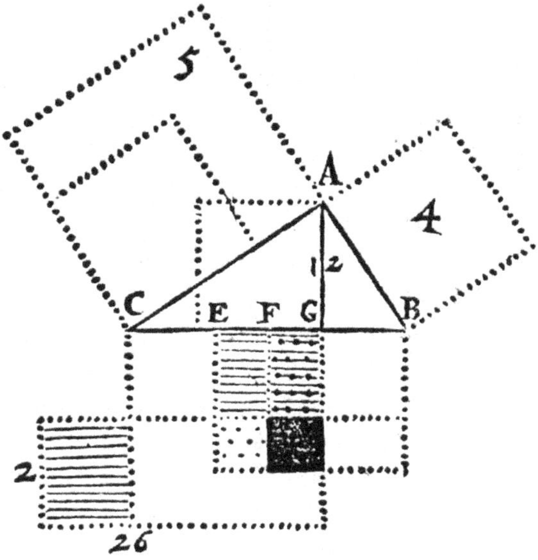

Figure 2.44: Question 32

Pose that GE is 2, so is GF 1, and as AC is to AB as 3 is to 2, the square of AC is to [39] the square of AB as 9 is to 4. Therefore, the square of AC is in the proportion of 5 more than the square of AB, as much as is then also the square of GC larger than the square of GB. So, then what the square of GC is larger than the of GB is of proportion 5 more if GE makes 2. Therefore say: if 5 is being changed to 52 (because 52 is two times 26) to what shall then be the square of AB, which is in proportion of 4, if the side AB makes 2 against AC 3, which is 5 more for the square of AC if AB and GE are 2. One shall then arrive at $41\frac{3}{5}$ for the square of AB.

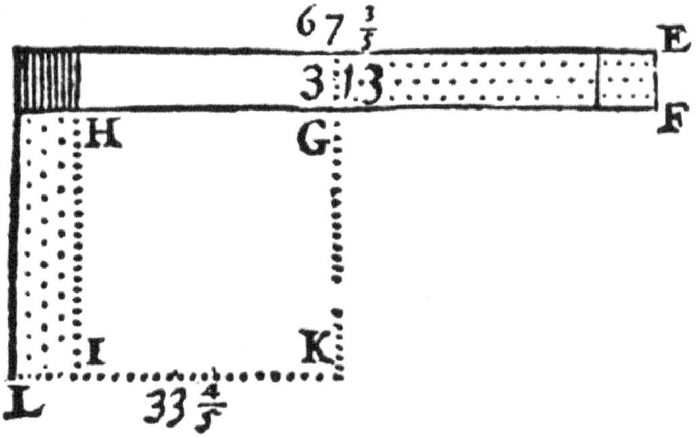

Figure 2.45: Question 32: solution

Furthermore, the square of AG is 144 and that of BF being 13, the half of BC is the square 169. These added together arriving at 313 for the squares of BF and of AG, and as much is also the square of AB when to this is added how much the square of BF is more than that of BG, which is proportion of 26 (if GF 1 and the square AB [40] is $41\frac{3}{5}$) minus as much as a small square of FE or FG. Therefore, to both sides adding such a small square, one arrives at 313, and the small square of FE is equal to the square of AB, and the square of AB is $41\frac{3}{5}$ times EF, and to this add 26 times EF, so much as the square BF is more than that of BG. Together is $67\frac{3}{5}$ times EF. At last take the half of $67\frac{3}{5}$, being $33\frac{4}{5}$, this multiplied as a square arrives at $1142\frac{11}{25}$. From this subtracted the 313, there remains $829\frac{11}{25}$ for the square $GHIK$. The $\sqrt{\mathfrak{z}}$ from this is $28\frac{4}{5}$ for KI. This subtracted from KL $33\frac{4}{5}$ there remains 5 for IL, or EF, or FG. So is BG 8 and GC 18 and AB will then be $\sqrt{208}$ / AC $\sqrt{468}$.

Question 33

From this triangle parts of its base are known, BD makes 80, DC 396 and the line DA makes 165. The question is: if the proportion of side AB against AC is as 25 to 33, then how much will the sides AB and BC be?

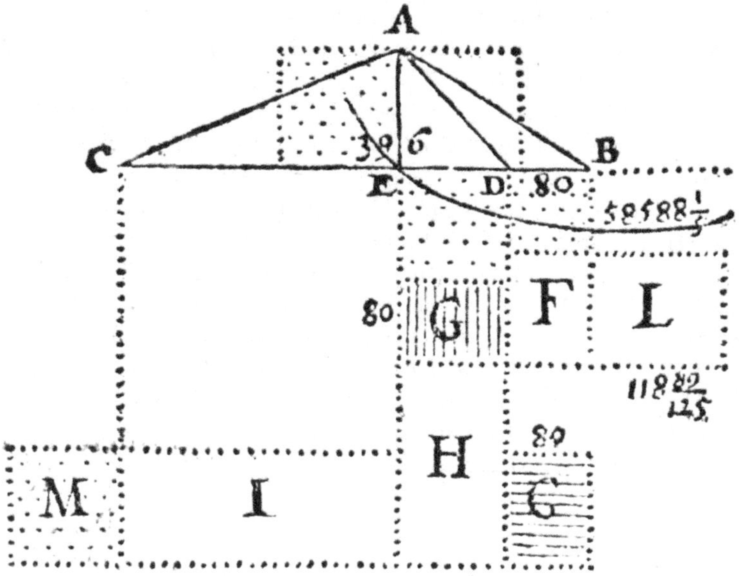

Figure 2.46: Question 33

The square of BD is 6400 and the square of DA is 27225. So the three squares of BD, DE and EA are together 33625, as BD is 6400 and DA 27225 [41] which is as much as [the squares of] DE and EA together. Thus [the squares of] BD, DE and EA are therefore together 33625. And if also the two parts designated by F and G are added to that, this together would be as large as the square of AB, as the square of BE and EA together are as much as [the square of] AB. Now side AB is to AC as 25 to 33. The square of AB thus stands to the square of AC as 625 against 1089, which is $1\frac{464}{625}$ times that the square of AC is greater than the square of AB. Let us then enlarge the square of BD, DE and AE (being together 33625) by $1\frac{464}{625}$ times. We shall arrive at $58588\frac{1}{5}$. Let us also enlarge both parts F and G which each 80 long, being as long as is BD. Together they are 160 long, which multiplied by $1\frac{464}{625}$ we shall arrive at a length of $118\frac{98}{125}$ for the part, here shown by the letter L, with the breadth remaining the same. So the enlargement of part L has been as much as that of the square AC. We further have to notice that the square of DC is 156816. So, if we add to this the squares of DE and EA which together are 27225 we arrive together with the square of DC at 184041 for the squares of EA, DC and M, with M being equal to the square of ED. So that the parts designated by G, H, I and N are what is more than the square of

AC. Therefore, of we also add the enlargement of AB to this, we shall arrive at $58588\frac{1}{5}$ for the parts L, F, G, D, E and the squares of G, H, I, M who all have the same breadth $1070\frac{98}{125}$ and are added together. Because L, F and G together are $278\frac{98}{125}$ and the square of DE to which the part G is added and H together have a length of 396, as long as DC. The three sums, $278\frac{98}{125}$, 396, 396 together makes $1070\frac{98}{125}$ long, as above. Also, the lengths $58588\frac{1}{5}$ and $1070\frac{98}{125}$ of which DE is the breadth are equal to the square of DC, EA and M or 184041, the area of DC, EA and M. At last, if we now subtract $58588\frac{1}{5}$ from each of both sides, there remain [42] the parts L, F, G, D, E, and the squares of G, H, I, M (whose lengths together are $1070\frac{98}{125}$ and the breadth DE) are still equal to $125451\frac{4}{5}$. Therefore, divide this by $1070\frac{98}{125}$ arriving at $117\frac{27}{169}$ for DE. So then BE and EC are known, BE makes $179\frac{27}{169}$ and EC $278\frac{142}{169}$. Also, now that BE and EC have been found artfully, so can AE be known by the 47th of 1 of Euclid (as DE is known). And one shall find for the sides AB and AC, AB $228\frac{11}{13}$ and AC $302\frac{1}{13}$, as one required, of which the proportion stands as 25 to 33.

Question 34

There is a triangle ABC of which AB makes 25 and AC makes 33 and in which a line AD is drawn cutting the base BC in D as such: that BD stands to DC as 1 to $4\frac{19}{20}$. Also, BD stands to DA as 1 against $2\frac{1}{16}$. The question is about the base BC and the line AD and also about the parts of the base BD and DC?

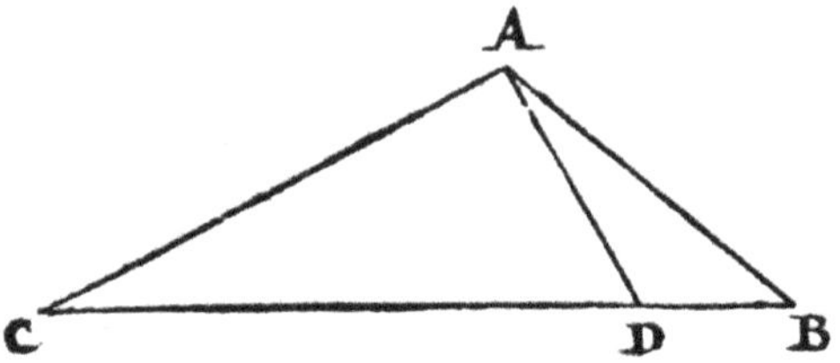

Figure 2.47: Question 34

To find these, proceed as taught in the previous 33rd example, and look for the proportion of AB to AC, as 25 to 33. How much shall AB and AC make when BD 1, DC $4\frac{19}{20}$ and DA $2\frac{1}{16}$ and one shall find for AB $2\frac{179}{208}$, or for AC $3\frac{807}{1040}$. Now say: AB $2\frac{179}{208}$ gives me 25, what shall give me 1? Arriving at BD $8\frac{88}{119}$. Further: what shall give me $4\frac{19}{20}$? Arriving at DC $43\frac{31}{119}$, and further: what shall give me $2\frac{1}{16}$? Arriving at AD $18\frac{3}{119}$. The whole of base BC then is 52. [43]

Question 35

I have a triangle ABC of which AB and AC each make 10, and the angle A is half as large as the angles B or C. Question: how much then makes side BC?

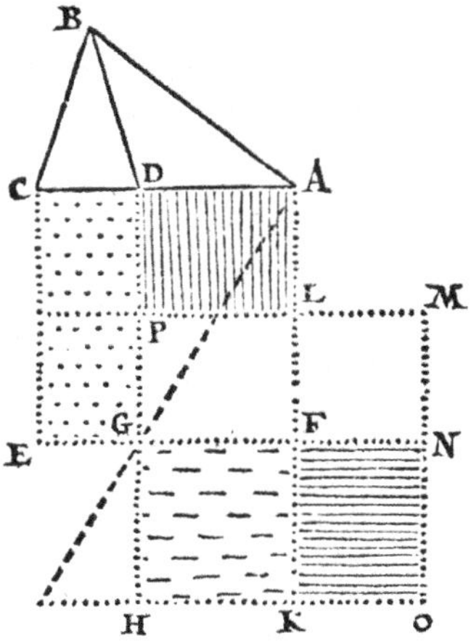

Figure 2.48: Question 35

As the angles B or C each are twice as large as A, draw a line from one of these angles (either B or C) (I will take B and draw BD) so that the angle B is divided into two equal angles, cutting the line AC in point D. So is AD equal to DB and A equal to DAB. Also, because angle BDC is equal to BCD, so is the small triangle BDC proportional to the large triangle ABC, as DC stands to DA as DA stands to the whole line AC. Hence, make a square of AC which here is $ACEF$, its area then makes 100. And as we draw DGH so will the parallelogram $DGEC$ be equal to the square $FGHK$, as EG stands to GH as AD to AF. Because the whole line is AC and GH or GF are the sides of the square $FGHK$, equal to AD. Now, halve $FADG$ by line LP, arriving at two equal parts $FLPG$ and $LADP$. This part [44] $LADP$ cut off and pasted to KF then becomes $FNOK$. So, the three parts included in $PLFNOKHP$ remain equal to the square $ACEF$ 100. To this add the square $MLFN$ 25, which is so much because LF makes 5. This adds up to 125 for the square $MPHO$. The $\sqrt{\text{ʒ}}$ of this is $\sqrt{125}$ for MP, or the side of the square. From this [subtract] LM 5, there remains for LP $\sqrt{125}-5$, which is equal to AD or BC, so that BC makes $\sqrt{125}-5$.

Problem 36

There is a battalion of musketeers arranged as a square of equal length and breadth, under which are 256 musketeers with 320 lancers. The question is: how many men thick are the musketeers encircled by the lancers?

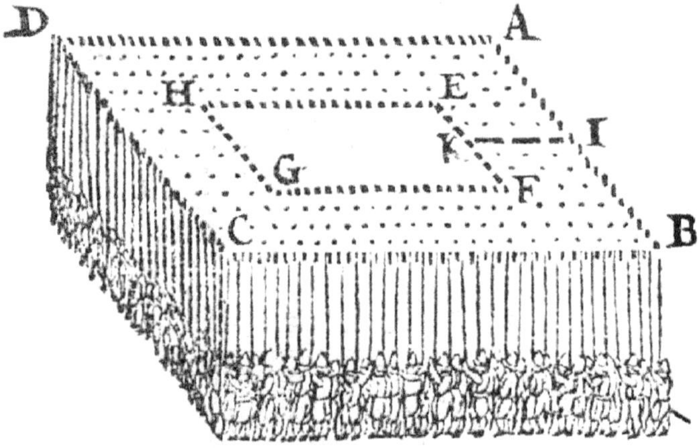

Figure 2.49: Question 36

Let the whole of the battalion of musketeers and lancers be the square $ABCD$ making in total 256 and 320 together, which is 576 men. And the inner square $EFGH$ of musketeers is 256 men strong. From this take the $\sqrt{\mathfrak{z}}$ arriving at 16 for EF or EH, the side of the square of musketeers, or as much as each row of the musketeers. Take also the $\sqrt{\mathfrak{z}}$ of the earlier 576, arriving at 24 for AB or AD, the side of the whole of the squadron. Subtracting 16 from this, the side of the musketeers, there remains 8. Its half is 4 for IK. And by so many men are the musketeers encircled by the lancers.

Question 37

A captain commands 120 lancers, to which he required to have as many musketeers as when arranged as a square battalion, that the lancers encircle the musketeers three men thick. Question: how many musketeers he therefore requires?

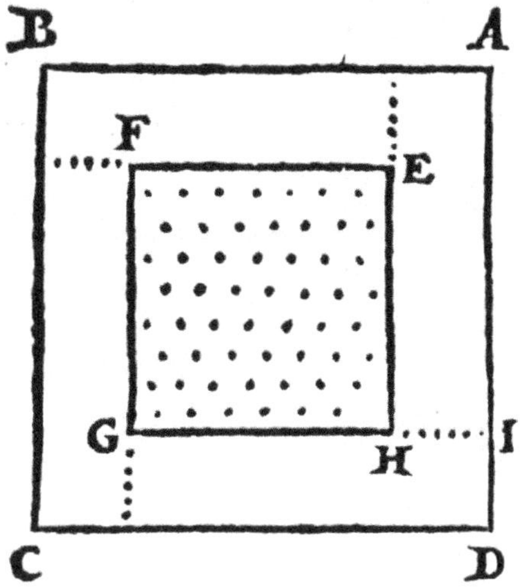

Figure 2.50: Question 37

Let $ABCD$ be the whole of the battalion and the square $EFGH$ the musketeers, encircled by the lancers three [rows] thick, as for example HI. Now, to find the number of musketeers, divide the number of lancers in four parts, arriving at 30 for each part. This, divided by HI 3, makes 10 for AI, from this 3 subtracted, there remains 7 for EH or EF, the side of the musketeers, multiplied as a square, one arrives at 49 for the square $EFGH$ or the number of musketeers, as one required to know.

Otherwise

Double 3 HI, arriving at 6, this multiplied as a square, arrives at 36, corresponding to the four squares as one on each corner as HI 3. This subtracted from 120, there remains 84. This divided by 4 times 3, which is 12, one shall arrive at 7 for EH or the side of the square $EFGH$, or the musketeers. Or one divides 84 in 4 parts, arriving at 21, which again divided by 3, arrives at 7 for EH, or 49 for the whole square as above. [46]

Question 38

There is a right-angled quadrangular piece of land $ABCD$, in which AD measures diagonally 16 rods to E, at which one finds that the perpendicular AE falls at the same point E. And if one measures E to F right angled, one finds EF connected by 2 rods. The question is: how many [square] rods large is such quadrangular piece of land?

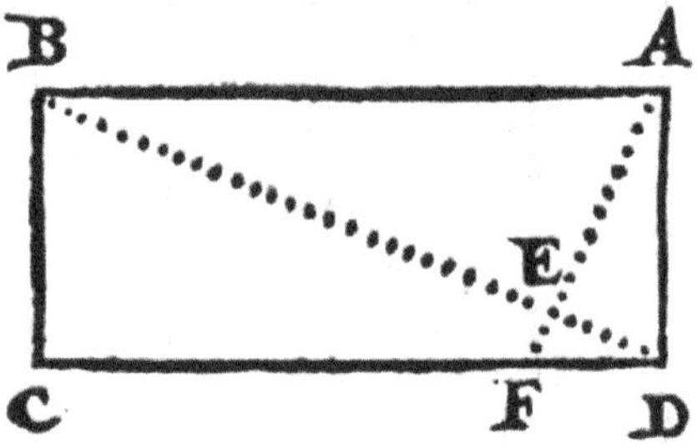

Figure 2.51: Question 38

As the four angles of this quadrangle are right angled, so EF stands to ED as ED stands to EA, and EA to EB. So, we now have to find how much are ED and EA. To do this, pose EF 2 and EB 16 in their smallest proportion, arriving at 1 and 8. Now, as EB is the third number of DE, take the $\sqrt{\mathfrak{z}}$ cube, because if the smallest proportion in DE was known, with EF 1 and multiplied as a square, this would bring forth the smallest proportion in AE and as a cube would bring forth EB, which is now known and makes 8, its cubic root makes 2 for DE in the smallest proportion. And if EF becomes smaller by 2, so DE 2 becomes larger again. So, the smallest proportion being 2 shall arrive at 4 for DE. So AE then is 8 and the whole of the line BD 20 and the area of the quadrangle $ABCD$ is 160.

Otherwise

Multiply FE 2 as a square, arriving at 4, this multiplied by EB 16 makes 64. The cubic $\sqrt{\mathfrak{z}}$ of this is 4 for BE as above.

Question 39

There is a heptagonal piece of land $ABCDE$ of which the angles B, C, D are right-angled and [47] if one extends the line AE towards angle C, also if one draws BD, then line DB will cut line AC right angled at point F. The question is: if AE makes 30 and EF 2 rods, how much then is the area of such heptagonal piece of land.

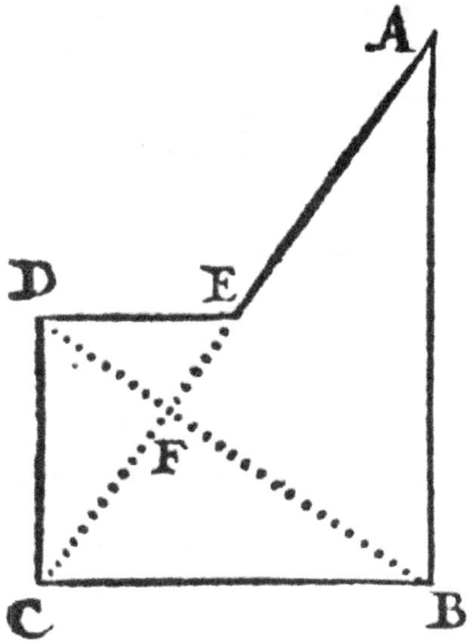

Figure 2.52: Question 39

Add EF 2 to EA 30, arriving at 32 for FA. This 32 multiplied by 2 arrives at 64. From this take the $\sqrt{\mathfrak{z}}$, arriving at 8 for FC, as FC is the mean proportional between FE and FA. Multiply this 8 again by 2, arriving at 16. The $\sqrt{\mathfrak{z}}$ taken from this arrives at 4 for FD as FD is the mean proportional between FE and FC. Now, that FC and FD both are known so can FB also be known by the same means. Or say: FE 2 gives me FD 4, what shall give me FC 8? Arriving at FB 16. So, multiply the half of FB by AC, and the half of FD by EC. Adding these products togther one arrives at 360 rods for the whole of the area of this heptagon $ABCDE$.

Otherwise

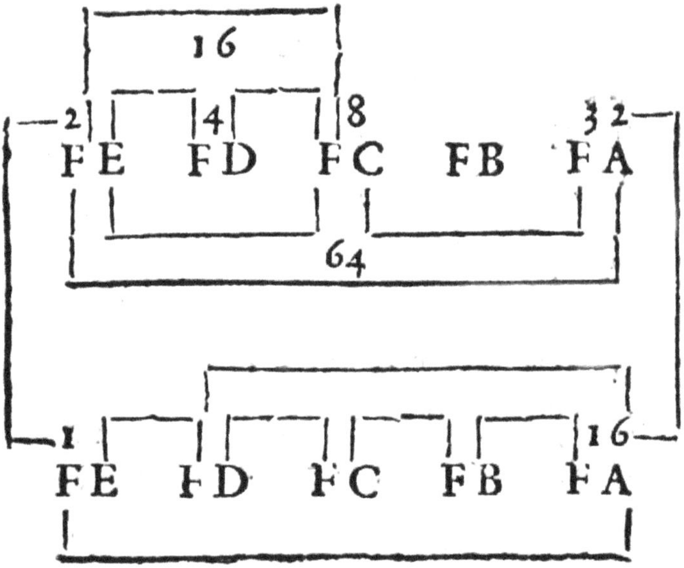

Figure 2.53: Question 39: alternative solution

Pose 2 FE and 32 FA in their smallest proportion, arriving at 1 and 16. From this take the $\sqrt{\text{ꝫꝫ}}$[2] arriving at 2 for FD, as FA is the fourth number in the progression from FD and now 2 subtracted from FE, so multiply [48] FD again by 2, arriving at 4 for FD as before. And now that FD is known, so can FC and FB also be known, as FE stands to FD as FD to FC, and FC to FB.

Question 40

Someone inherited a rectangular quadrangle[3] $ABCD$ which is 3 times as long as its breadth and 12 rods more in length? The question is: when the area of the whole of the quadrangle is 180 rods, how much is the breadth of AB and the length of BC?

[2] A curious notation for the cubic root.

[3] Cardinael has 'square'.

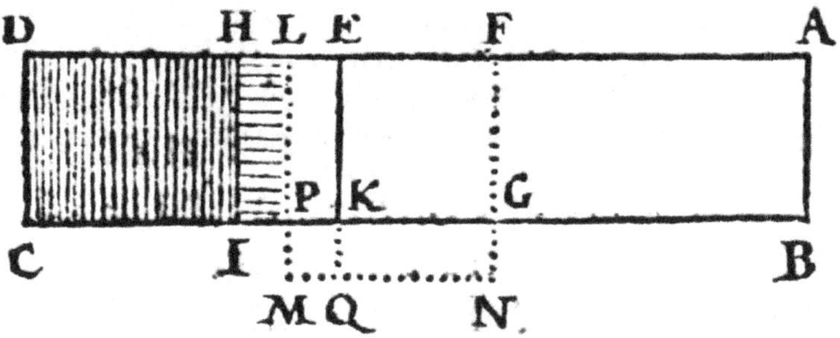

Figure 2.54: Question 40

Let AE stand to AB as 3 to 1. So, ED 12 and take $\frac{1}{3}$ from $AEKB$, which is $FEKG$, also $\frac{1}{3}$ from $EDCK$, which is $EHIK$. So, then $FHIG$ is $\frac{1}{3}$ from the whole of the quadrangle $ABCD$ and makes 60. Furthermore, as EH is $\frac{1}{3}$ from ED, so EH or KI makes 4. Its half is 2 for KP and is the square $KPMQ$ 4. This added to 60, one arrives at 64 for the square $FLMN$. Its $\sqrt{\mathfrak{z}}$ is 8 for FL or LM. Now EL 2 subtracted from FL 8, there remains 6 for EF, and so much is also AB, the breadth. And AE 3 times 6 is 18, to which added ED 12, one arrives at a length of 30 for AD.

Question 41

I have a right-angled quadrangle $ABCD$ of which the length AB is 3 times as much as its breadth, and if the length was 18 rods less, the area would be 120 rods. The question is to the length and the breadth of this quadrangle. [49]

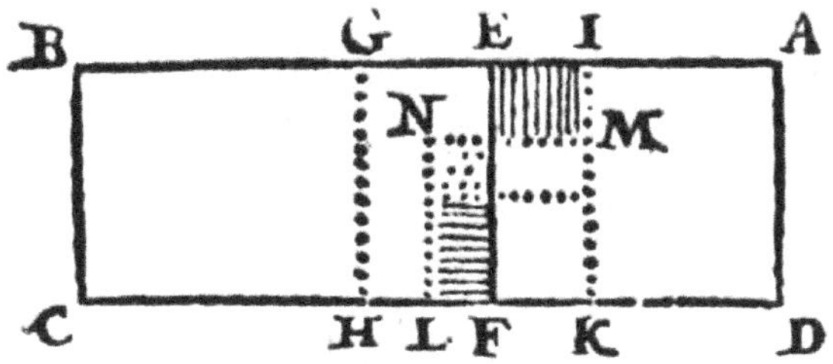

Figure 2.55: Question 41

The proportion of the length against the breadth is 3 to 1. Therefore, divide 18 in three parts, arriving at 6 for each part, or for EG. Now, let EI be $\frac{1}{3}$ of AE, so is IG $\frac{1}{3}$ of the whole line AB, or the square $IGHK$, being the

$\frac{1}{3}$ of whole quadrangle $ABCD$, and $IEFK$ then makes 40. The half of GE or HF is 3 for FL, its square is 9, this added to 40, arriving together at 49 for the square $MNLK$. Take the $\sqrt{\mathfrak{z}}$ of this, arriving at 7 for KL. This added to FL 3 is together 10 for HK or IK, which is equal to the breadth AD. So, AB must be three times as much, which is 30.

Question 42

There is another right-angled quadrangle $ABCD$ of which the length AB makes 40 and [which] is divided into two parts by the line EF, so that the area of part $EFCB$ is 192 [square] rods. But EB is longer than EA and EA is twice as long as AD. The question: how much then will be the breadth AD?

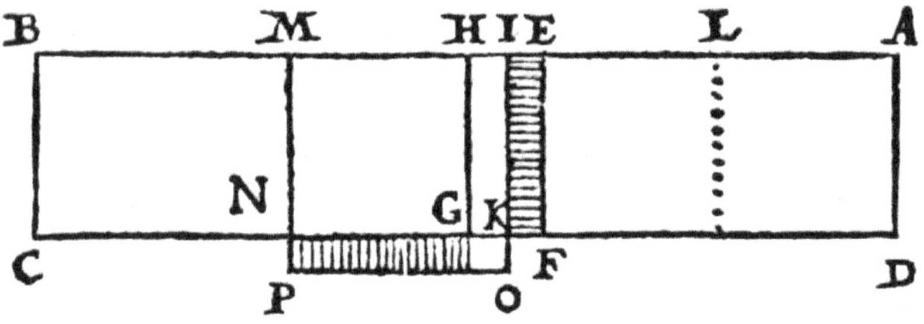

Figure 2.56: Question 42

Take the half of AE and EB, together being 40, arriving at 20 for LM, as AE is twice as long as AD. Also, take the half of 192, arriving at 96 for $EMNF$ and the half of LM (being 20), is 10 for IM. The square of IM is 100 for the square $IMPO$. From this subtract the 96 of before, there remains 4 for the square of OK, or of KG, which are equal to [50] IH. The $\sqrt{\mathfrak{z}}$ of 4 is 2 for KG, or IH. This subtracted from IM 10, there remains 8 for HM. And so much is also the breadth AD, as one required.

Question 43

When, from the aforementioned right-angled quadrangle $ABCD$, the length AB is known, being 40, and divided into two parts by the line EF, such that part $EBCF$ has an area of 192 [square] rods, and EB is shorter than AE, and

AE is twice as much as the breadth AD, the question is, how much then is the breadth AD?

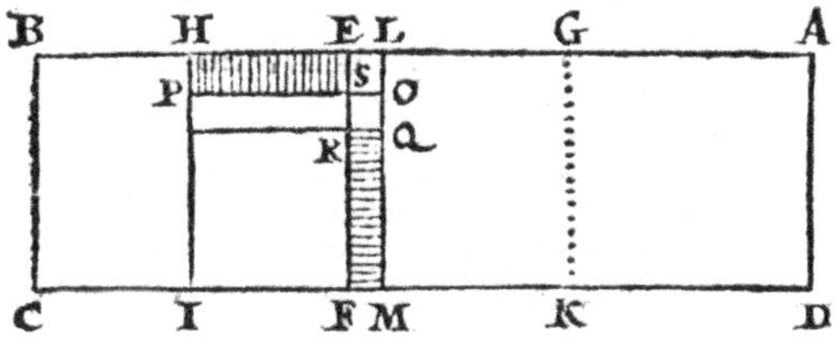

Figure 2.57: Question 43

The half of AB 40 is 20 for GH. So, $GHIK$ is the half of the whole quadrangle and $EHIF$ makes 96, because $EHIF$ is the half of $EBCF$. Furthermore, taking the half from GH, which is LH or MI 10, multiplied as a square, arrives at 100 from the square $MIPO$. Of this subtract 96, so there remains 4 for the square $OQRS$. The $\sqrt{\mathfrak{z}}$ is 2 for OQ or LE, which added to MO 10, which is equal to MI or LH, is together 12 for LM. And so much is also the breadth AD, as one required to know.

Question 44

There is given a right-angled quadrangle $ABCD$ of which each of the sides make 70 feet, in which there are made two other squares, as $EBOF$ and $HFGD$, having together 2500 [square] rods. The question [51] is how much the squares $EBOF$ and $HFGD$ make separately?

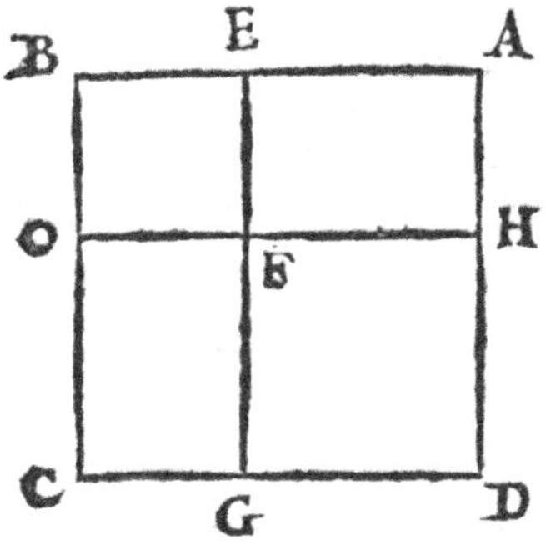

Figure 2.58: Question 44

Multiply 70 as a square, arriving at 4900, from this subtract 2500, remaining 2400. The half is 1200 for one of the parts $AEFH$ or $FOCG$. Further, halve AB, as in the second figure, arriving at 35 for KB, the square of which $KBLI$ then makes 1225.

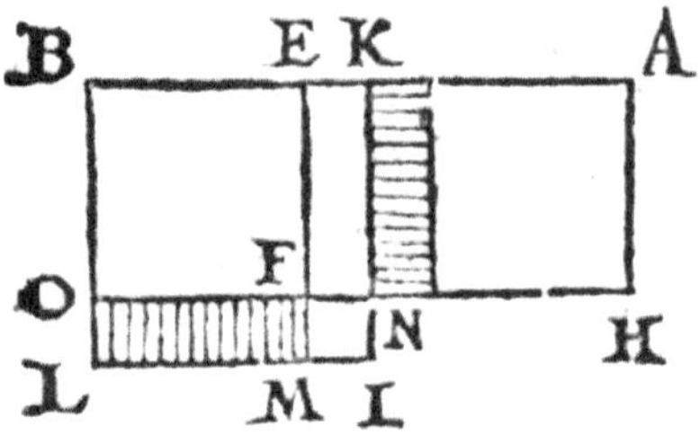

Figure 2.59: Question 44: solution

From this subtracted 1200 $AEFH$, there remains 25 for the small square $NFMI$, the $\sqrt{\mathfrak{z}}$ from this is 5 for MI, which is equal to KE. This subtracted from 35 KB, there remains 30 for EB, or for the side of the square $EBOF$. Or add KE 5 to 35 AK, arriving at AE 40 for the side of the square $HFGD$. The area of $EBOF$ then is 900, [that] of $HFGD$ is 1600.

Otherwise

The part $AEFH$ being 1200 is the mean proportional between both squares $EBOF$ and $HFGD$, which together make 2500. As from our fifth example each particular square shall be known. Multiplying the half of 2500 by itself, arriving at 1500125, from this [subtract] the square of 1200 (being 1440000), remaining 122500. The $\sqrt{\mathfrak{z}}$ of this is 350. Added to the half of 2500 arrives at 1600 for the square $HFGD$, and subtracting the aforementioned 350 from half of 2500, there remains 900 for the square $EBOF$, as above. [52]

Question 45

There are two right-angled quadrangles $ABCD$ having 96 and $EBGF$ having 24, which both are of the same proportion and the length of AB is three times as much as the breadth BE of the other quadrangle. Question: the length and breadth of these two quadrangles in particular.

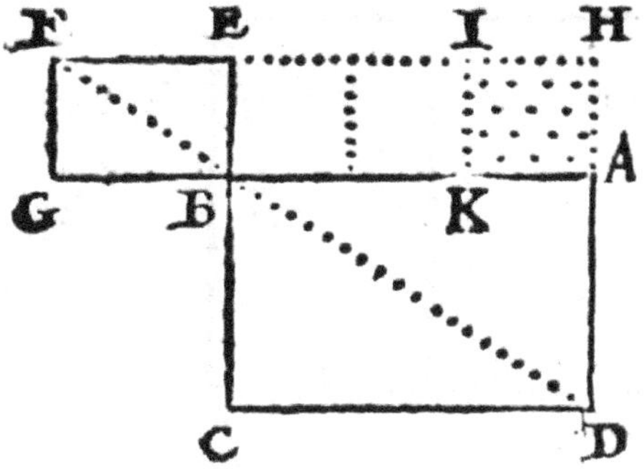

Figure 2.60: Question 45

Look for the mean proportional between both areas of $ABCD$ 96 and $EFGB$ 24, and you shall arrive at 48 for the part $ABEH$, of which the length of part AB stands to the breadth of BE as 3 to 1. Therefore, divide 48 by 3, arriving at 16 for the square $HIKA$. The $\sqrt{\mathfrak{z}}$ of this is 4 for AH or the breadth BE. So the length BG is 6 and the length AB 3 times as much as BE. That is then 12 for AB. The breadth BC then is 8.

Question 46

We have a right-angled quadrangle $ABCD$, the breadth AD of which gives 24 rods and if one cuts of part $FBCE$ parallel to AD, or BC, with an area of [square] 336 rods, there will remain $AFED$ which is of the same proportion as the whole of the quadrangle $ABCD$. I then ask: how much will the side be of the aforementioned quadrangle AB?

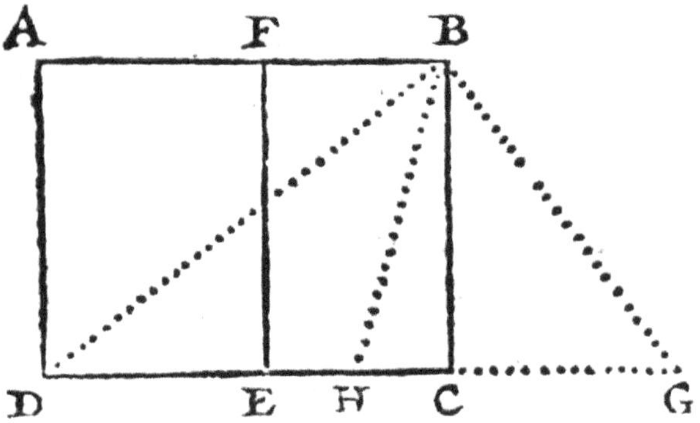

Figure 2.61: Question 46

Divide 336 by 24 arriving at 14 for EC. Now, make CG as much as DE, so then the mean of EC, which is also the mean of DG, which is point H, and HC will be 7, because EC is 14. Now, as CG stands to CB, so stands CB to CD. Therefore DBG is a right angle. Now look for HB, multiply HC 7 as a square, arriving at 49, and CB 24, squared as AD arrives at 576. These squares together are 625 for the square of HB. The line HB then is 25, as much is also HD. To this add HC 7 arriving at 32 for DC, or the length AB, as was asked for.

Question 47

There is a quadrangular garden $ABCD$, whose area makes 180 [square] rods, in which garden is made a quadrangular field, as here $EFGH$, with the same proportion as the garden $ABCD$, the field having an area of 80 rods, of which its length EF is 5 rods less than the length AB of the whole garden, so that IK or LM each makes $2\frac{1}{2}$. The question is to the length and breadth of such a garden?

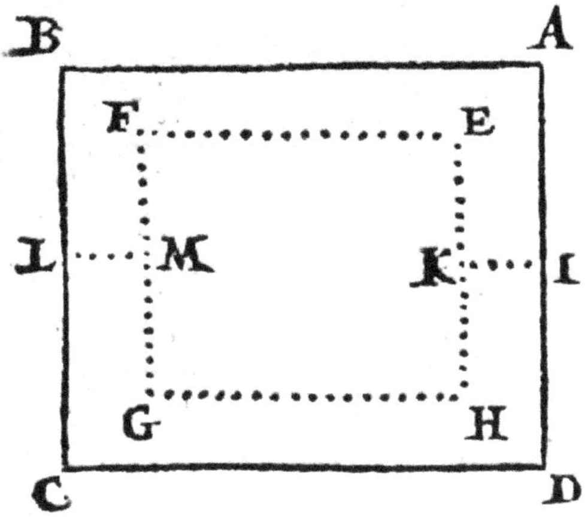

Figure 2.62: Question 47

Divide 180 by 80 arriving at $2\frac{1}{4}$. So, take the $\sqrt{\mathfrak{z}}$ of this, arriving at $1\frac{1}{2}$. Now, subtract 1 from $1\frac{1}{2}$, there remains $\frac{1}{2}$. As $\frac{1}{2}$ stands to 5, so stands 1 to HG, the length of the inner field. Also, as $\frac{1}{2}$ stands to 5, so stands $1\frac{1}{2}$ to DC, the length of the garden. Therefore say: $\frac{1}{2}$ gives me 5, what shall give me 1? Arriving at 10 for the inner length GH. The inner breadth then is 8 for EH. Also say: $\frac{1}{2}$ gives me 5, what shall give me $1\frac{1}{2}$, arriving at 15 for the length AB of the garden. The breadth BC then is 12. [54]

Question 48

One has a right-angled quadrangular farmstead $ABCD$, with length AB 16 and breadth BC 10, on which land he wants to delve a ditch, starting from D to A and from A to B such that, the ditch here drawn as $DABGFED$ with an area of 120 [square] rods, and EF being parallel to DA and FG parallel to AB, and also with ED being equal to BG. The question then is: How many rods must DE or BG be so that the ditch will be 120 [square] rods of area?

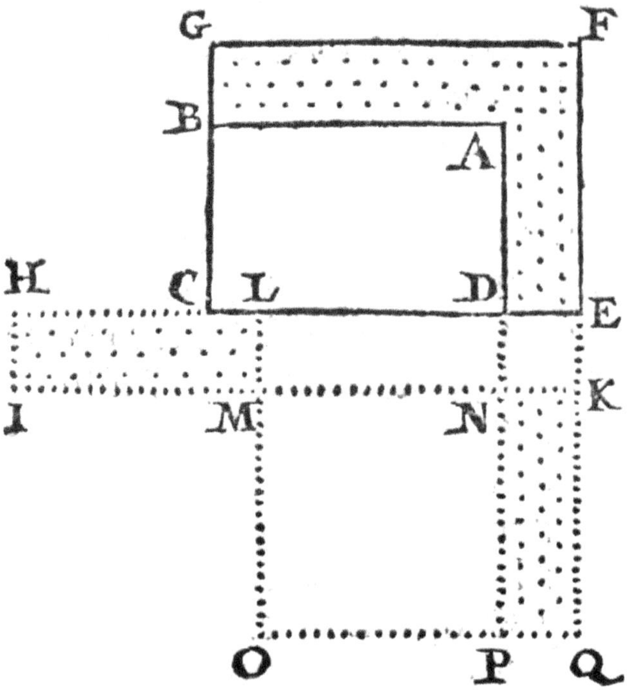

Figure 2.63: Question 48

Extend DC so much longer as AD, so CH becomes equal to AD 10, and DH then is 26. Also, let EK be equal to ED so that the parallelogram $EHIK$ is equal to the ditch $EDABGFE$, so that parallelogram $EHIK$ also makes 120 rods. Now, divide $DHIN$ into two equal parts by line LM. So is LD or LH each 13, as the whole line DH is 16 and 10, together 26. Its half is 13 for LD or LH. Now, make part $KNPQ$ equal to $LHIM$, so will gnomon $QPNMLDEKQ$ also be 120. And now that QK is equal to LM, making also 13, so is the square $NPOM$ 169. This added to 120, the area of the aforementioned gnomon, arrives together at 289 for the square $ELOQ$. The $\sqrt{\text{ʒ}}$ of this is 17 for EQ or for the sides. From this [55] subtract KQ 13, there remains 4 for EK, as much as is ED and BG, as one requires.

Question 49

There is a quadrangular[4] battalion, three times as long as its breadth, of which the lancers are covered by the musketeers four men thick. The question is: if the musketeers are 567 men strong, how many lancers were there?

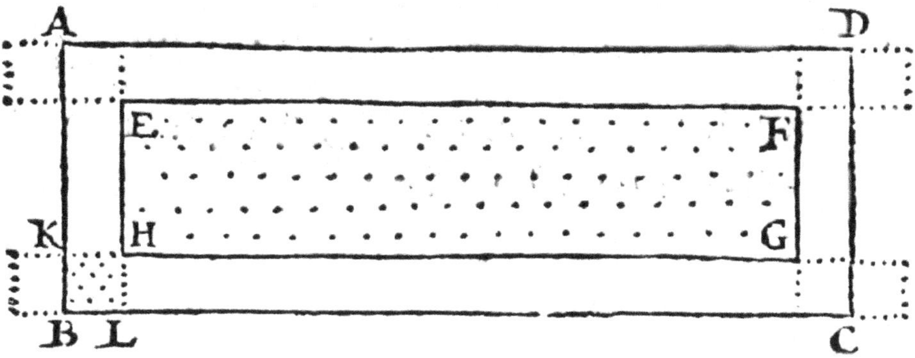

Figure 2.64: Question 49

Add to the 576 men musketeers four times as much as $KHLB$, this being 64, as KH or HL are equal, each 4, the square $KHLB$ then is 16, this times 4, one arrives at 64 as above. As there is such a square at each of the corners and added to the number of musketeers, one arrives at 640. This divided by 4, the thickness BK or KH, one arrives at 160. Its half is 80 for the breadth AB and the length BC together. Also, as AB stands to BC as 1 to 3, so add also 3 and 1, arriving at 4 and say: 4 gives me 80, what shall give me 1? Arriving at 20 for AB. And what shall give me 3? Arriving at 60 for BC. The length and breadth of the battalion being known so is the area of the whole battalion 1200 men. From this subtracted 576, the number of musketeers, there remains 624 men as the number of lancers. [56]

[4]Cardinael has 'square'

Otherwise

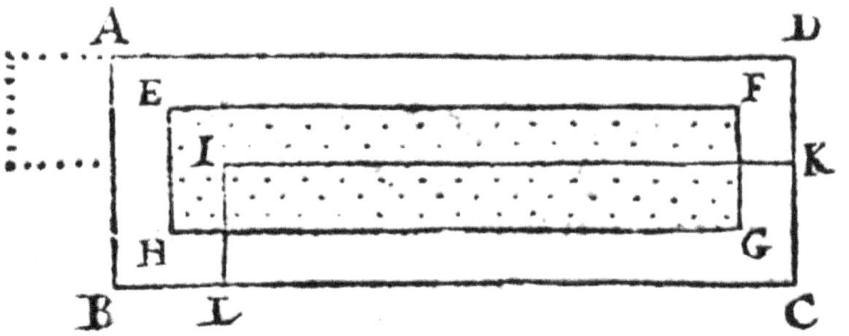

Figure 2.65: Question 49: alternative solution

Double the thickness 4, arriving at 8 as will be for example BL or KD in the figure shown. And as $EFGH$ is equal to $IKCL$, so will be $ABLIKDA$ 576. This divided by 8, arriving at 72, to this added 8, arriving at 80 for AB and BC together. This divided into two parts, as we learned before, so that one part will be to the other as 1 to 3. Arriving at 20 for AB and 60 for BC, as before.

Question 50

A commander has set up a battalion in a rectangle[5], three times as long as its breadth, of which the lancers are covered by the musketeers 4 men thick. The question is: if the lancers are 624 men strong, how many musketeers must there have been?

[5]Cardinael has 'square'.

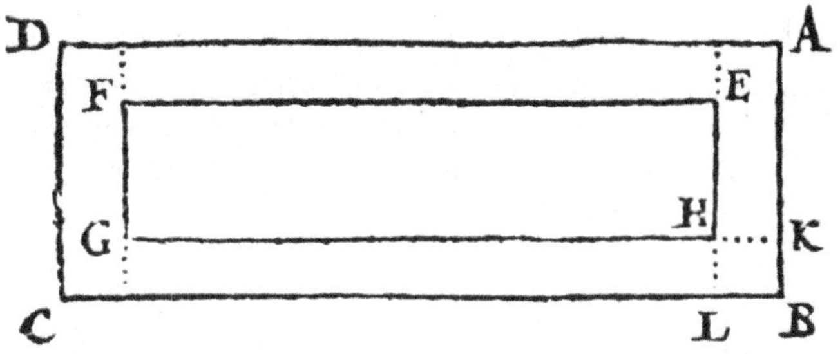

Figure 2.66: Question 50

Let $ABCD$ be the battalion, the inner area $EFGH$ be the lancers, which are covered by the musketeers four men thick, here represented by HK or HL. We thus have to look for AB and BC as the whole of the battalion is known to us, and as such also the number of musketeers. Now, to find this: take $\frac{1}{3}$ of 624 $EFGH$, which is 208, let this be $EHNM$ in the second figure. [57]

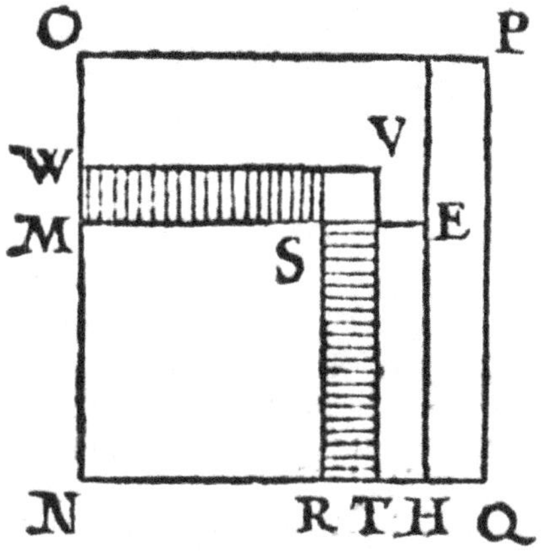

Figure 2.67: Question 50: solution

Also, take $\frac{1}{3}$ of the whole of the quadrangle $ABCD$ which here is $QPON$, so that MO is 8 and QH $2\frac{2}{3}$, as MO is twice as much as BK or HL. Now make that square $RNMS$, so is HR $5\frac{1}{3}$ and TH or TR is the half $2\frac{2}{3}$. And if one multiplies this $2\frac{2}{3}$ of TR as a square, one arrives at $7\frac{1}{9}$. This added to 208 (the area of $EHNM$) one arrives at $215\frac{1}{9}$ for the square $TVWN$. The $\sqrt{\mathfrak{z}}$ of this is $14\frac{2}{3}$ for TN. To this added $2\frac{2}{3}$ HT and $2\frac{2}{3}$ HQ, arrives together at 20 for QN, or the $\frac{1}{3}$ of BC, which is equal to AB. So AB makes 20 and BC 60.

Question 51

When the number of 567 musketeers were known to me and that they encircle the number of lancers aforementioned (being 624) four men thick, and that the proportion of the breadth AB against the length AD were now unknown to me, then to find AB and AD.

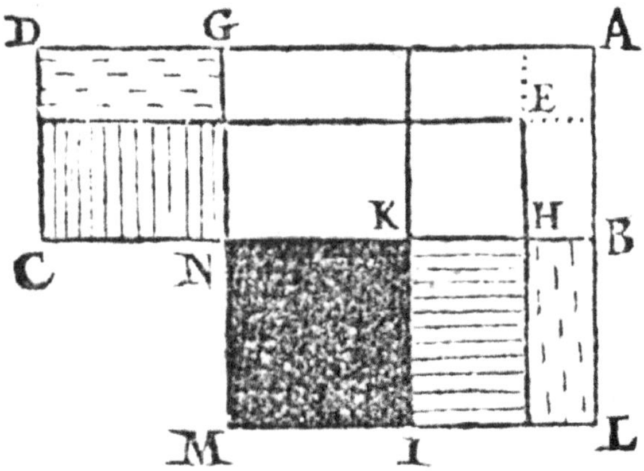

Figure 2.68: Question 51

So, cut off of the quadrangle $ABCD$ as much as $GDCN$ and paste it to BK which makes $BKIL$. So, then $GALM$ is a square, making AG or AL 40, as BAD is equal to LAG, the square $AGML$ then is 1600. Now, add the number of musketeers (being 576) to 624, the number of lancers, the sum making 1200. This subtracted from 1600 there remains 400 for the square $KNMI$. The √ȝ from this is 20 [58] for IK, which is equal to BL. So BL makes 20, this subtracted from LA 40, there remains 20 for AB. So, AD must be 60, as GD is equal to BL.

Question 52

There is a right-angled quadrangle $ABCD$, whose area is 98 [square] rods, in which is made a separating ditch FE being 13 rods of length. Also, DF and BE each make 2 rods. One asks: how much is breadth AB and length BC?

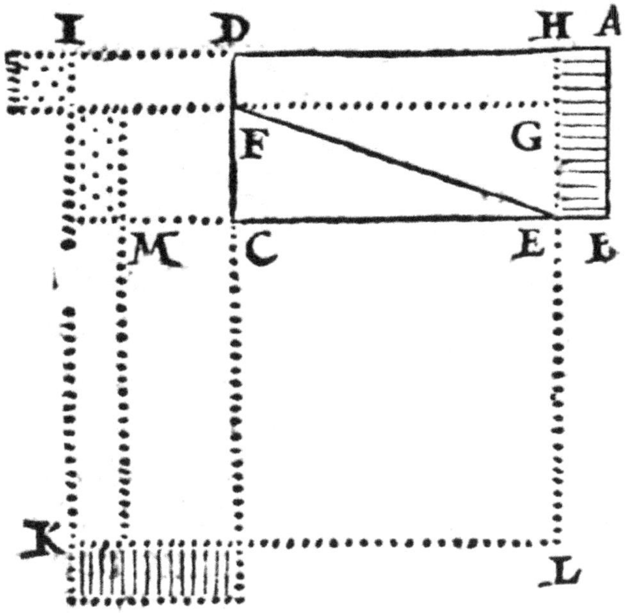

Figure 2.69: Question 52

To find these: so multiply 13 as a square, arriving at 169. This added to two times 98, arriving together at 365. Also, multiply BE or DF 2 as a square, arriving at 4. This subtracted from 365, there remains 361 for the square $HIKL$. The $\sqrt{\mathfrak{z}}$ from this is 19 for HI or the sides. From this subtract 2, there remains 17 for ECM or for EC and CF together. Furthermore, now to find EC and CF in particular, this can be done as EF is known, as taught in our 14th example. And we shall arrive at 5 for CF and 12 for EC. So the breadth AB is 7 and the length BC 14. Concerning the demonstration of this exercise you will have enough with the figure shown. [59]

Question 53

There is a right-angled quadrangle $ABCD$ in which are drawn two lines, one EH parallel to AD, the other GK parallel to DC, which cut each other in F. Also, that FG stands to FH as FE to FK. The question is: how much is the area of this quadrangle $ABCD$ when FA is $\sqrt{90}$, FD 5, FC $\sqrt{160}$ and FB 15? Also, [what is] the length and breadth of such quadrangle?

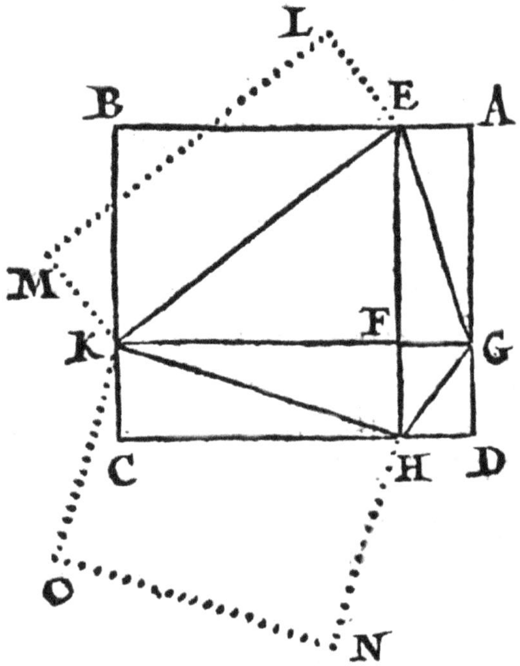

Figure 2.70: Question 53

Draw parallelogram $EKML$ on line EK of equal proportion as $AEFG$ or $FKCH$. So, [the area of] quadrangle $EKML$ is as much as both quadrangles $AEFG$ and $FKCH$ together, by the 31st of 6 of Euclid, as the square of EK is a much as both the squares of EF and FK together. Also, now that EK is being known, is 15, so then is also EL or KM known, and it makes 5, equal to GH, because the lengths of the lines FE and FK make EK 15 in the length. Also, the breadths of the lines FG and FH give HG in the same proportion. Also, now that $EKML$ is made of the same proportion as $AEFG$ or $FKCH$, so must EL be equal to GH, and therefore makes 5. The area of $LMKE$ thus is 75 and so much are also both quadrangles $AEFG$ and $FKCM$ together. In the same way can the two other quadrangles $GFHD$ and $EBKF$ be known, $HKON$ being as much as both squares $GFHD$ and $EBKF$ together, as $HKON$ [60] is of the same proportion as $GHFD$ or $EBKF$, and HN or KO is being made equal to GE, making HN or KO $\sqrt{90}$. The area of $HKON$ then is 120, and as much are also both the quadrangles $GFHD$ and $EBKF$ together, which added to the aforementioned 75, the area of the quadrangles $AEFG$ and $FKCH$, arriving together at 195 for the area of the whole of the quadrangle $ABCD$, as one required to know.

Now, to find the length AB and the breadth BC, so multiply EK and GH each as a square, arriving at 25 and 225, making 250 together. Now, divide the area of both quadrangles $EBKF$ and $GFHD$, together being 120, into two parts such that the one part stands to the other as 25 to 225. Therefore: 250 gives me 120, what shall give me 25? Arriving at 12, and so much is the area of the quadrangle $GFHD$. Also: what shall give me 225? Shall arrive at 108 for the area of the quadrangle $EBKF$ or the square GE 90, and HK 160. Is together also 250. Say: 250 gives me 75, the area of both quadrangles $AEFG$ and $FKCH$, what shall give me 90? One shall arrive at 27 for the area of $AEFG$ and what shall give me 160? One shall arrive at 48 for the area of $FKCH$. Now, let FG be 1, so FE is 3. Divide 27 by 3 arriving at 9. The $\sqrt{\mathfrak{z}}$ from 9 is 3 for FG. So, FH is 4, FK 12 and FE 9, the length AB 15 and the breadth BC 13.

Otherwise

Pose that FG be 1, so FE is 3, FH $1\frac{1}{3}$, FK 4, the breadth EH then $4\frac{1}{3}$, the length GK 5. Multiply EH $4\frac{1}{3}$ by GK 5, arriving at $21\frac{2}{3}$. Also, multiply each square, arriving at $18\frac{7}{9}$ and 25. Now say: $21\frac{2}{3}$ gives me 195, the area of the whole of $ABCD$, what shall give me $18\frac{7}{9}$ and 25? So arriving at$18\frac{7}{9}$ for the square of the breadth EH 169. And for 25 at the square of the length GK 225. Taking the $\sqrt{\mathfrak{z}}$ from each of these [61] arriving at 13 for EH or the breadth BC and 15 for GK, or the length AB, as above.

Question 54

In this circle shown, whose diameter be AB, is drawn another circle whose diameter is CB, their difference being that AB is 10 longer than CB and the difference of their areas is $361\frac{3}{4}$. Question: the diameters AB and CB?

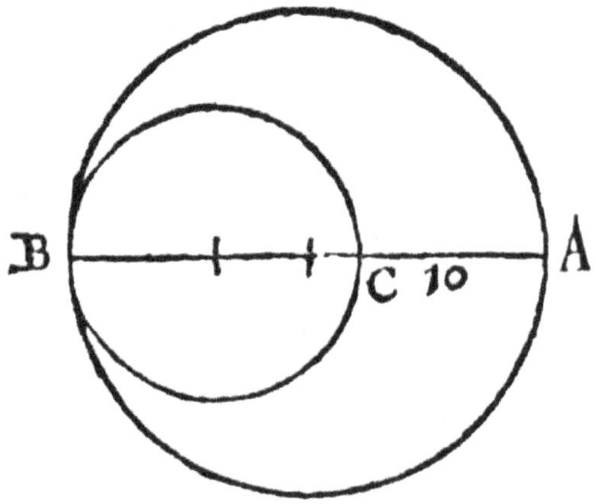

Figure 2.71: Question 54

It is known from the instructions of Archimedes that the proportion of the diameter to the circumference is as 7 to 22. So let for this purpose DEF be a right-angled triangle, its [side] DE is the half of the diameter of AB and EF the circumference. The proportion of DE to EF then is $3\frac{1}{2}$ to 22. The difference between the halves of the diameters be EL 5 and $ELHF$ $361\frac{3}{7}$ be the difference of their areas.

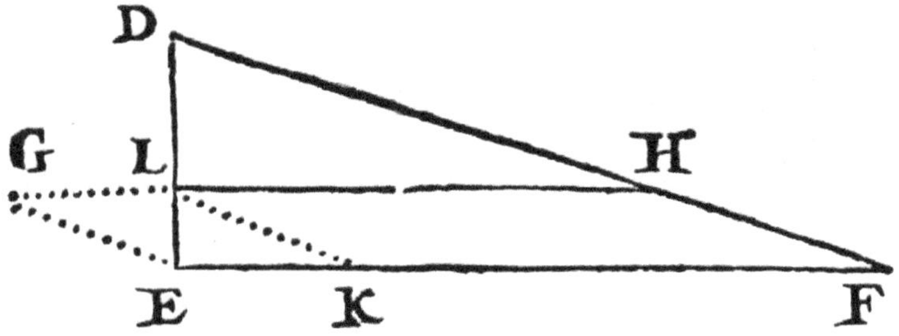

Figure 2.72: Question 54: solution

Now say: DE $3\frac{1}{2}$ gives me EF 22, what shall give me EL 5? One shall arrive at $31\frac{3}{7}$ for EK. This multiplied by the half of EL, being $2\frac{1}{2}$, one shall arrive at $78\frac{4}{7}$ for the area of LEK. So much is also ELG. This added to $361\frac{3}{7}$ together is 440 for the area of $EGHF$. Divided by EL 5, one arrives at 88 for EF, and so much is the circumference of the larger circle with diameter AB. The diameter AB then is 28, and so is the diameter of the smaller circle CB 18. [62]

Question 55

There is a circle whose diameter is AB in which is drawn a smaller circle whose diameter is CB touching each other at point B, by which that what makes one circle longer than the other is shaped a moon. Also that AC, the difference of their diameters, is 9. And perpendicular from the center E is drawn ED, so that FD then is 5. The question is the area of such moon?

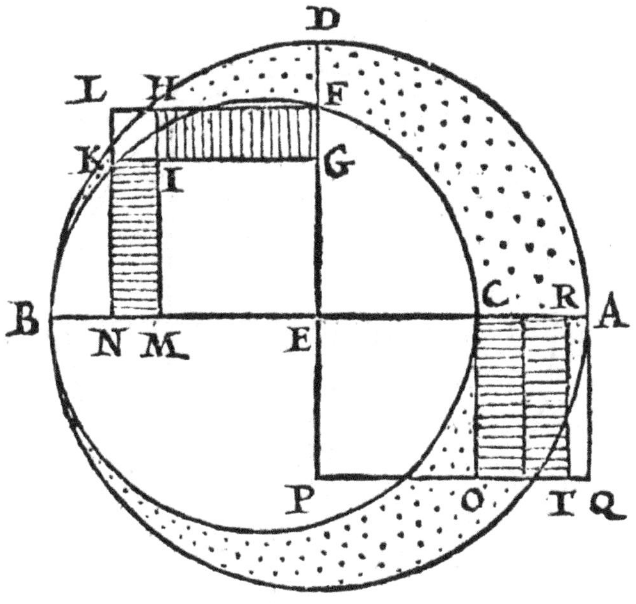

Figure 2.73: Question 55

Subtract 5 FD from 9 AC, there remains 4 for FG. And as FE is the mean proportional between CE and EB: therefore is the square $EFLN$ equal to that parallelogram that can be made from AE and EC, being $AEPQ$. Now,

AQ is equal to EG or EM. Therefore add 4 GF to 4 IK, which is together 8 for TD or RC. This, subtracted from 9 AC there remains 1 for AR, which part of $ARTQ$ is equal to that of the smaller square included in $IHLK$. And as also GF is known to me, being 4, so is KL or LH also 4. The area of this small square $IHLK$ then is 16. The part $ARTQ$ then is also 16. This divided by AR 1, arrives at 16 for AQ, and so much is also EC, to which is then added 9, which is 25 together for AE, the half the diameter of the larger circle. The half the diameter of the smaller circle then is $20\frac{1}{4}$. Now look [63] for the area of both circles, following the instructions of Archimedes, subtracting the smaller from the larger leaves the area of the moon, as one required.

Question 56

In a circle, whose diameter in the figure is shown by AB, are drawn the two perpendiculars DC and FE on the diameter. Also that CE makes 42. Now one asks, if DC makes 24 and FE 18, how much will the diameter be of such a circle?

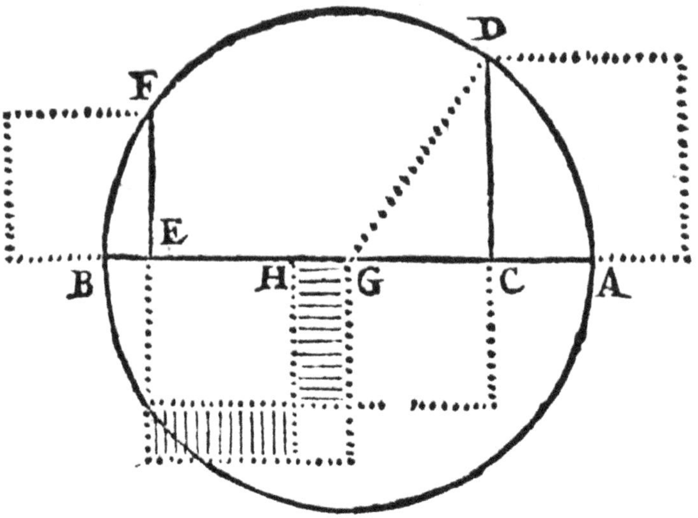

Figure 2.74: Question 56

Multiply DC 24 as a square, arriving at 576. From this subtract the square of EF 18, being 324, there remains 252. Also, if 42 is multiplied as a square, arriving at 1764, of which is subtracted the aforementioned 252, there remains

1512. This divided by the double of 42, which is 84, one shall arrive at 18 for CG. This squared (being 324) and added to the square CD 576 together is 900. The $\sqrt{\mathfrak{z}}$ from this is 30 for the half of the diameter. So the whole diameter AB is 60. [64]

Otherwise

Subtract the square of EF from the square of CD, there remains 252. This divided by CB 42, arrives at 6 for GH. This subtracted from CB 42, there remains 36, its half is 18 for CG, as before. The reasons for these operations can be easily deduced from the figure shown.

Question 57

There is a semicircle whose diameter is AB making $\sqrt{424}$, in which is inscribed a triangle EFD of which side ED makes 10 and FD 12. The question is: if side EF is parallel to diameter AB, how much then is this side of the triangle?

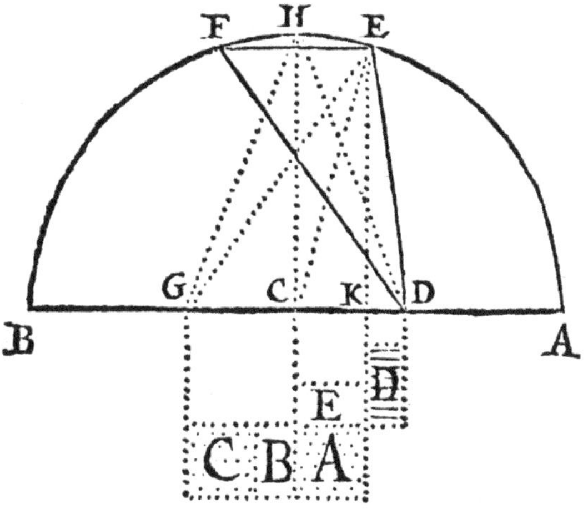

Figure 2.75: Question 57

Draw a line CH from the center C perpendicular on AB, also line CE and the perpendicular EK on AB. And CE is $\sqrt{106}$, equal to CH or the half of the diameter.

Further, draw line EG so that CG is equal to CD. So is EG equal to DF, and EG thus makes 12. Now, multiply DE 10 as a square, arriving at 100. Also [multiply] 12 EG as a square, arriving at 144. These squares are together 244, and as much are also both the squares of DH and GH. Therefore halved, arriving at 122 for the square of DH or GH. From this subtracted, the square of HC, being 106, there remains 16 [65] for the square of DC or CG. The $\sqrt{\mathfrak{z}}$ from this is 4 for DC or CG. And also known is DG, GE and ED, so KD can be found as in our first example, and one shall arrive at $1\frac{1}{4}$ for KD. So is KC $2\frac{3}{4}$ and EF $5\frac{1}{2}$ as one required. To prove that both squares of DE and EG are equal to [the squares of] DH and HG make of DK a square and also of KC and KG, also of DC and CG. Now, the square of DK and the square of KG are together with two squares of EK, as much as the squares of ED and EG together. So, if we subtract from these squares of ED and EG as much as two squares of EK and two squares of KC, so shall remain as much as the square of DK, of KC, of CG and the parts E and B. When we diminish the squares of A and C, which are equal, each with the square of KC, the remaining parts make precisely two squares, one of DC, the other of CG, as the area D is equal to part B. At last, if one subtract also two squares of HC from the two squares HD and HG, there shall remain also the squares of CD and CG. And as we have subtracted the squares of ED and EG equally from the squares of HD and HG, their remainders are the same. Hence, the squares of ED and EG together are equal to both the squares of DH and GH.

The reasoning behind this question depends on the tenth proposition of the second book of Euclid and can also be made by our twentieth question.

Question 58

Look at the picture at the following page.

There is a triangle ABC, whose sides AB and BC have the same length, in which there are inscribed two circles, one with diameter 8, the other with diameters 12. [66] Question: [what are] the sides of such a triangle?

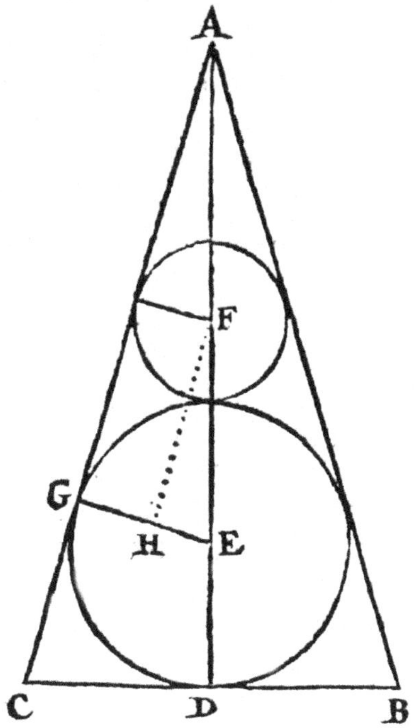

Figure 2.76: Question 58

Subtract half of the diameter 8 from half of the diameter 12, which is 4 from 6, there remains 2 for EH.[6] Also add the two half diameters, arriving at a sum of 10 for FE. Now say: EH 2 gives me EF 10, what shall give me EG 6? One shall arrive at 30 for EA. Adding to this ED 6, is together 36 for AD, the perpendicular of this triangle. Now, subtract the square of EG from the square of EA, there remains 864 for the square of AG. So AG is $\sqrt{864}$. Furthermore, also say: AG $\sqrt{864}$ gives me AE 30, what shall give me AD 36? One shall arrive at $\sqrt{1350}$ for AC and as much is also AB. And as AD has also be found before, being 36, then BD or DC is known by the 47th of 1 of Euclid. One shall arrive at both or for the base BC $\sqrt{216}$. And as such the three sides of the triangle are known as required.

[6] Cardinael has 3. Corrected in both translations.

Question 59

The triangle ABC here shown is given to me, of which the sides AB and BC are equal, each making 40. A circle is inscribed in the triangle so that the circle touches all three sides of the triangle. And if one draws a line through the center D, parallel to base BC, as here is EDF, so [67] that such line cuts the circle into two equal parts, but [cutting] from the triangle the part AEF, which is $\frac{25}{64}$ of the area of the triangle ABC. The question is: how much shall make the base BC?

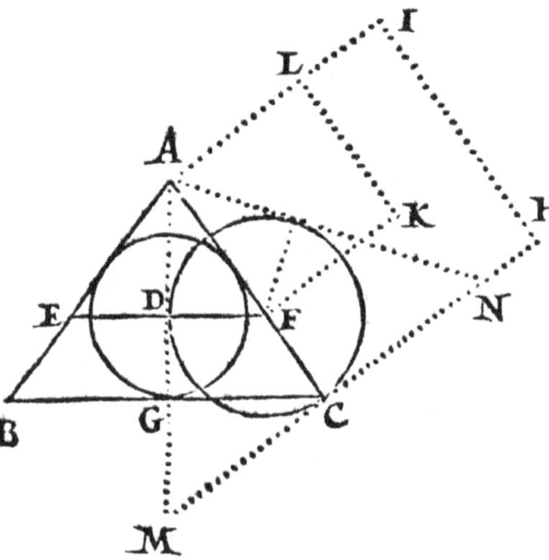

Figure 2.77: Question 59

Multiply AC 40 as a square, arriving at 1600 for the square $ACHI$. Cutting off from this the $\frac{25}{64}$, being 625 for the square $AFKL$, as the given triangle ABC stands to the square $ACHI$ as the triangle AEF stands to the square $AFKL$. So, now take the √ȝ of 625, arriving at 25 for AF, and FC then is 15. Furthermore, note that GD stands to DA as CF stands to FA, which makes it known that FC then is the half of the diameter of the circle which can be inscribed in triangle AMN (as GD is the half of the diameter of the circle that can be inscribed in ABC). Therefore, FD is also equal to FC. So then FD makes 15. Now say: AF 25 gives me FD 15, what gives me AC 40? One shall arrive at 24 for GC. So, the whole of base BC is 48, as one required to know.

Question 60

Look at the picture at the following page.

In a circle is inscribed a square with equal sides, as large as possible, such as here $ABCD$ [68] with segments ABF or BCG being found 224. Question: how much then is the diameter of the circle?

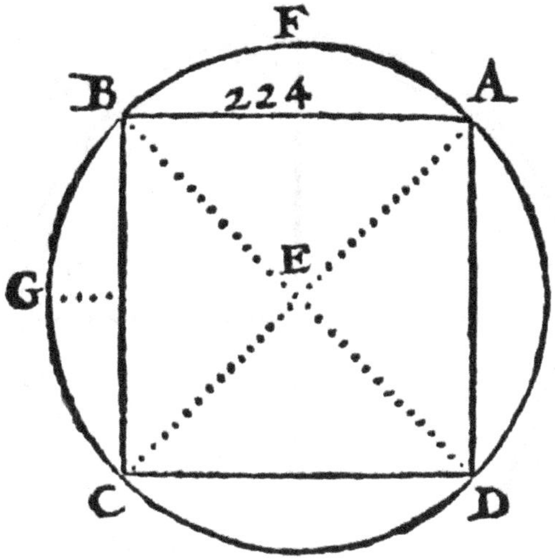

Figure 2.78: Question 60

Pose that AB, the diameter of the circle, makes 1 and look for the area of the circle, following the teachings of Archimedes, and you shall arrive at $\frac{11}{14}$ of which $\frac{1}{4}$ is $\frac{11}{56}$ for the area $AEBF$. Now, also look for the area of the triangle AEB, and you shall arrive at $\frac{1}{8}$. Subtract this from the $\frac{11}{56}$ aforementioned,[7] as the area of the fourth part of the circle $AEBF$, there remains $\frac{1}{14}$ for the segment ABF. And, as $\frac{1}{14}$ stands to the square of 1, also 224 stands to the square of the real diameter AC. Therefore say: $\frac{1}{14}$ gives me 1, what shall give me 224? Arriving at 3136 for the square of the diameter AC. The $\sqrt{\mathfrak{z}}$ from this is 56 for the diameter AC, as in the question and, following the teachings of Archimedes, as the proportion of the diameter to the circumference of the circle, being 7 to 22.

[7] Cardinael has $\frac{11}{15}$, Rudd has $\frac{11}{14}$ and Kurz corrects the misprint.

Question 61

Look at the following picture.

In the circle shown is inscribed an octagon whose sides make $\sqrt{72-\sqrt{2592}}$. The question is how much then is diameter AD?

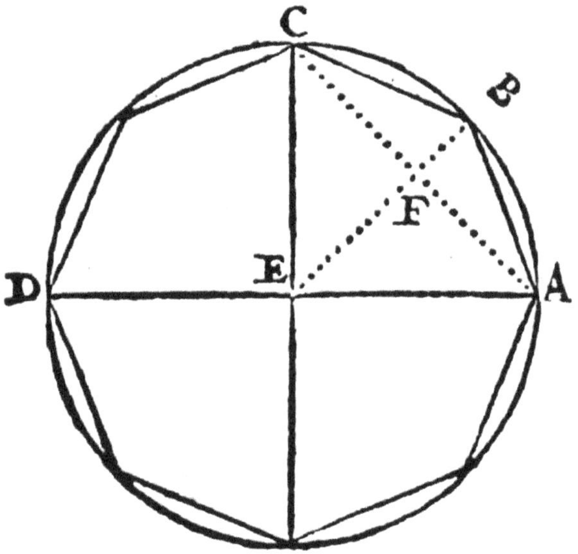

Figure 2.79: Question 61

Pose that half of the diameter of AE or EC be 1. So is AC $\sqrt{2}$ and FC or FA then is $\sqrt{\frac{1}{2}}$. Also, FE is as much as FC or FA. Now, subtract FE $\sqrt{\frac{1}{2}}$ from BE 1, there remains $1-\sqrt{\frac{1}{2}}$ for BF. The square of BF added to the [69] square of AF arrives together at the square of AB $2-\sqrt{2}$, or for both sides of the octagon, when half of the diameter AE is 1. Therefore, divide $\sqrt{72-\sqrt{2592}}$ by $\sqrt{2-\sqrt{2}}$ and you shall arrive at 6 for half of the diameter AE, the whole of diameter AD is 12.

Question 62

In the circle below is inscribed a square $ABCD$ (as large as possible) such that HG (falling perpendicular at G, the half of AD) makes $\sqrt{32}-4$, in which square is also inscribed a triangle AEF with equal sides, as large as is possible. Question to the diameter of the circle that can be inscribed in the triangle and to the diameter of the circle in which the triangle is inscribed . [70]

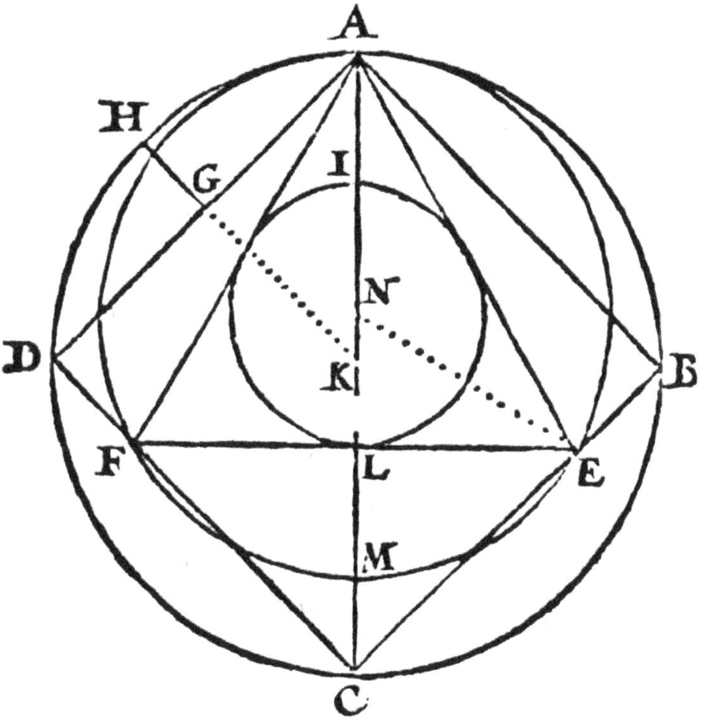

Figure 2.80: Question 62

Pose that each of the sides of triangle AEF makes 2, so the perpendicular AL will be $\sqrt{3}$. Also, as LF is equal to LC, so is LC 1 also equal to LF, as we posed 2 for EF and AC, then is $1+\sqrt{3}$. This multiplied as a square arrives at $4+\sqrt{12}$. Its half is $2+\sqrt{3}$ for the area of the square $ABCD$. So its sides each are $\sqrt{2+\sqrt{3}}$. Furthermore, pose that AG be 1, so is GK also

1, or AD 2, and KA or KH then is $\sqrt{2}$ and GH $\sqrt{2}-1$. Now say: $\sqrt{2}-1$ gives me $\sqrt{32}-4$, what shall give me 2? One shall arrive at 8 for AD. Also say: $2+\sqrt{3}$ gives me 8, what shall give 2? Arriving at $\sqrt{512-\sqrt{196608}}$ for the sides AE, EF or FA of the triangle. As AD is known, being 8, so shall AC, the diagonal line of the square $ABCD$ make $\sqrt{128}$. And, as we posed 2 for EF, the side of the triangle AEF (as we did before), so shall we find for AC $1+\sqrt{3}$. Therefore say: $1+\sqrt{3}$ gives me $\sqrt{128}$, what shall give 2, the side EF of the triangle? And you shall arrive at $\sqrt{512-\sqrt{196608}}$ for the side EF (as above). Now take the half of $\sqrt{512-\sqrt{196608}}$ which is $\sqrt{128-\sqrt{12288}}$ for LF. The square of this [subtracted] from the square of AF, there remains $384-\sqrt{110592}$ for the square of AL. So line AL is $\sqrt{384-\sqrt{110592}}$. From this take $\frac{1}{3}$, arriving at $\sqrt{42\frac{2}{3}-\sqrt{1365\frac{1}{3}}}$ for NL, the half of the diameter of the circle that can be inscribed in the triangle. The $\frac{2}{3}$ is $\sqrt{170\frac{2}{3}-\sqrt{21845\frac{1}{3}}}$ for the half of the diameter AN or NM of the circle in which such triangle can be inscribed . Or, one poses 1 for NL, so NE is 2 and EL $\sqrt{3}$. Side EF then is $\sqrt{12}$ and say: $\sqrt{12}$ gives me $\sqrt{512-\sqrt{196608}}$, what shall give me 2 and what shall give me 1? And out of 2 comes for EN or AN $\sqrt{170\frac{2}{3}-\sqrt{21845\frac{1}{3}}}$, and for 1 for NL one shall arrive at $\sqrt{42\frac{2}{3}-\sqrt{1365\frac{1}{3}}}$ as above. [71]

Question 63

There is a semicircle whose diameter is AB, being 50, in which there is drawn a line AF, with length $\sqrt{250}$ and also a line in the semicircle repeated four times and a little more, as much as CB. So the question is how many times this will be the line CB.

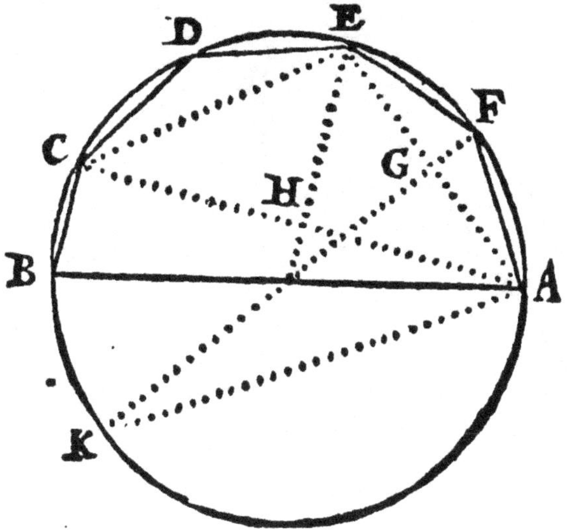

Figure 2.81: Question 63

Multiply AF $\sqrt{250}$ with FE $\sqrt{250}$ arriving at 250. This divided by the diameter AB 50, one shall arrive at 5 for FG. As KF stands to FA as AF to FG, hence subtracting the square of FG from the square of FA leaves the square of AG, 225. The square root from this is 15 for AG, or GE. Therefore AE is 30 as is also EC. In the same way, look for EH and AH, and AC shall be known, on shall arrive at 48 for AC. Its square [subtracted] from the square of AB 50, there remains the square of CB 196. The square root of this is 14 for the line CB, as one required.

Question 64

Look at the following page.

There is a quadrant or a fourth part of a circle $ABCD$, in which is drawn a quadrangle $EBCD$ so that ED runs parallel with BC and ED and BC are together of equal length as DC alone. One asks when DC makes $10 - \sqrt{20}$, how much the sides of the quadrant will be, AB or BC?

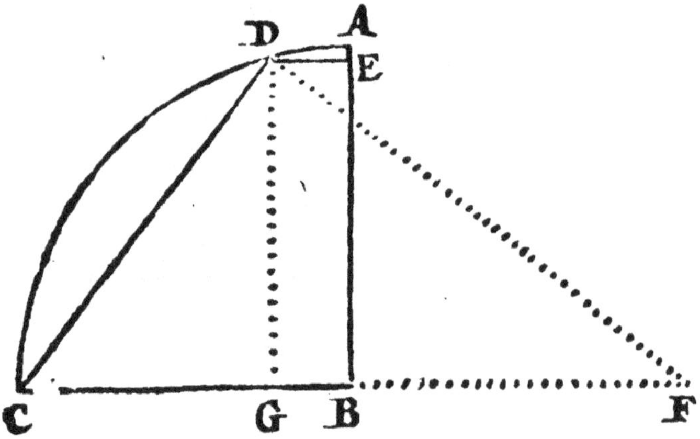

Figure 2.82: Question 64

[72] As ED and BC are as much together as CD, then FG is therefore equal to CD. As FB is equal to BC, and B is the center of the quadrant ADC, FDC is therefore a right-angled triangle, with D being the right angle, in which CG stands to GD as GD to GF, and as FG is equal to CD, so stands CG to GD as GD to CD, also CG to CD or FG as CD to FC. From this one notices that the line FC is divided into two parts, being FG and GC, such that CG stands to FG as FG to whole of the line FC. And also, as FG (being equal to DC) is now known, being $10 - \sqrt{20}$, so one finds (as in the 25th problem) GC, and one shall arrive at $\sqrt{180} - 10$ for GC and for the whole of the line FC $\sqrt{80}$. Its half is $\sqrt{20}$ for BC or for AB which one required to know.

Question 65

Look at the following figure.

Also, there is a circle, in which is inscribed the triangle ABC, whose sides are AB 13, BC 14, CA 15 and in the same triangle is also inscribed a circle, as large as possible. The question is how much the diameter of this circle is.

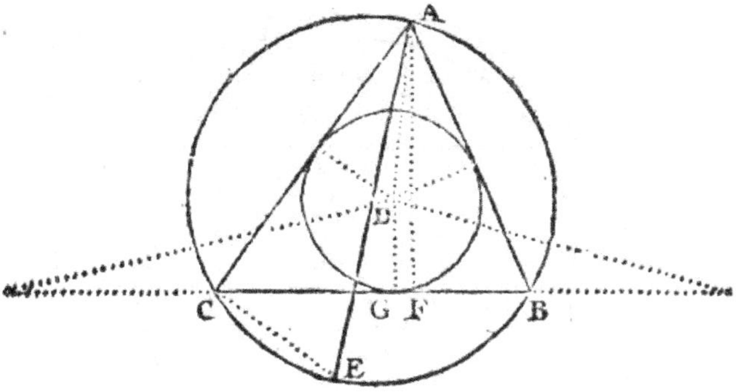

Figure 2.83: Question 65

Firstly, to find the diameter AE, for constructing the triangle, one first looks as in [73] our first example for the perpendicular AF, which will be 12, as the triangle EAC stands on same base AC as triangle BAC. Therefore, the angles AEC and ABC must be equal (because of prop. 21 of the third book of Euclid) and also the angle ACE is a right angle, as is AFB. So the angle BAF is equal to EAC, from which follows that the triangle ACE is of the same proportion as the triangle AFB. Therefore AF stands to AB as AC to the diameter AE. Therefore say: AF 12, gives me AB 13, what shall give me AC 15? On shall arrive at AE $16\frac{1}{4}$, the circles diameter, to be able to construct the triangle.

To find the diameters of the circles, so the triangle can be constructed, one first looks for the area of the triangle ABC which here is 84, doubling this gives 168, now add all three sides, [which] are together 42, and divide 168 by 42, arriving at 4 for GD, the half of the diameter of the inscribed circle. The whole diameter thus is 8. [74]

Question 66

Of the circle here shown, the diameter is AB 65, in which is constructed a triangle ACD of which the side AC is 25 and AD is 33. Question is what is the unknown side of the triangle CD?

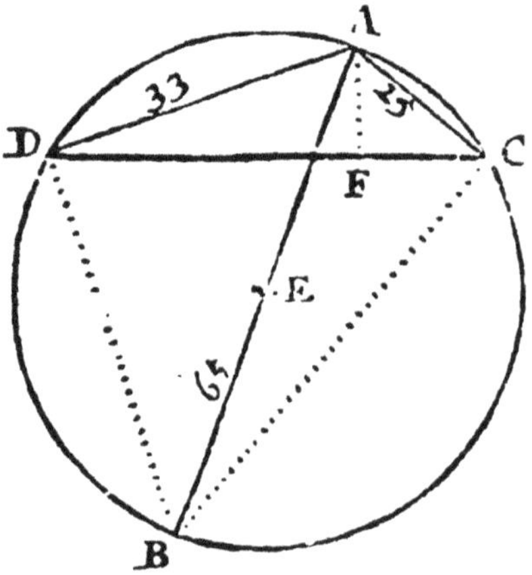

Figure 2.84: Question 66

Say: AB 65, gives AD 33, what shall give me AC 25? On shall arrive at $12\frac{9}{13}$ for AF. Furthermore, if one subtracts the square of AF from the square of AC, there remains the square of CF $463\frac{153}{169}$. From this, $\sqrt{\mathfrak{z}}$ is $21\frac{7}{13}$ for CF. Also, when looking for FD one shall find FD $30\frac{6}{13}$. This added to FC makes together 52 for CD. If one desires to know the reason why one arrives at such operations for finding AF, this is learned from the previous sixty-fifth problem.

Otherwise [1]

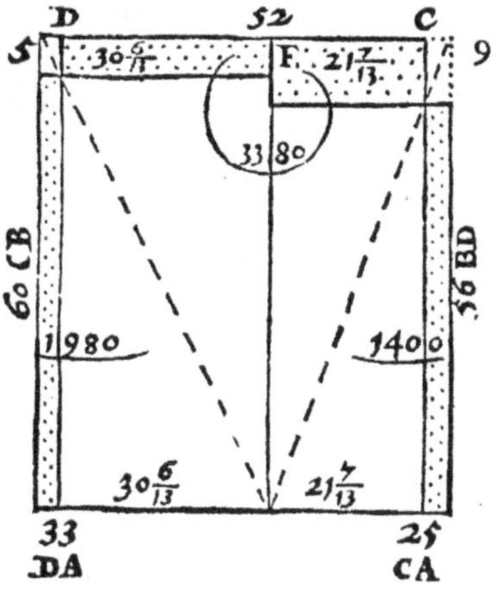

Figure 2.85: Question 66: alterative solution

The triangle ABD is in equal proportion to the triangle ACF, and the triangle ACB in equal proportion to AFD. Therefore, look first for BD and BC, which can be known because the angles ACB and ADB are right angled and AB, AD and AC are all three known. One shall arrive at BC 60, and 56 for BD. Say as follows: AB 65, gives me CB 60, what shall give me AD 33? Arriving at FD $30\frac{6}{13}$. Also say: AB 65, gives me BD 56, what shall give me AC 25? Arriving at CF $21\frac{7}{13}$. To this add FD $30\frac{6}{13}$, [which] is together 52 (as above) for CD.

Otherwise [2]

[75] Now that CB and BD are found, CB being 60 and BD 56, so multiply AC 25 by BD 56, arriving at 1400. Also multiply AD 33 by CB 60, arriving at 1980. These two products added together gives a sum of 3380. This divided by AB 65, arriving at 52 for CD, the side of the triangle, as above. The reason for this operation is easy to deduce from the two ways of solving the sixty-sixth question, because as the right-angled quadrangle, constructed by CB with AD, is equal to the one constructed by AB with FD, and also, as the right-angled quadrangle of BD with AC is equal to the one of AB with FC. This makes that the quadrangle of FD and FC together with the line AB, being CD with AB is as much as the aforementioned two right-angled ones together. As is to be seen in the accompanied figure, because they both are brought to the same length (as there is line AB 65).

Question 67

[76] There are three hunters A, B, C each lying at a particular place to shoot the parrot, however each of them at the same distance from the parrot, and B lies at 66 feet from C, B 50 feet from A and A 104 feet from C. I ask, when the parrot was raised 156 feet high above the earth how many feet have the hunters to shoot to reach the parrot?

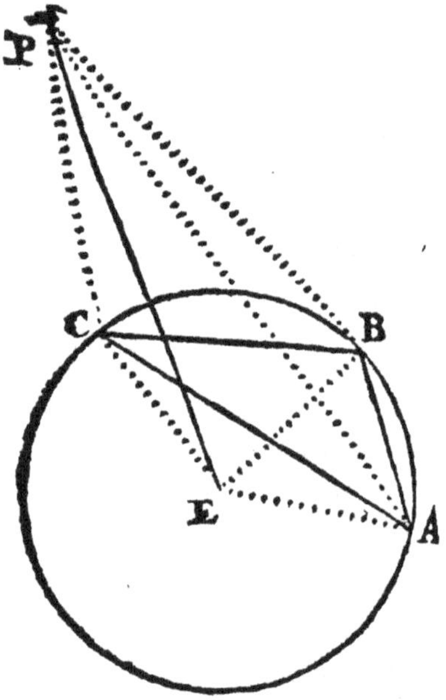

Figure 2.86: Question 67

As the hunters are at the same distance away from the parrot, one can understand that their places (as there are A, B and C) construct a triangle inscribed in a circle. Therefore, as above in our sixty-fifth question, first find how much the semi-diameter AE, BE or CE are. And one shall arrive at the semi-diameter AE 65, and that is as much feet the hunters are distant from the place where the parrot was raised, as in E. And now, to know how many feet the hunters have to shoot, one adds the square of AE to the square of EP, the height to which the parrot was raised above the earth. And extracting the $\sqrt{\mathfrak{z}}$ of this, one shall arrive at 169 feet for AP, BP or CP. And so much the hunters will have to shoot to hit the parrot. [77]

Question 68

In a circle is inscribed a quadrangle with unequal sides *ABCD*, *AB* being 25, *BC* 33, *DC* 60 and *AD* 16. I ask, how much then will be diameter *DE*, also how much then will be the diagonal lines *AC* and *BD*?

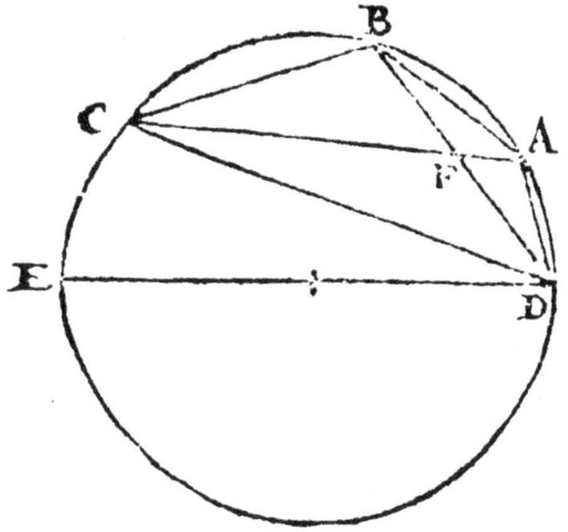

Figure 2.87: Question 68

The angle *DAC* is equal to the angle *BDC*, because of the 21st proposition of the third book of Euclid, as the triangles *DAC* and *DBC* stand on the same base *DC*, and also because the angle *ADB* is equal to *ACB* and the angle *AFD* is equal to *BFC*. From this follows that the triangle *AFD* is of equal proportion as the triangle *BFC*, so that *FA* stands to *FD*, as *FB* to *FC*, or also that *FA* [stands] to *FB*, as *FD* to *FC*. Therefore, let us pose for *AF* 1, so is *FB* $2\frac{1}{16}$, *FD* $2\frac{2}{5}$, *FC* $4\frac{19}{20}$. The diagonal line *AC* thus $5\frac{19}{20}$, and *BD* $4\frac{37}{80}$. Multiplying these diagonal lines by each other, this amounts to the multiplication of *AD* with *BC*, also of *AB* by *DC* and adding the products together. Now *AC* $5\frac{19}{20}$ multiplied by $4\frac{37}{80}$, arriving at $26\frac{883}{1600}$, and *AB* 25 multiplied by *DC* 60 is 1500. And *AD* 16 multiplied by *BC* 33 arrives at 528, together is 2028. Hence, this stands in proportion to $26\frac{883}{1600}$ as the square of *AF* stands to 1. Therefore, divide 2028 by $26\frac{883}{1600}$, thus $76\frac{5364}{14161}$ by the square for *AF*. The square root is $8\frac{88}{119}$ for *AF*. So the diagonal line *AC* is 52 and *DB* 39. [78]

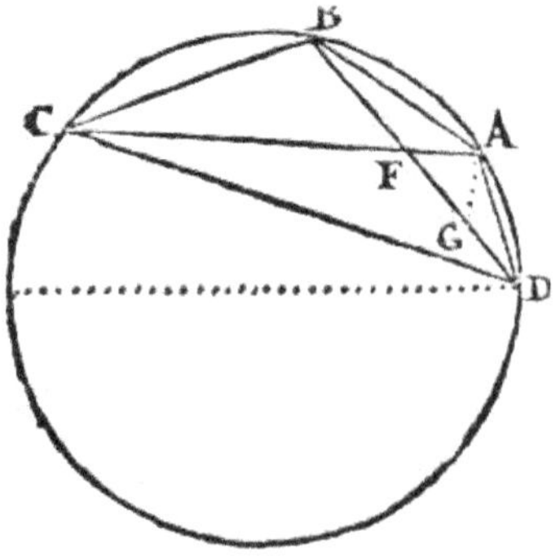

Figure 2.88: Question 68: alternative solution

Now that these diagonal lines are known, the triangle ABC becomes known, or the sides of the other triangles. The sides of the triangle ABC will be AB 25, BC 33, AC 52. We then find (by the sixty-fifth question) 65 for the diameter DE.

The diagonal lines can also be found in a very artful way by the 33rd and 34th problem of our questions, where in triangle ABC the proportion is known of the line BF to the parts AF and FC of base AC. And the two sides AB 25 and BC 33 are also known. So is determined by our 33rd and 34th question how much the base AC and the segments AF and FC each are in particular. And AC being known, you then have (as seen above) the diameter DE, which makes it possible to know the diagonal BD and its parts DF and FD.

And concerning the demonstration that the opposite sides of the quadrangle, as AB against DC and AD against BC, multiplied together amount to as much as multiplying the diagonal lines, as here AC multiplied by BD, so is done by Ptolemy and Copernicus. Next, we will demonstrate the same.

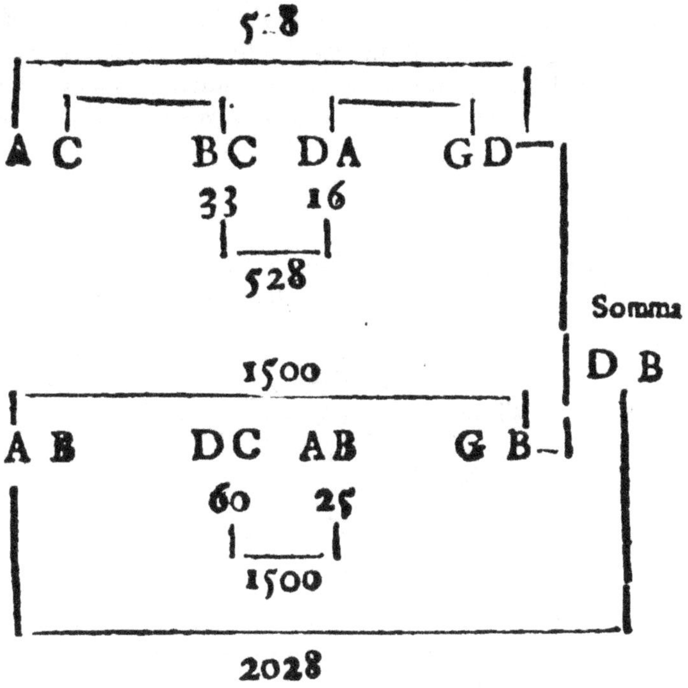

Figure 2.89: Question 68: alternative solution

Let us draw the line AG so that the angle DAG is equal to the angle FAB. Now by the 21st proposition of the third [book] of Euclid, the angles ADG and ACB will be equal to each other, because the triangles ADB and ACB both stand on the same base AB. Therefore the angle AGD must be equal to the angle ABC, and also both the triangles AGD and ABC must have equal angles, so that the sides [79] are proportional to each other. Furthermore, also the angles DAC and GAB are equal, and the angle ABG is equal to the angle ACD, for the same reason as before. Therefore, also the third angles AGB and ADC must be equal to one another. The triangles ABG and ACD are therefore of the same proportion. So stands AC to BC as DA to GD. Therefore is the right-angled quadrangle of BC and DA equal to the right-angled quadrangle of AC and GD. Also, because AC stands to DC as AB to GB, the right-angled quadrangle of DC and AB thus are equal to the right-angled quadrangle of AC and GB. And if one now adds the right-angled quadrangle of AC and GD to the right-angled quadrangle of AC and GB then that will make together a right-angled quadrangle of AC and DB, because GD and GB together make the whole line DB. Which then makes the right-angled quadrangle of AC with DB equal to the right-angled quadrangles of BC and DA together with DC and AB, which we had to prove. As is here pictured by numbers and also by geometrical figures in our 66th question which serves as an explanation of the same. [80]

Question 69

There is a circle whose diameter AB makes 20. Question: how much will make the sides of the pentagon and the decagon that can be inscribed in the circle?

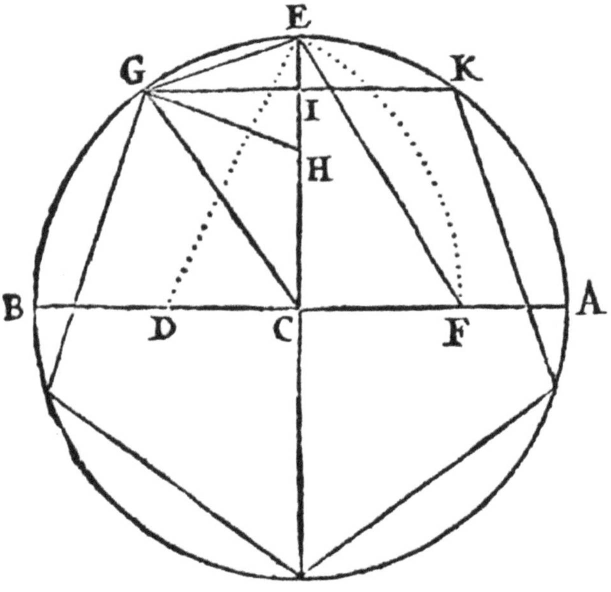

Figure 2.90: Question 69

Multiply the half of the semi-diameter CD 5 as a square, arriving at 25. Also [multiply] the semi-diameter CE 10 as a square, arriving at 100. This [added] to 25 is together 125 for the square of DE. So line DE is $\sqrt{125}$ and so much will also be FD. From this subtracted CD 5, there remains $\sqrt{125} - 5$ for FC, or the side EG of the decagon.

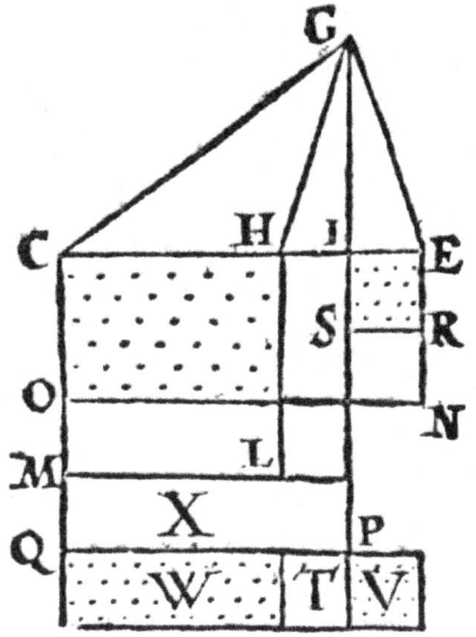

Figure 2.91: Question 69: demonstration

The square of FC $\sqrt{125} - 5$ is $150 - \sqrt{12500}$, added to the square of the semi-diameter CE 10, being 100 arrives together at $250 - \sqrt{12500}$ for FE or the side KG of the pentagon. The demonstration of such calculation is (according to the 30th proposition of the eleventh book of Euclid) that when the semi-diameter is divided into two parts, such that the smaller is in proportion to the larger as the larger part is to semi-diameter, so the larger part will be the side of a decagon, as here EG, so the larger part FC will be equal, as the isosceles triangle ECG is such a triangle as the angles CEG or CGE are each [81] twice as much as the angle ECG, as demonstrated in our thirty fifth question. Because if ECG be one, so the angles CEG and CGE will each be 2. The sum of such angles of such triangle will therefore be 5 as in a semi-circle. By the 32nd proposition of the first book of Euclid, the whole circle then will be 10. Angles such as ECG will therefore be the arc EG which is $\frac{1}{10}$ of the circle, and line EG the side of the decagon.

And to prove that the square of the semi-diameter CE and the square of [the side of] the decagon FC (being together the square of FE) are as much as the square of [the side of] the pentagon, or that line FE be equal to the side KG of the pentagon. So, consider from the figure of the pentagon aforementioned the triangle EGC, of which EG is the side of the decagon and IG half the side of a pentagon, and CG or EC each is a semi-diameter. Now, we must prove that the squares of GC and GE together are as much as the square of GK, or that both squares of GE and GC, be four times as much as the square of GI, as GI is the half of GK. So let us now subtract the square of IG from [the square of] EG, so that there remains the square $EISR$. Also, [subtracting the square of] GI from the square of GC, there remains the square $ICQP$. So, these two remainders should be together equal to the two squares of GI, as from each of the squares of EG and CG are subtracted the square of GI. And as the angles CEG and CGE from this triangle are equal, being each twice as much as the angle GCE, so stands EH to HC as HC to the whole [82] of the line EC, as is proved in our thirty fifth question. Therefore then, the square of HC, being $HCML$ is equal to the quadrangle that can be constructed from the whole line EC and EH, as is here $ECON$. Now, the line HC and EG or HG are equal. So, the [area of] quadrangle $ECON$ is so much larger than the square of GI, as is the square $EISR$. So let us cut off this, there remains part $NRSICON$, equal to the square of GI.

Furthermore, as we have said before, that the square $EISR$ and $ICQP$ together are two squares of GI, so let us cut off the square $HCML$ from this, so remains the square $EISR$ with gnomon $IPQMLH$. But, as $HCML$ is as larger than the square of GI as the square $EISR$, so we must paste to the remainder such square $EISR$, here drawn as T. Such gnomon $IPQMLH$ together with the two squares of $EISR$, together equal to the square of GI, which I have here drawn as V, T, W, X, are equal to the part $NRSICON$ aforementioned. And as these two parts each are equal to a square of GI, together they must also be equal to the squares of EI and IC. From this follows that both the squares of EI and IC together are then also two squares of GI. And such four squares of GI make as much as both the squares of EG and CG, as the square of EI and GI together with the square of IC and GI make as much as both the squares of EG and CG, by the 47th proposition of the first book of Euclid. Or, both these squares of EG and CG are as much together as the square of GK only, as GK is twice as much as GI, which we had to prove. [83]

Question 70

There is a shifted quadrangle [i.e. parallelogram] $ABCD$ of which sides AB or DC each make $\sqrt{106}$ and the diagonal line AC makes 12, and BD 10. The question is: when side AD equals side BC, then how much will AB or BC be?

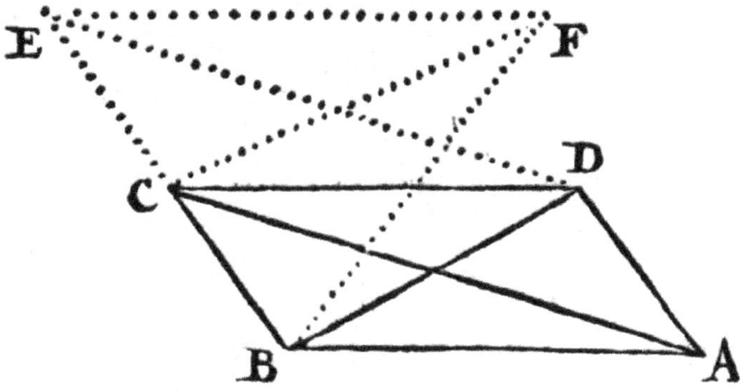

Figure 2.92: Question 70

To find this, notice that the squares of BD and DE together are always as much as the squares of BF and FC together. So, when line DC stands perpendicular on base BE at point C, being at the half of BE, as here is line FC, it also makes $\sqrt{106}$, as DC, so BF and FE are also equal. So, this being the case, add both squares of BD 10 and DE 12 together, arriving at 244. Its half is 122 for one of the squares of BF or FE. From this, subtract the square of FC being 106, so there remains 16 for the square of BC or CE. So BC is 4 as is AD.

This question can also be solved in the manner of our twentieth question. The demonstration of these operations is contained in our fifty seventh question.

Question 71

Look at the following page.

There is an irregular quadrangle $ABCD$ of which side AB makes $\sqrt{192\frac{1}{4}}$, BC 13, CD 15, the [84] diagonal BD makes 14 and AC $19\frac{1}{2}$. Question: how much will be side AD?

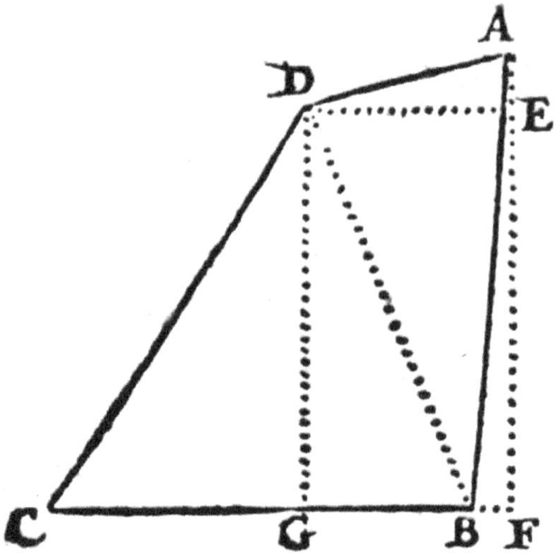

Figure 2.93: Question 71

First look for *BG* and the perpendicular *DG* and one shall arrive at *DG* $12\frac{12}{13}$ and for *BG* $5\frac{5}{13}$. Also look for *AF* and *FB* and one shall arrive at $13\frac{11}{13}$ for *AF* and $\frac{19}{26}$ for *FB*. Add $\frac{19}{26}$ to $5\frac{5}{13}$, which together is $6\frac{3}{26}$ for *FG* or *ED*. Now, subtract $12\frac{12}{13}$ *DG* from $13\frac{11}{13}$ *AF*, there remains $\frac{12}{13}$ for *EA*. This square of *EA* added together with [the square of] *ED* is $38\frac{1}{4}$ for the square of *AD*. So line *AD* is $\sqrt{38\frac{1}{4}}$ as one required.

Question 72

From the irregular quadrangle *ABCD* here shown, the sides are known, *AB* makes 26, *BC* 40, *CD* 20 and *AD* 34. Also, the diagonal line *AC* makes 42. Question: how much then will make the other diagonal line *BD*?

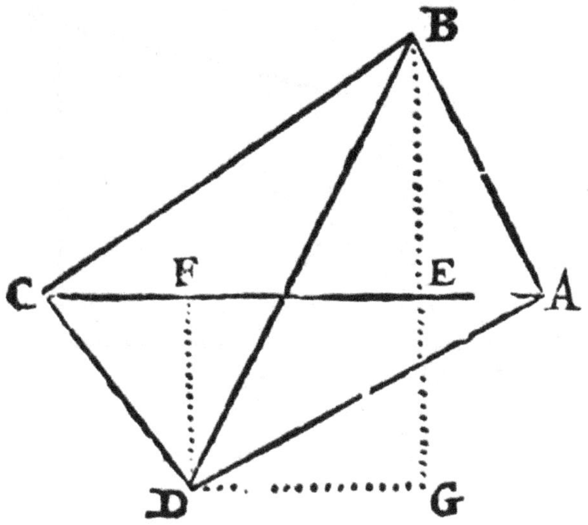

Figure 2.94: Question 72

To find this diagonal line BD, first look for AE and FC as in our first example, arriving at AE 10 and for FC 12. After that, look for both the perpendiculars, as EB and FD, and one shall arrive for EB at 24 and for FD 16. So BG is 40. Now, add AE and FC, [which] together is 22. This [subtracted] from AC 42, there remains EF 20, and so much is also [85] GD. At last, the square GD stands to the square GB, arriving at 2000 for the square of BD. So the diagonal line BD is $\sqrt{2000}$.

Question 73

There is a quadrangle $ABCD$ of which angle A is right angled and AB makes $33\frac{3}{5}$, BC 39, CD 25 and AD $44\frac{4}{5}$. One desires to know how much perpendicular EC will be?

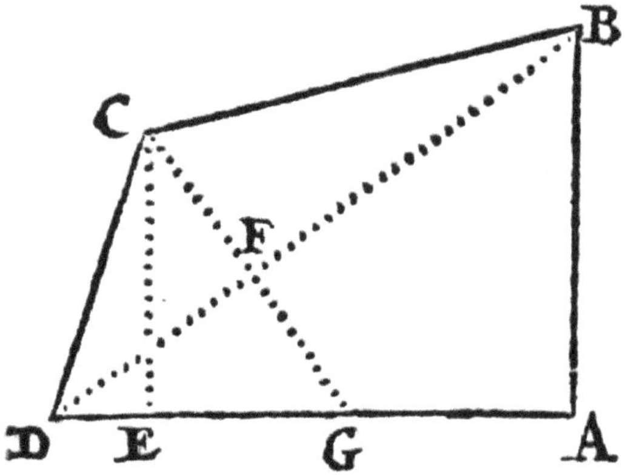

Figure 2.95: Question 73

First, look for DF which can be done as AB and AD are known, and therefore is also BD known (being 56), by the 47th proposition of the first book of Euclid. So, one shall find 20 for DF. So, also FC is known, making 15. Now say: DA $44\frac{4}{5}$ gives me AB $33\frac{3}{5}$, what shall give me DF 20? One shall arrive at 15 for FG. This added to FC is together 30 for CG. Also say: BD 56 gives me AD $44\frac{4}{5}$, what shall give me GC 30? One shall arrive at EC 24, as one required.

Question 74

Look at the following page.

There is another quadrangle $ABCD$ of which is known to me that B is right angled and that AB makes 48, BC 14, CD 30 and AD 40. The question is the perpendicular DE?

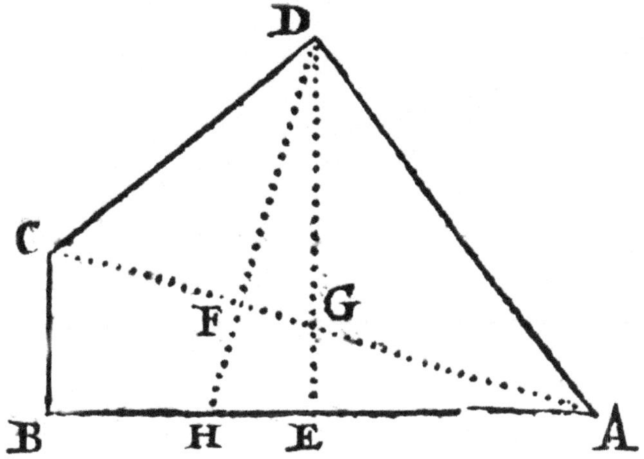

Figure 2.96: Question 74

First, look for AC, arriving at 50 for AC. Then, [look] for how much AF and DF are, and one shall arrive at 32 for AF [86] and 24 for DF. Say: AB 48 gives me BC 14, what shall give me DF 24? One shall arrive at 7 for GF. Also: AB 48 gives me AC 50, what gives me 24 DF? One shall arrive at 25 for GD. Subtract GF 7 from AF 32, there remains 25 for AG. And say more: AC 50 gives CB 14, what shall give AG 25? Arriving at GE 7. This added to GD 25 is together 32 for the line ED, as one required.

Otherwise

With FD and FA being found, so say: 48 AB gives 14 BC, what shall give me 32 AF? One shall arrive at FH $9\frac{1}{3}$. This added to FD is together $33\frac{1}{3}$ for HD. Now say: AC 56 gives me AB 48, what shall give me HD $33\frac{1}{3}$? One shall arrive at 32 for DE. This line can also be found otherwise as is been taught in our 72nd question.

Question 75

Look at the following page.

One has a quadrangular piece of land $ABCD$ with an area of 576 rods, of which C is a right angle. The line AB makes 15, BC 48 and CD 20. The question is to the side AB of such quadrangle.

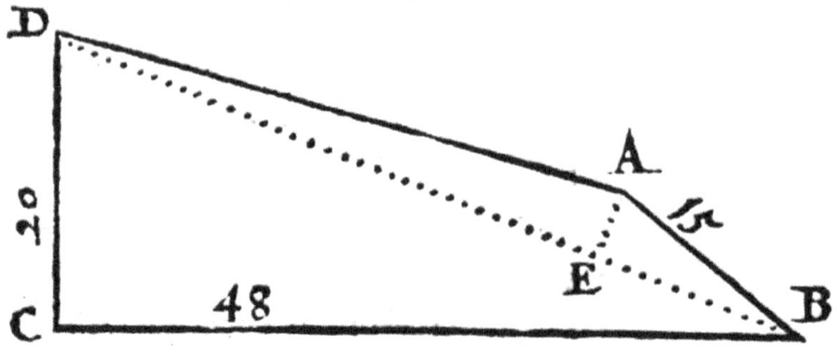

Figure 2.97: Question 75

To find AD, first look for the area of the triangle BCD, and one shall arrive at 480. This subtracted from 576, there remains [87] 96 for the area of ABD. And as C is a right angle, so is the diagonal line BD also known (by the 47th proposition of the first book of Euclid), which shall be found to be 52. Its half 26, dividing 96, the area of the triangle ABD, arriving at $3\frac{9}{13}$ for the perpendicular AE. Now that AE is known, so is also BE known (by the 47th proposition of the first book of Euclid aforementioned) being $14\frac{7}{13}$, subtracted from the whole of BE, there remains ED $37\frac{6}{13}$, and (by the 47th proposition of the first book of Euclid) as ED and EA are known, so one arrives also at AD $\sqrt{1417}$, as one required to know.

Question 76

Look at the following page.

A parrot is above a triangular piece of land ABC and at each angle there is a hunter to shoot the parrot. It has been found that the hunter at A has to shoot 140 feet to reach the parrot, at B 150 feet and at C 130 feet. Now the question is, if the triangular piece of land measures 130 feet from A to B, 140 feet from B to C and 150 feet from C to A how high is then the parrot D raised above the horizon, also how many feet each of the hunters is away from the pole on which the parrot is above.

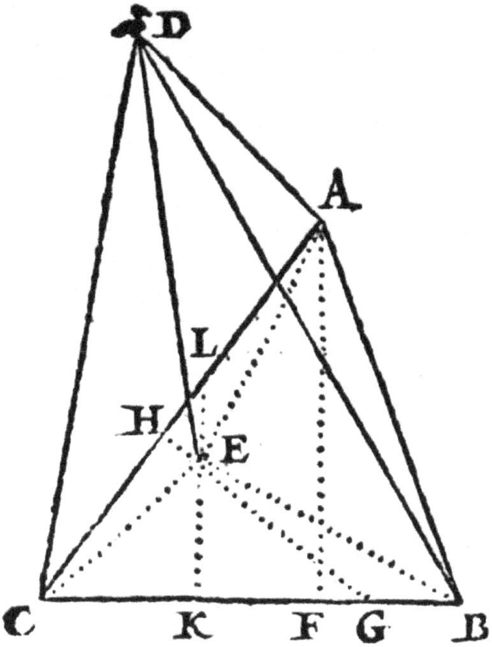

Figure 2.98: Question 76

First, look where the perpendicular falls down of the triangle [88] BDC from angle D, on line BC, which is at K, KC making 50, and KB 90. Also, look where the perpendicular falls down of triangle ABC on the plane field at line BC, which is at F, FB making 50 and FC 90, and the perpendicular FA 120. Now say: FC 90 gives me FA 120, what shall give me KC 50? One shall arrive at KL $66\frac{2}{3}$. Also, look where the perpendicular falls down of triangle ADC, from angle D on line AC, which is at H. So is CH 66 and say: FC 90 gives me FA 120, what shall give me CH 66? And one shall arrive at 88 for GH. Furthermore: FC 90 gives me CA 150, what shall give me CH 66? And one shall arrive at 110 for GC. So is KG 60. Also: AF 120 gives me FG 90, what shall give me KG 60? One shall arrive at 45 for EK, so EG shall be 75 and EH 13. The square of EK [added] to the square of KB and from this taken the $\sqrt{\mathfrak{z}}$, arriving at BE $\sqrt{10125}$. Also, look for CE, arriving at $\sqrt{4525}$. For AE 85 feet, for the distance that the hunters are away from the pole E.

Now, to find the height of the parrot, as here ED, so subtract the square of AE from the square of AB. There remains 12375 for the square of ED. The $\sqrt{\mathfrak{z}}$ from this is $\sqrt{12375}$ for the height ED, or how much the parrot is above the horizon. Or one subtracts the square of BE from the square of BD, or the square of CE from [the square of] CD, and the $\sqrt{\mathfrak{z}}$ taken from this, shall arrive at ED $\sqrt{12375}$ feet (as above), this is approximately $111\frac{1}{4}$ feet. [89]

Question 77

There is a triangular piece of land ABC, of which the sides make AB 13, BC 15 and CA 14, which one wants to divide into three equal parts, from side BC at point D. Question: where will the lines DE and DF cut the lines AB and BC if DC makes 6 rods?

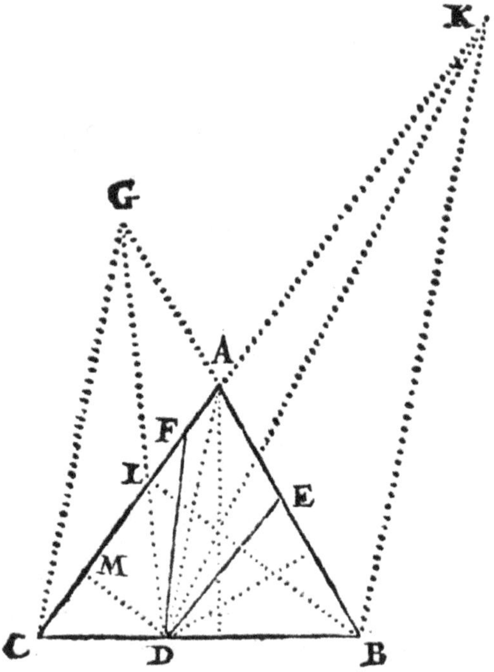

Figure 2.99: Question 77

As BC makes 15 and DC 6, so is BD 9. Now say thus: BD 9 gives me BA 13, what shall give me BC 15? One shall arrive at $21\frac{2}{3}$ for BG. And triangle BDG is now equal to ABC, as AD and GC are parallel and the triangles ACD

and AGD stand on the same base AD, which makes them equal according to the 37th proposition of the first book of Euclid, and hence BDG and ABC are equal, as stated above. So, divide base BG into three equal parts, as one wanted to divide the triangle ABC into three equal parts. Arriving at $7\frac{2}{9}$ for BE. So is triangle EDB, being cut by line ED, the $\frac{1}{3}$ as one required.

Now say: DC 6 gives me AC 14, what shall give me BC 15?[8] Arriving at 35 for KC. This being cut into three equal parts, as KDC is also equal to the [90] triangle ABC. Arriving at $11\frac{2}{3}$ for the $\frac{1}{3}$ FC. So is triangle FDC the $\frac{1}{3}$ of the whole of triangle ABC, cut by line DF. The remainder $DEAF$ then also is $\frac{1}{3}$ of the whole triangle. And as such the whole triangle ABC has been divided into three equal parts as one required, and with BE and CF known, being cut by lines DE and DF.

Otherwise

First look for the area of triangle ABC, and one shall arrive at 84, the $\frac{1}{3}$ of this is 28. Now look for the perpendicular BL as in our first question, so which will fall from angle B on AC in L, one shall arrive at 12 for BL. Say: CB 15 gives me BL 12, what shall give me DC 6? One shall arrive at DM $4\frac{4}{5}$, its half is $2\frac{2}{5}$. The $\frac{1}{3}$ of the area, being 28, divided by $2\frac{2}{5}$ arrives at $11\frac{2}{3}$ for CF. And look for BE in the same way, and one shall arrive at $7\frac{2}{9}$ for BE (as above).

Question 78

Look at the following figure.

Two brothers own together a quadrangular piece of land $ABCD$ to be divided, of which AB is parallel to DC, AB making 24 and DC 36, and the angles B and C are right angles. They want this piece divided into two equal parts, by line EF, such that EB and EC are each the proper half of BC, knowing that each gets 8 feet of breadth. Question: how much breadth will each receive from the side AD. That is: how much will be DF or FA, so that each has the half of the land described.

[8] Cardinael omits BC for the value 15.

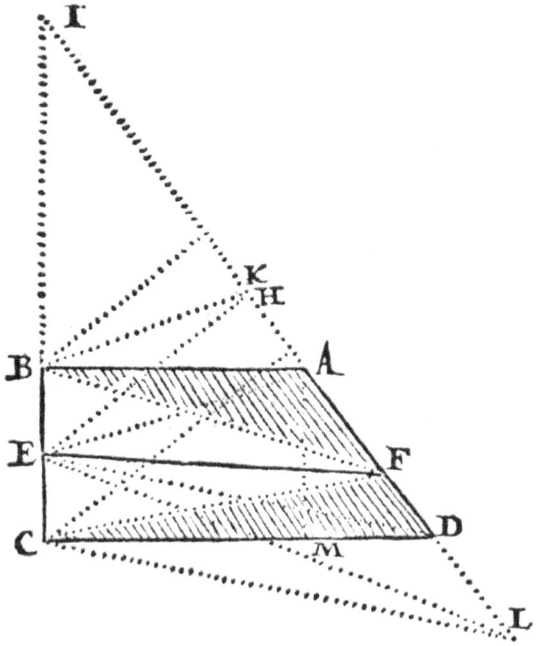

Figure 2.100: Question 78

First, look how much side BI will be, and to find this, say as such: DM 12 gives me MA 16, what [91] shall give me AB 24? One shall arrive at 32 for BI. So EI is 40. Now, as AD 20 stands to DM 12, also [stands] EI 40 to EH, and one shall arrive at 24 for EH.

Furthermore, multiply EB 8 by AB 24, arriving at 192 for double the area of triangle ABE, as ABE is right angled. This divided by EH 24 arrives at 8 for AK, as triangle AKE becomes equal to triangle ABE, knowing that double the area of AKE also be 192. Now that HE is also being known, being the perpendicular on AH, so can AK be known, arriving at 8 for AK. Also, look for DL by the same means. Multiply EC 8 by DC 36, arriving at 288 for double the area of triangle ECD, as C is a right angle. Now, also triangle LDE becomes equal to DEC, and therefore double its area is 288. This also divided by HE, the perpendicular aforementioned, being 24, one shall arrive at 12 for DL. At last, add DL 12, DA 20 and AK together, arriving at the sum 40 for LK. The half of this is 20 for LF or FK. From this LD 12 subtracted, there remains 8 for DF. Then FA is 12 as one required. [92]

Otherwise

Add AB 24 to DC 36, arriving together at 60. And say: 60 gives me AD 20, what shall give me AB 24? One shall arrive at 8 for DF. Or, what shall give me 36? One shall arrive at 12 for FA, as above.

Question 79

A farmer owns a quadrangular piece of land $ABCD$, of which AB runs parallel to DC, and the angles B and C are right angles, AB making 24, BC 16 and CD 36. He wants to divide the land into two equal parts, such that FE divides the land, starting from the half of AD. Question: how much will EC and EB be?

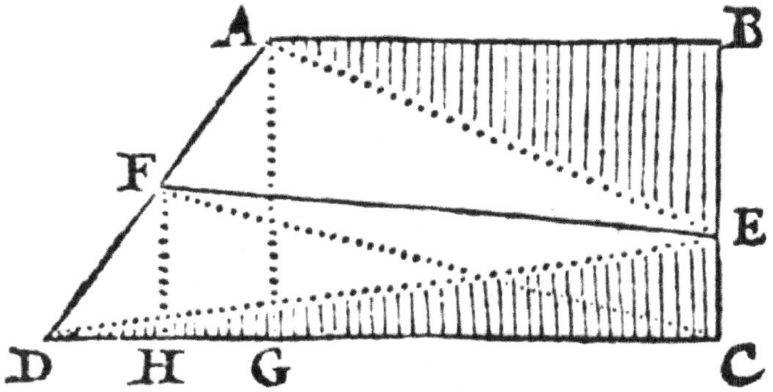

Figure 2.101: Question 79

First look for FH and HD, which can be known as AG and AD are known, and one shall arrive at FH 8 and HD 6. Now, if the half of FH 8, being 4, is multiplied by CD 36, one arrives at 144 for the area of triangle CFD. This subtracted from 240, the half of the area of the whole quadrangle, there remains 96 for the area of ECF. Also, subtract HD 6 from CD 36, there remains 30 for CH. Its half is 15. So, now divide 96 by 15, arriving at $6\frac{2}{5}$ for EC. So, EB is $9\frac{3}{5}$.

This question can also be solved in a way which was taught in our seventy eight question, in which the quadrangle is transformed into a triangle with the extended line BC as its base, etc. [93]

Otherwise

Add AB 24 to DC 36, arriving together at 60. Say: 60 gives me BC 16 what shall give me 24. One shall arrive at $6\frac{2}{5}$ for EC, or what shall give me 36? One shall arrive at $9\frac{3}{5}$ for EB.

Question 80

There is a quadrangle $ABCD$, of which AD runs parallel to BC, AB making 25, BC 18, CD $21\frac{2}{3}$ and AD 8, in which is drawn the parallel line EG such that [the area of] triangle ABC is equal to [the area of] quadrangle $EGCB$. Question: the length of parallel line EG and also how much that AE and DG will be?

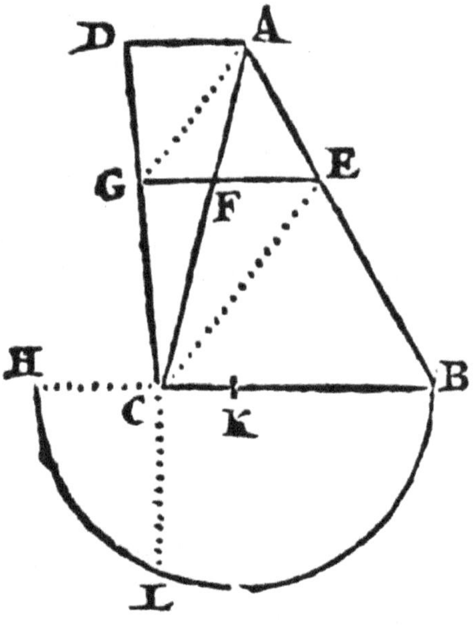

Figure 2.102: Question 80

To do this, let use make CH equal to AD, so that BH becomes as much as BC and AD together. The center point of this line BH be K, on which is drawn a circle, and if one draws the perpendicular CL from C to the circumference, so will CL be the mean proportional between BC and CH. Therefore, multiply CH 8 by BC 18, arriving at 144. The $\sqrt{\mathfrak{z}}$ from this is 12 for CL and so much is then also EG. That EG is the mean proportional between AD and BC can be proven as such: as [the area of] triangle ABC has remained the same as [that of] quadrangle BEG so let us draw the diagonal line EC. This cuts off triangle CGE which must be equal to triangle CAE. It is therefore that we draw line AG, so that there are 2 triangles standing on the same base EC, which [94] have an equal area. From the 37th proposition of the first book of Euclid then follows that the line AG is parallel to EC and AG then is the diagonal line of quadrangle $ADGE$, as line EC is the diagonal line of quadrangle $EGCB$. Furthermore, as AD runs parallel to EG and AG to EC, the triangles ADG and EGC are therefore of the same proportion. Also, as EG runs parallel to BC and EC to AG, triangles AGE and ECB are of equal proportion. Now, these triangles, both of the same proportion, make $ADGE$, as they both make quadrangle $EGCB$. From this follows that quadrangle $ADGE$ is of the same proportion as quadrangle $EGCB$. So, BC stands to EG as EG stands to AD, and hence EG is the mean proportional between BC and AD, which we had to prove.

Further notice that as BC stands to CG, also EG [stands] to GD, or [as] BC [stands] to BE, as GE [stands] to EA. From this AE and DG become known, being EG 12 and BC 18. They are together 30, to be divided by AB 25, what shall give me EG 12 and BC 18? One shall arrive from 12 at AE 10 and from 18 at EB 15. Also, look for GD and GC and one shall arrive at GD $8\frac{2}{3}$ and GC 13, as one required.

Question 81

Look at the following figure.

Someone owns a right-angled quadrangular piece of land $ABCD$, with length 16 and breadth 12 rods. He wants to have this divided into two equal parts, such that one part will be the right-angled quadrangle $DGFE$ and the other the border around that quadrangle, such that AE and CG (the breadth of the border) [95] remain everywhere equal and that EFG be parallel to ABC.

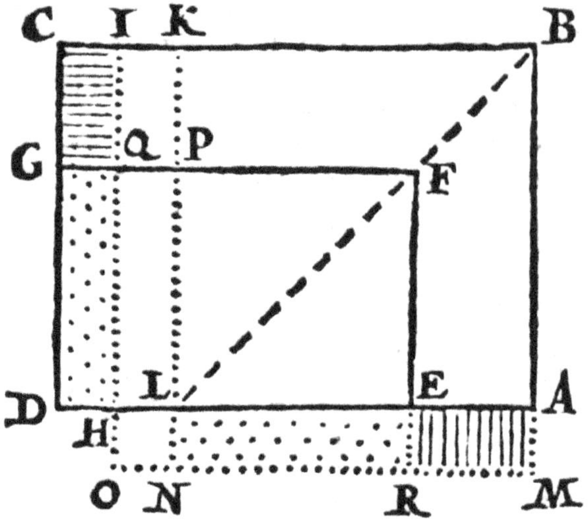

Figure 2.103: Question 81

To do [so], let us draw the square $ABKL$ whose side will be 12. So, KC and LD will be 4. Now, let us remove half of the cut-off part $KLDC$, which then becomes $IHDC$ and paste it to AL, so that $ALNM$ becomes equal to the given half $IHDC$. So, to complete square $BIOM$, there will be missing a small square, being $LHON$. Furthermore, given that the quadrangle $AFGD$ is the half of the area of $ABCD$, the area of $EFGD$ will thus be 96. Now, from $EFGD$ has been removed a part $HQGD$ and added to EL, with $ENLR$ being equal, so that the area of $RFQHLNR$ must have remained the same as before $EFGD$ 96. And that we now complete this by the small square $LHON$ with an area of 4 (as KI or NO make 2). So such square $RFQO$ will make 100. Now take the $\sqrt{\mathfrak{z}}$ from 100, arriving at 10 for RO or EH. To this added HD, which is 2, arriving together at 12 for ED. Also, subtract LH 2 from EH 10, there remains 8 for EL, which is equal to EF. So, EF or GD then is 8. As such is found (as required) 12 for the length ED or FG and 8 for the breadth EF or DG, making 96 for the area $EFGD$, being the half of the whole quadrangle $ABCD$, so that AE and GC are equal and each makes 4. [96]

Question 82

I have a quadrangular piece of land $ABCD$ of which sides AD and BC run parallel, the angles C and D are right angled and AD makes 38, DC 16 and BC 50. I want to divide this piece into two equal parts, parallel to ABC in such a way that AI or KF shall be equal to FS or CG. I then ask: how much will AE and CG be?

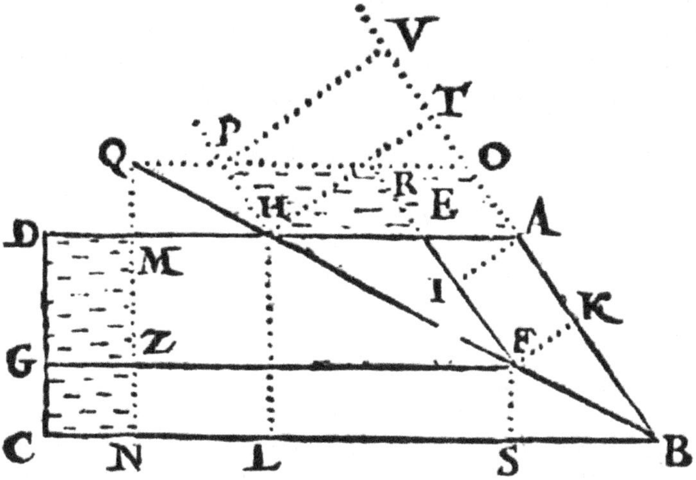

Figure 2.104: Question 82

Make AH equal to AB. Now, as CD, AD and BC are known, so can AB also be known and one shall arrive at 20. And draw line BHQ such that the angle B is divided into two [equal parts]. So, AH then is also equal to AB 20 and HD 18. This HD we now must divide into two parts HM and MD such that HM is in proportion to MD as BL is to BA, and as LC is being made equal to HD, making 18. So is LB 32. Therefore say: 20 and 32 are together 52. This gives me 18, what shall give me 20? And what shall give me 32? From 20 one shall arrive at MD $6\frac{12}{13}$ and from 32 at $11\frac{1}{13}$ for MH. Furthermore, let us draw the shifted parallelogram $AOPH$, which is equal to $NMDC$, as $THPV$ is equal to $NMDC$, this so as TH is equal to HL and HP equal to MD, as follows from previous division. And as the triangles HTB and HLB have equal angles and as $THPV$ stands between two parallel [lines] AV and

HP, therefore [97] is $AOPH$ equal to $THPV$ or $NMDC$. So $OBNMHPO$ then is as much as the quadrangle $ABCD$, with an area of 704. Add to this the area of $HMQP$, which can be known because the area of $ABLH$ is known, as AH, HL and BL are known, and one shall arrive at 416 for the area of $ABLH$. Therefore say: the square of AB 20 gives me 416, what shall give me the square of HP $6\frac{12}{13}$? One shall arrive at $49\frac{11}{13}$ for the area of $HMQP$. This added to 704, arrives at $753\frac{11}{13}$ for the area of $OBNQ$. From this subtracted the half 352, there remains $401\frac{11}{13}$ for $RFZQ$. Also say: AB 20 gives me CD 16, what shall give me HP $6\frac{12}{13}$? Arriving at $5\frac{7}{13}$? for MQ. So QN is $21\frac{7}{13}$. This multiplied as a square arriving at $463\frac{153}{169}$. And again say: $753\frac{11}{13}$ gives me $463\frac{153}{169}$, what shall give me $401\frac{11}{13}$? One shall arrive at $247\frac{49}{149}$ for the square of QZ. So line QZ is $\sqrt{247\frac{49}{149}}$. This subtracted from QN $21\frac{7}{13}$, there remains $21\frac{7}{13} - \sqrt{247\frac{49}{149}}$ for ZN, and so much is also CG.

At last say: 16 gives me 20, what shall give me $21\frac{7}{13} - \sqrt{247\frac{49}{149}}$? Arriving at $26\frac{12}{13} - \sqrt{386\frac{66}{169}}$ for AE.

Question 83

Look at the following side.

There is a quadrangle $ABCD$ (as above), of which AD runs parallel to BC and the angles C and D are right angles, AD making 38, DC 16, BC 50. One wants to divide this quadrangle into two equal parts, parallel to lines BC and CD, such that FI shall be equal to FH. Question: how much then have to be GD and BE?

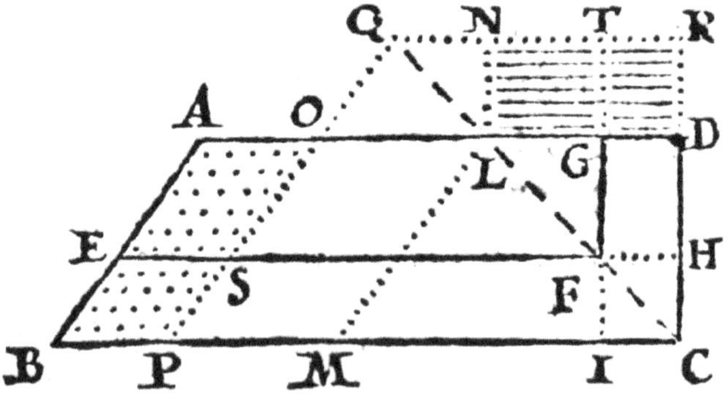

Figure 2.105: Question 83

To find this, subtract DC 16 from DA 38, there remains 22 for LA or MB. So MC is 28. This 28 [98] MC added to DC 16 makes together 44. And say: 44 gives me 22, what shall give me 16? Also: what shall give me 28? One shall arrive at 16 for OA 8 and from 28 shall arrive 14 for OL. So, line AL is divided into two parts OA and OL in such a way that OA stands to OL as MC to CD. Therefore, extend line CD as long as OA, so that CR or DR becomes equal to OA, so CR then is 24. Now, also add MP 14 to MC 28, which is together 42 for PC. This being done, let us now find the area of the whole quadrangle $ABCD$, and from this the half, which shall be 352. We also have to find the area of $CRQP$ and shall find 792. From this subtracting 352, being the half of the aforementioned quadrangle $ABCD$, there remains 440 for the area of $SQTF$. Furthermore, multiply CR 24 as a square, arriving at 576. And say: 792 gives me 576, what shall give me 440. And one shall arrive at 320 for the square of FT. So, TF is $\sqrt{320}$. This subtracted from TI, being 24, there remains $24 - \sqrt{320}$ for FI. So much is also FH or GD which we had to find.

Now, to find also EB, say: CD 16 gives me HC $24 - \sqrt{320}$ which is equal to FI or IC, what shall give me AB 20? And one shall arrive at $30 - \sqrt{500}$ for BE.

Question 84

Look at the following figure.

One owns an irregular quadrangular piece of land, as here $ABCD$, of which AB makes 65 [99], BC $\sqrt{605}$, CD $\sqrt{1800}$ and DA 20, the perpendicular EC 22. One wants to have this piece divided into two equal parts, such that KFL runs parallel to DAB and with FG and FH being equal. Question: how much will be FG or FH, also DK and LB?

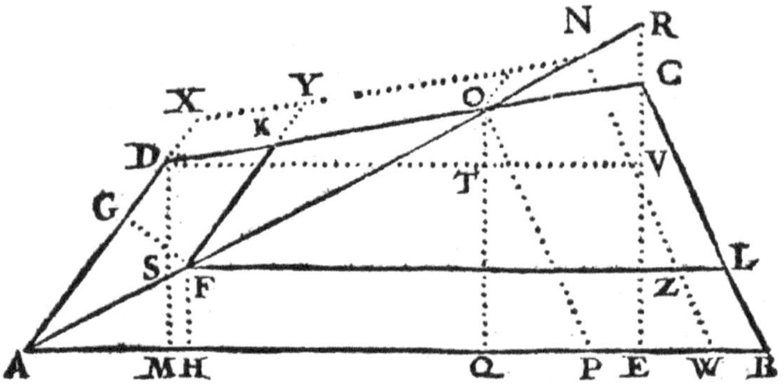

Figure 2.106: Question 84

First look how much BE will be (as CE is known), which can be found by the 47th proposition of the first book of Euclid, and one shall arrive at 11 for BE. So EA then is 54. Also, look for the perpendicular DM as is taught in our 73rd question. One shall find 16 for DM. The area of the quadrangular $ABCD$ shown is then known. Now, add AM and AD together, is 32. And say: 32 gives me DM 16, what shall give me AD 20, and AM 12? One shall arrive at DS 10 from 20 and SM 6 from 12, by the third proposition of the sixth book of Euclid. Now say: AM 12 gives me MS 6, what shall give me AE 54? One shall arrive at 27 for ER. From this subtract EC 22, there remains 5 for CR, and add SD 10 to CR 5, which together is 15. And say: 15 gives me CD $\sqrt{1800}$, what shall give me 10, and 5? And one shall arrive at DO $\sqrt{800}$ from 10 and at CO $\sqrt{200}$ from 5. Subtract DM 16 from CE 22, there remains 6 for CV. And say again: CD $\sqrt{1800}$ gives me CV 6, what shall give me OD $\sqrt{800}$? Arriving at OT 4. This [added] to TQ 16, being equal to DM is [100] together 20 for OQ. Furthermore, let OP be parallel to CB and say: EC 22 gives me EB 11, what shall give me QO 20? Arriving at 10 for PQ. And [adding] the square of PQ to the square of QO arriving at the square of OP, so line OP then is $\sqrt{500}$. Also TD is known, being equal to QM 28. So PA is 50. Now that the four lines are known AP 50, PO $\sqrt{500}$, OD $\sqrt{800}$ and DA 20, we then can find BL and DK as follows. First, look for the area of $ADOP$, and this shall be 700. Also, the area of the whole quadrangle $ABCD$ is 1015. The half is $507\frac{1}{2}$. Now, divide BP 15 into two parts such that the proportion of the one against the other is as AD against AP, arriving at WB $4\frac{2}{7}$ and for WP $10\frac{5}{7}$. So, let us add WB $4\frac{2}{7}$ to AD, then AX becomes $24\frac{2}{7}$ and AV is $60\frac{5}{7}$. Also, now that $AXNW$ is of the same proportion as $ADOP$, so can be known NX and NW by saying: the square of AD (being 400) gives me $589\frac{39}{49}$ for the square of AX, what shall give me 800, the square of OD, and 500, the square of OP? From 800 one shall arrive at $1179\frac{29}{49}$ for the square of XN, and from 500 at $737\frac{12}{49}$ for the square of NW. So, line XN is $\sqrt{1179\frac{29}{49}}$ and NW $\sqrt{737\frac{12}{49}}$. The area of $AXNW$ is thus known, say: 400, the square of AD gives me 700, the area of $ADOP$, what shall give me $589\frac{39}{49}$, the square of AX? And one shall arrive at $1032\frac{1}{7}$ for $AXNW$. From this, subtract half of the area of the whole quadrangle $ABCD$, being $507\frac{1}{2}$, there remains $524\frac{9}{14}$ for the area of $FYNZ$. The lines NY and NZ are then also known to me, as the proportion of these lines is known. Say again: 700 gives me 800 for the square of OD, and 500 for the square of OP, what shall give me the area $524\frac{9}{14}$? One shall arrive at $599\frac{29}{49}$ for the square of NY and $374\frac{73}{98}$ for the square of NZ. Now, take the $\sqrt{\mathfrak{z}}$ arriving at NY $\sqrt{599\frac{29}{49}}$ and $\sqrt{374\frac{73}{98}}$ for NZ. Subtract this NY $\sqrt{599\frac{29}{49}}$ from NX [101] $\sqrt{1179\frac{29}{49}}$ and there remains $\sqrt{1179\frac{29}{49}} - \sqrt{599\frac{29}{49}}$ for YX. So much is also KD. Also, subtract NZ $\sqrt{374\frac{73}{98}}$ from NW $\sqrt{737\frac{12}{49}}$, there remains $\sqrt{737\frac{12}{49}} - \sqrt{374\frac{73}{98}}$ for ZW, and so much is also LB. And thus KD and LB are also found as required. Now, to find FH, say so: CB $\sqrt{605}$ gives me 22 for EC, what shall give me LB $\sqrt{737\frac{12}{49}} - \sqrt{374\frac{73}{98}}$? One shall arrive at $24\frac{2}{7} - \sqrt{299\frac{19}{49}}$ for FH or FG.

Question 85

There is a quadrangular piece of land $ABCD$, right-angled in B, which one wants to divide into two equal parts, such that the line which divides it into two, as here EF, runs parallel to DC. Question: how much will be ED or FC if AB makes 10, BC 32, CD 25 and AD 13?

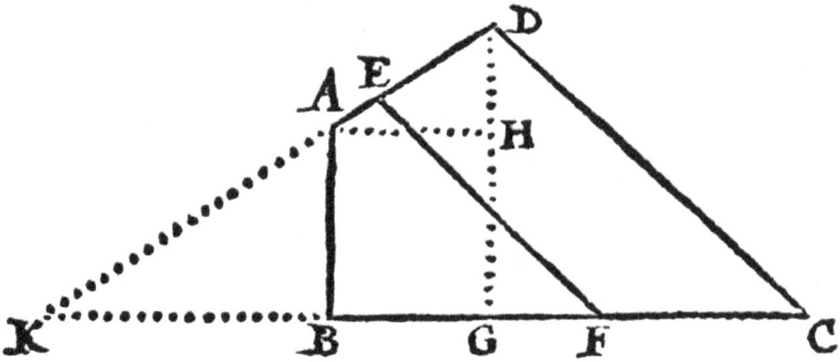

Figure 2.107: Question 85

First, look how much the perpendiculars DG and AH will be, as previously taught as in our 74th question, and one shall find 15 for DG, 5 for DH and 12 for AH. Now say: HD 5 gives me AH 12, what shall give me DG 15? And one shall arrive at 36 for KG. So GC is 20 and KC 56. These lines being known, look also for the area of the quadrangle $ABCD$, arriving at 300. Its half is 150, the area of triangle KDC is 420. From this subtracted 150, there remains 270 for the triangle KEF. Now say: [102] 420 gives me a square of KC being 3136, what shall give me 270? Arriving at 2016 for the square of KF. So line KF is $\sqrt{2016}$. This subtracted from KC 56, there shall remain for FC $56 - \sqrt{2016}$. Also, look for ED and one shall arrive at ED $39 - \sqrt{977\frac{11}{14}}$.

Question 86

One wants to divide quadrangle $ABCD$ here shown into two equal parts, such that line HK falls perpendicular on line BC. Question: if BC makes 50, BF 10, FE 30, EC 10 and the perpendicular DE 30, FA 40, how much will be the length of line HK (dividing the land in two), also how much will be KC and HD?

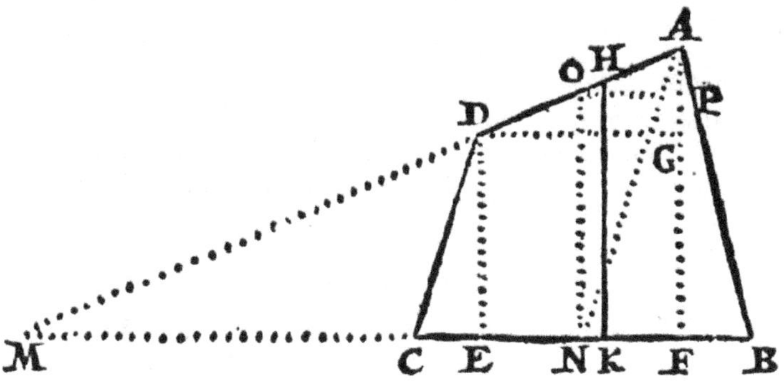

Figure 2.108: Question 86

Say: GA 10 gives me GD 30, what shall give me ED 30? Arriving at 90 for EM. Multiply the half of ED (being 15) by 90, arriving at 1350 for the area of DEM. To which must be added the area of $EDHK$, which is 550. Is together 1900 for the area of HKM. One shall arrive at $1266\frac{2}{3}$ for the square of HK, so line HK is $\sqrt{1266\frac{2}{3}}$. Now, to find HD and KC say as such: 1350 gives me the square of EM 8100, what shall give me 1900? One shall arrive at 11400 for the square of KM. So line KM is $\sqrt{11400}$. As CM is known to me, as EM and EC are known to me, being CM 80, so then KC is $\sqrt{11400} - 80$. [103] Now that HK is known to me and KM, as also DE and EM (by the 47th proposition of the first book of Euclid), so will HM and DM also be known to me. Arriving at HM $\sqrt{12666\frac{2}{3}}$ and DM $\sqrt{9000}$, so then HD is $\sqrt{12666\frac{2}{3}} - \sqrt{9000}$.

Otherwise

Divide the whole of the area, being 1400, by 40 AF, arriving at 35 for BN. So, FN is 25. Now say: GD 30 gives me GA 10, what shall give me OP 25, which is equal to FN. Arriving at AP $8\frac{1}{3}$. This subtracted from AF 40, there remains $31\frac{2}{3}$ for PF, as much is also ON. Multiply ON $31\frac{2}{3}$ by AF 40, arriving at $1266\frac{2}{3}$. The $\sqrt{\text{ℨ}}$ from this is $\sqrt{1266\frac{2}{3}}$ for HK, as HK is the mean proportional between AF and ON.

Question 87

Look at the following figure.

To extend a given quadrangle (as here shown $ABCD$ of which AD runs parallel to BC) into a longer one, as long as line BG given, such that the lines of the extended quadrangle are all parallel to the quadrangle aforementioned.

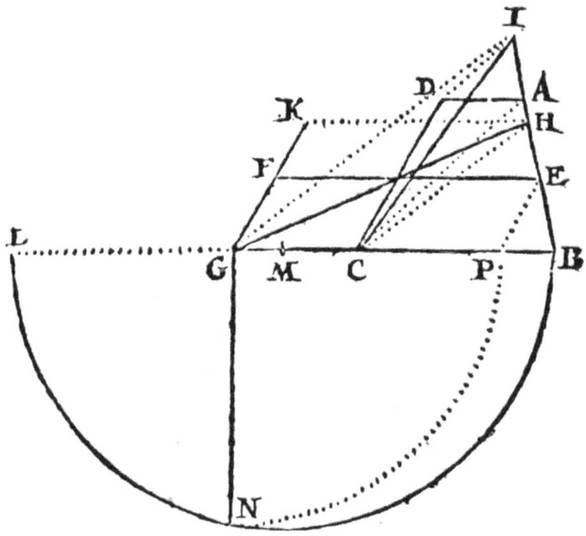

Figure 2.109: Question 87

To do this, bring the whole of the quadrangle into a triangle, arriving at BIC with angle ABC and base BC. Then, extend the triangle such that its base BC becomes as long as the given line BG. And one shall arrive at BHG hold in angle ABC. This being done, draw line GK from point G, parallel to CD. Also draw line HK parallel to BG and make GL equal to HK, such that BG is extended to L, and at the center [104] of BL, as in M draw a semicircle. From point G draw the perpendicular GN to the circumference and make PG equal to GN. Also, draw a [line] parallel to GK which cuts AB at point E. So, line EF, parallel to BG will be equal to GN, and EF is the mean proportional between HK and BG, as has been taught in our 80th question. Triangle BHG is equal to the quadrangle $BEFG$, and therefore also equal to triangle BIC. Thus, the given quadrangle $ABCD$ is equal to $BEFG$, constructed on given line BG, which sides run parallel to those of $ABCF$, as one required.

Question 88

Look at the following figure.

How one can construct a triangle from a given quadrangle $ABCD$ (of which AD runs parallel to BC) [105], of which two angles of the triangle will be equal to the two angles B and C of the quadrangle. As for example, as here shown, that triangle BFE be equal to quadrangle $ABCD$ and EF runs parallel to CD, and that part $CEGD$ be equal to triangle AGF.

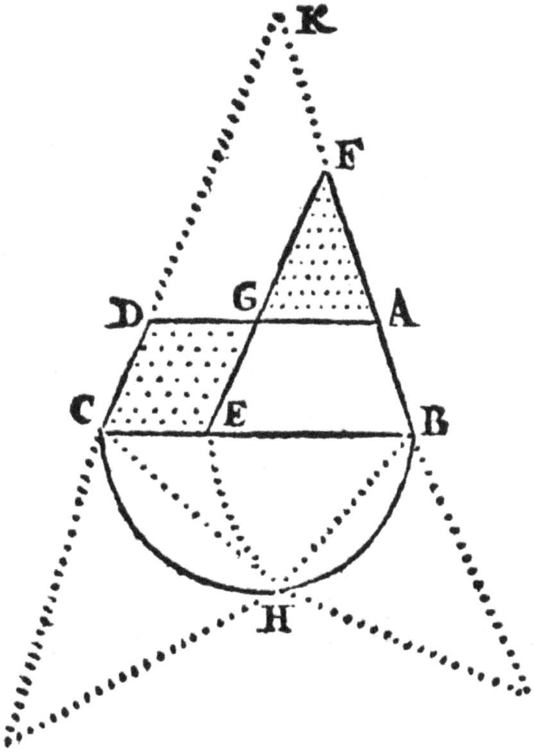

Figure 2.110: Question 88

Draw a semicircle on BC and make CH equal to AD. Then draw line BH, and make BE equal to BH. Now, if one draws a line from point E parallel to CD and also extend line BA until both cross each other at point F, so will triangle BFE be equal to quadrangle $ADCB$, as one required. Or, that triangle AGF is equal to quadrangle $ECDG$. Because, if we subtract triangle BKC from triangle AKD, there remains quadrangle $ABCD$. Now that HC is made equal to AD, therefore if the triangle of the same proportion is subtracted from BCK, there remains on BH a triangle in the same proportion as triangle BCK or ADK, by the 31st proposition of the sixth book of Euclid, as BHC is a right angle. And also, as BE is made equal to BH, so is triangle BEF (which is made of the same proportion as BCK or ADK) equal to quadrangle $ABCD$. [106]

Question 89

There is a given triangle ABC in which one wants to construct a right-angled square of equal sides, as large as possible.

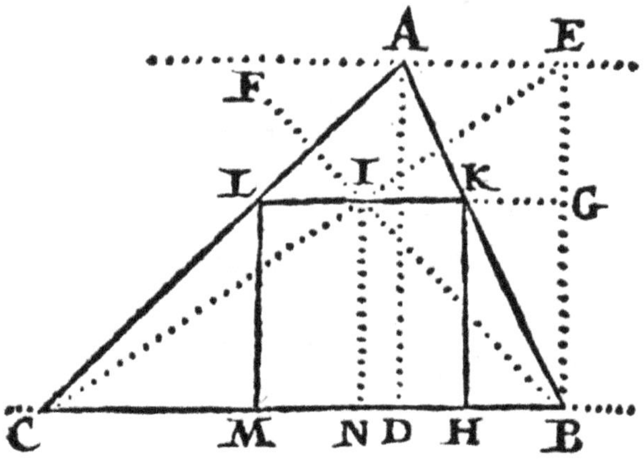

Figure 2.111: Question 89

To do this, construct a right-angled triangle on the same base BC as here EBC, such that the perpendicular EB is equal to the perpendicular of the given triangle AD. Then draw line BF which will divide the right angle[9]

[9]Cardinael has 'rectangle'.

EBC into two equal parts, such that angle *EBF* equals angle *FBC*, which line *BF* crosses line *EC* at point *I*. So, *IN* is a side of the square. So, if one then draws line *GL* trough *I*, parallel to *BC* and *HK* and *ML* parallel to *AD*, so *HKLM* will be the square inscribed in triangle *ABC*, as large as possible. Because line *BI* divides the right angle[10] *EBC* into two equal parts, so is *GINB* a square and as this right triangle *EBC* also stands on the same base as *ABC*, therefore also *GI* is equal to *KL* and as *GI* is equal to *GB*, so is *KL* also equal to *GB* or *KL* equal to *KH*, or the square *HKLM* equal to *BGIN*.

Otherwise

To perform this by means of numbers, let the base *BC* be 14, the perpendicular *AD* 12. This multiplied by each other arrives at 168 for double the area of triangle *ABC*. Now, add *BE* 12 to *BC* 14, arriving together at 26. And 168 divided by 26 arrives at [107] $6\frac{6}{13}$ for *GI* or *IN*, and so much will also make the side of the inscribed square *HKLM* as one required.

Question 90

One wants to divide this quadrangle *ABCD* with unequal sides into two parts at point *E*, such that the part *ABEF* stands in proportion to the other part *FECD* as given line *GL* stands to *LM*.

[10] Cardinael has 'rectangle'.

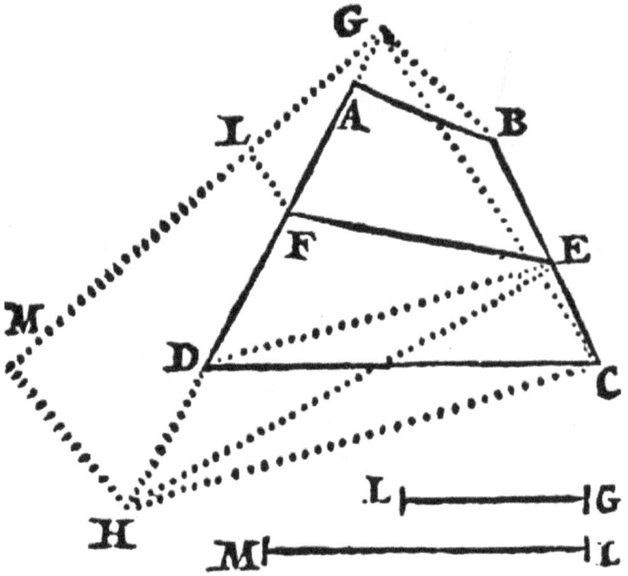

Figure 2.112: Question 90

To do this, draw BG parallel to EA and extend DA to G, so that triangle EAG is equal to EAB. As they stand on the same base EA, by the 37th proposition of book 1 of Euclid, draw CH parallel to ED and extend AD to H, so is triangle EDH equal to triangle EDC and the whole triangle GHE then is equal to the quadrangle $ABCD$. Furthermore, draw an arbitrary line from point G as required, as long as is GL and LM [together], the two given lines, which will here then be GLM. So, if we now draw lines MH and LF parallel to MH, then line GH shall be divided by point F into two parts, GF and FH which will be in proportion to each other, being GF to FH as GL to LM. Therefore, line FE being drawn shall divide triangle GHE into two parts such that part GEF stands in proportion to the other part EHF as line GL to LM. Also, as GEF is equal [108] to $ABEF$, and also triangle EHF is equal to part $EFDC$, so then is the given quadrangle $ABCD$ divided into two parts, where part $ABEF$ stands to the other part $EFDC$ as the given line GL stands to LM, as one required.

Question 91

There is given a hexagon with unequal sides, as here $ABCDEF$, from which one wants to cut a belt, as large as the quadrangle with [as side] the perpendicular on side AB, shown by G, in such [109] a way that the parts that remain at both sides from the belt, part $AVWEF$ being twice as large as the other part $TXCB$. The question is how one shall do this geometrically?

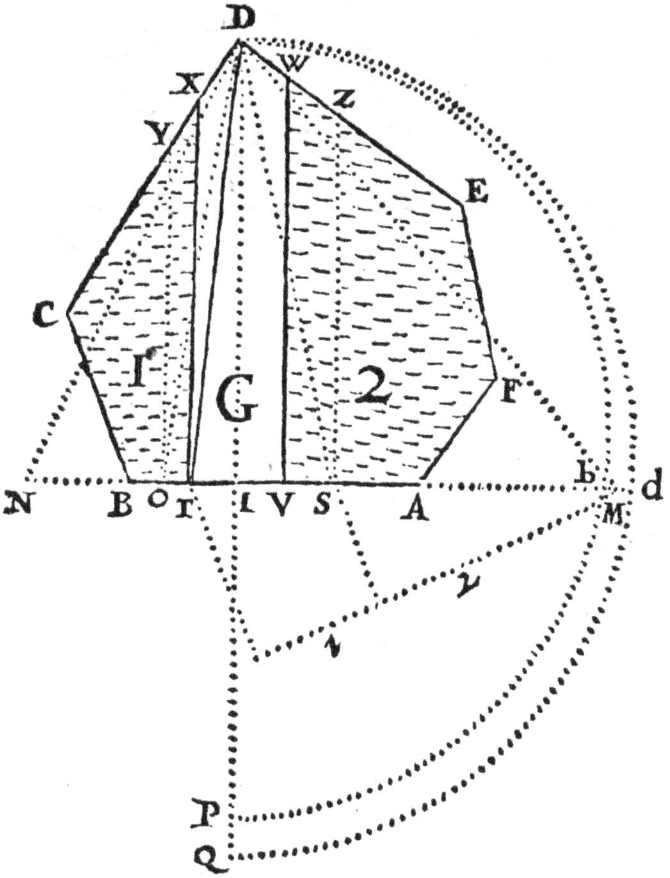

Figure 2.113: Question 91

First, transform the hexagon $ABCDEF$ to a triangle, from angle D with DI as perpendicular and this shall be triangle DMN. Also, transform quadrangle G to a triangle with its perpendicular as long as DI, which shall be triangle HKL, right-angled in K, and KL being made equal to DI.

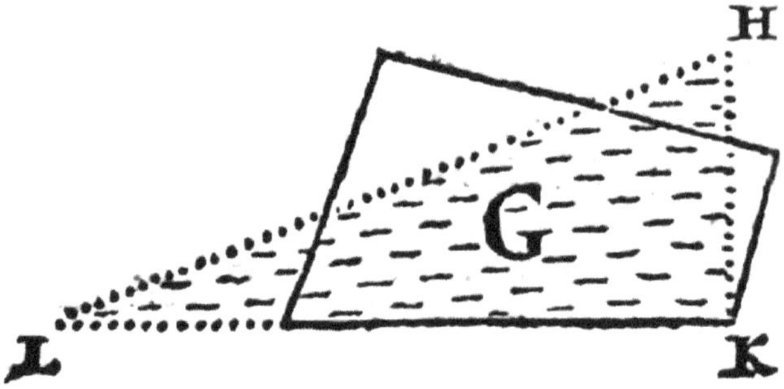

Figure 2.114: Question 91: transforming a quadrangle to a triangle

This being done, make TN equal to HK. Then MT will be divided into two parts, as here MS and ST, so that MS stands to ST as 2 to 1. Now, make NO equal to ST, so is SO equal to TN and MS also stands to ON as 2 to 1. Triangle SOD is equal to the given quadrangle G because base HK is equal to SO and the perpendicular DI is equal to KL. Also, triangle DMS stands to triangle DON as 2 to 1. Furthermore, let us now draw OY parallel to the perpendicular ID and make IP equal to OY so that DIP is a right line. On this we draw a semicircle so that Ib be the mean proportional between IP and ID, making TX equal to Ib. This being done, also draw SZ parallel to perpendicular DI and make IQ as long as SZ so that DIQ becomes a right line, with Ib being the mean proportional between DI and SZ. Then also make VW equal to Id so that hexagon $ABCDEF$ becomes divided into three parts of pieces by two lines TX and VW, of which the middle part $VWDXT$ be the belt [110] or a part as large as the given quadrangle, being drawn as G, of which sides WV and XT fall perpendicular on AB. Hence, part $AVWEF$ stands to part $BTXC$ as 2 to 1, as one required.

Question 92

One has a triangular piece of land, of unequal sides and angles such as ABC, in which is given a point D, through which one desires to dig a ditch, such that the triangle ABC will be divided into two equal parts. Question: how shall this be done geometrically?

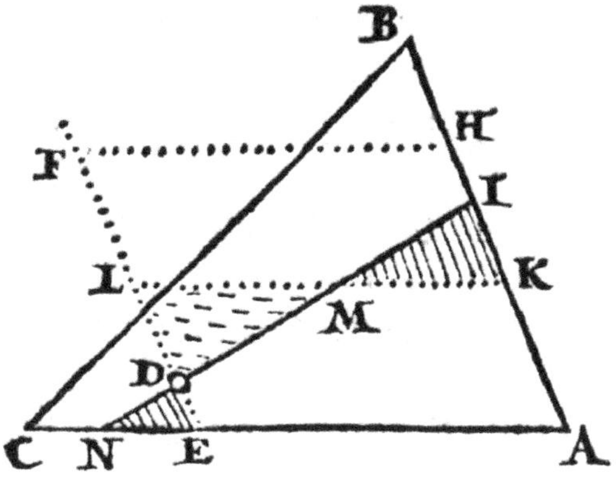

Figure 2.115: Question 92

I first draw a right line through point D parallel to AB, as EDF, as it happens to be the case that I noticed that the separation shall originate from a point on side AC, through given point D, to some point on line AB. Then, I construct quadrangle $HAEF$ equal to triangle ABC. From this, the half subtracted, this is quadrangle $KLEA$. And if I now draw a line through given point D, such that triangle MDL is cut off from quadrangle $KLEA$, of equal size as triangles KMI and DEN together, so will triangle IAN be equal to quadrangle $KLEA$. Hence also [equal] to half of the given triangle ABC, as quadrangle $KLEA$ is the half of the given triangle ABC. So, line IDN will then divide triangle ABC into two equal parts, through [111] point D as required.

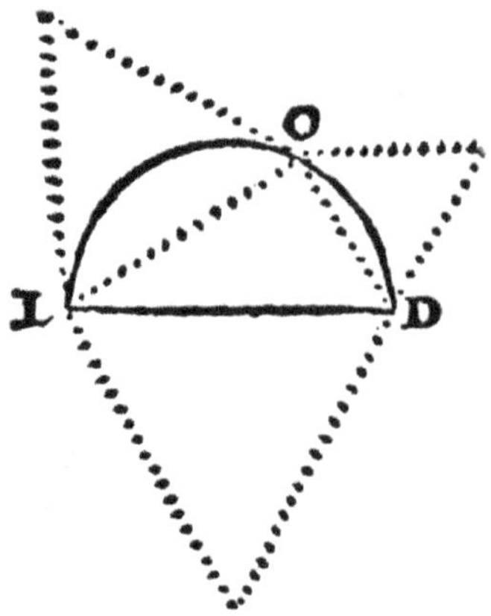

Figure 2.116: Question 92: solution

Now, to draw such line NDI, I draw a semicircle on DL. From the end of this line at D I draw the line DO, equal to ED. Then I make KI equal to OL. This being done, I draw line IDN from I through given point D, which is the line we required, cutting AB at I and AC at N. As all three triangles IKM, MLD, DEN of the same proportion, which can be easily noticed by the angles, and triangle DEN together with triangle IKM is of the same size as triangle MLD by itself, therefore the square of ED subtracted from the square of DL, there remains the square of KI and hence (by the 31st proposition of the sixth book of Euclid), KI can be known. So, then also AI and AN are known, which make up together with IN the separation required, with triangle AIN half the size of given triangle ABC.

Question 93

Look at the following page.

I have an irregular quadrangle $ABCD$, in which I place point E through which I desire to draw a line, which shall divide this quadrangle into two equal parts.

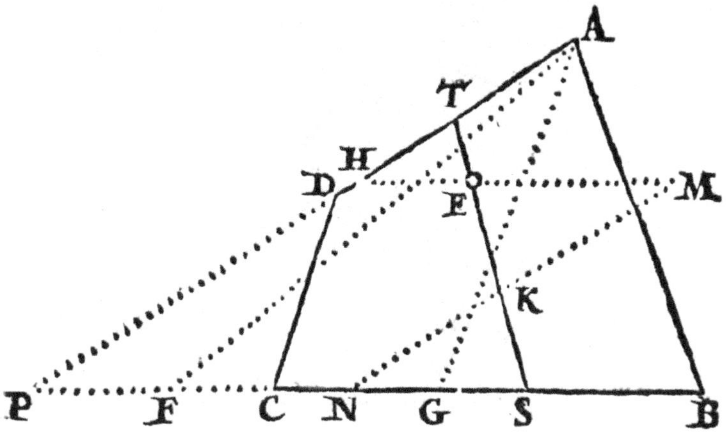

Figure 2.117: Question 93

To do this, extend lines AD and BC so that they cross at P and transform quadrangle $ABCD$ into a triangle at angle ABC, which will be triangle ABF. This triangle ABF will be divided into two equal parts, as here at point G by line AG. This being done [112] so draw line HM through given point E, parallel to base BP, and construct quadrangle $HMNP$ equal to triangle AGP, such that MN be parallel to HP, for which there are several ways to do so. Then, also draw line TS through given point E, as we have been taught before in our 92nd question, such that triangles TEH and SKN together are as much as triangle EMK by itself. So, then triangle TSP is equal to quadrangle $MHPN$ or equal to triangle AGP. Therefore, if DCP is subtracted from triangle TSP, then $TSCD$ will be the one half and $TSBA$ the other half. And TS is the line required, so that the given quadrangle has been divided into two equal parts.

Question 94

Look at the following figure.

Two farmers each have a certain piece of land bordering each other. One has piece $ABCD$, the other piece $ABCEF$, the pieces being separated by an angled ditch ABC. Now, they desire to have their ditch delved straight, such [113] that the straight ditch runs parallel to BC, the straight ditch here drawn by GH, cuts off as much land from one as from the other, that what one loses the other will win, knowing that triangle GAI is as much as quadrangle $ICHB$. Question: how shall this be done geometrically?

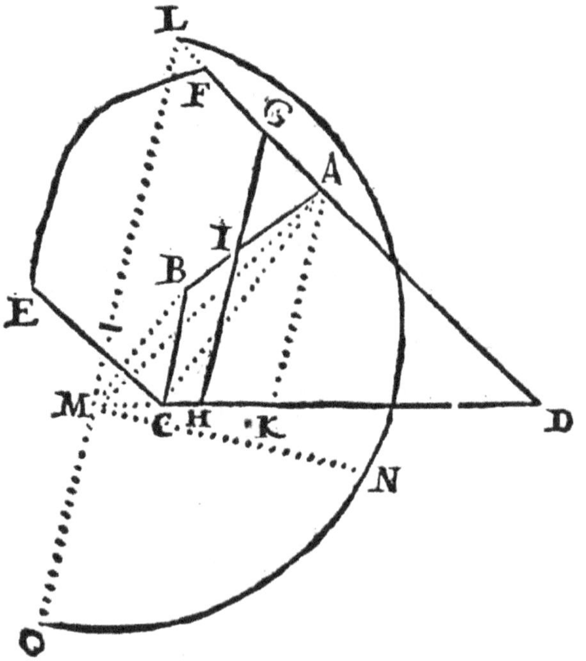

Figure 2.118: Question 94

First, transform the quadrangle $ABCD$ to a triangle, and to do this, draw BM parallel to AC, hence a line from A to M so that AMD becomes a triangle which is equal to the quadrangle $ABCD$. This being done, draw a line from A, parallel to BC, which is line AK. Also, draw a line from the point parallel to BC or AK, which is here line ML, and extend DA to L. Furthermore, when LM is extended to point O, as much as AK, then MO is equal to AK and MN is the mean proportional between MO and ML. At last, when GH is made equal to MN such that GH be parallel to AK or LM, then quadrangle $GHKA$ will be equal to triangle KAM, as being taught in our 80th example. Therefore, quadrangle $GHKA$ is also equal to $ABCK$. The large triangle DGH is then also as large as the quadrangle $ABCD$, or triangle GAI is equal to $IHCB$ as [114] one required. And that what one wins by the cutting line GH, he will lose again, so that part $GHCEF$ remains equal to the part $ABCEFA$, as before. One could also solve this by other means, as the way that is taught in our 88th question.

Question 95

Given an irregular quadrangle $ABCD$ which one desires to have geometrically divided into two equal parts, such that line KLM (the dividing line) runs parallel to ADC and LP be equal to LQ, so that parts $BKLM$ and $ADCMLK$ each are the half of the given quadrangle $ABCD$. Question: the length of AK and CM.

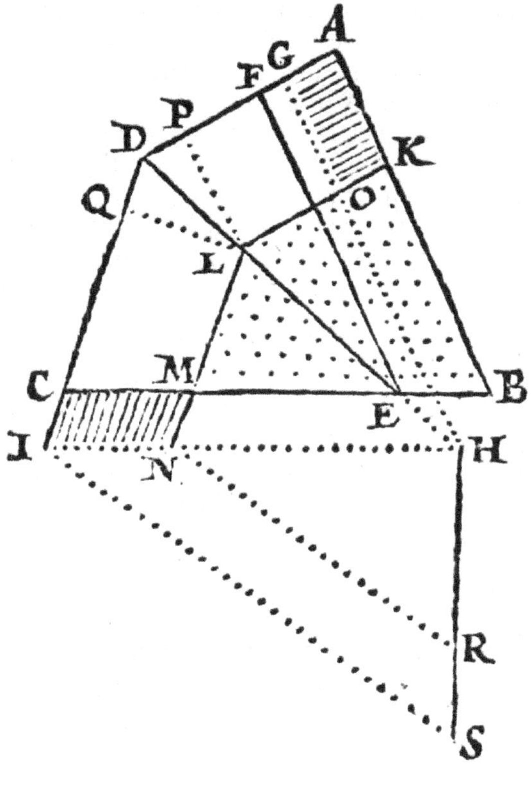

Figure 2.119: Question 95

First, draw the line ED such that the angle ADC is divided into two equal parts, so that line BC will be cut at point E. Then, also draw line EF parallel to AB, which will cut line AD into two unequal parts AF and FD at point F. Now, divide part AF into two parts such that part GA stands to the other

GF as line DC to FD. Then extend DC to I such that CI becomes equal to AG. So, quadrangle $GHID$ is of the [same] proportion as quadrangle $FECD$. Furthermore, make the half of the given]115] quadrangle $ABCD$ into a right-angled square. Also, [make] a square of quadrangle $GHID$, the side of the square being HS. And if one subtracts the square of $GHID$ from the given square of half the quadrangle $ABCD$, there shall remain a square of which the sides are HR. Now, draw SI and RN, parallel to SI. So HI stands to HN as HS to HR. Also draw NL parallel to ID and LK to DA, so that quadrangle $OLNH$ is equal to the square of HR, as the quadrangle $GHID$ is equal to the square of HS and because the square of HS is so much larger than that of HR as half of the given quadrangle $ABCD$. So, quadrangle $GHID$ is also so much larger than $OLNH$ as half of the given quadrangle $ABCD$ or $OGDINLO$ being the half of $ABCD$, and because line ED divides the angle ADC into two equal parts, so is LP equal to LQ. Also, as CI is equal to AG so will the parts $CINM$ and $AGOK$ are of equal size, as they stand between two parallels and part $KADCMLK$ is equal to $OGDINLO$. Hence, also half of the given quadrangle $ABCD$ and the other half $BKLM$ have been divided by line KLM into two equal parts, so that LP be equal to LQ, which are the lengths of lines AK and CM, as one required.

Question 96

Look at the following page.

The three lines here shown AB, BC and CD, are given, which are the three perpendiculars of an unknown triangle. One desires to establish geometrically the length of the side of such a triangle.

Figure 2.120: Question 96: the given catheti

To find the sides of such a triangle, take a line of any length (I take for example EK) and make EH as long as AB and [draw] the right-angled quadrangle $EHIK$.

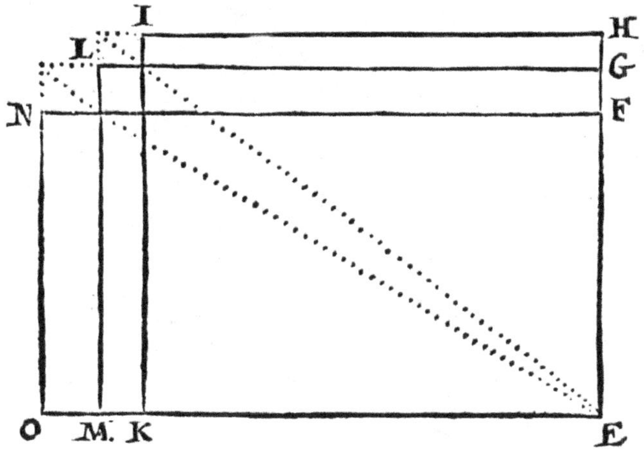

Figure 2.121: Question 96: the construction

Also, make EG as long as BC and EF equal to CD. On these two lines EF and EG draw on each a right-angled quadrangle, as large as quadrangle $EHIK$ was drawn, which will be quadrangle $EGLM$ on EG and quadrangle $EFNO$ on EF. This being done, one arrives at three lines EK, EM and EO, in proportion to each other as the sides are to each other of the triangle one required. Which will be so because the quadrangle constructed by the perpendicular of a triangle and the side on which this falls will always be equal to [the quadrangle constructed from] any other perpendicular and the side on which it falls, which will have an area of double the triangle.

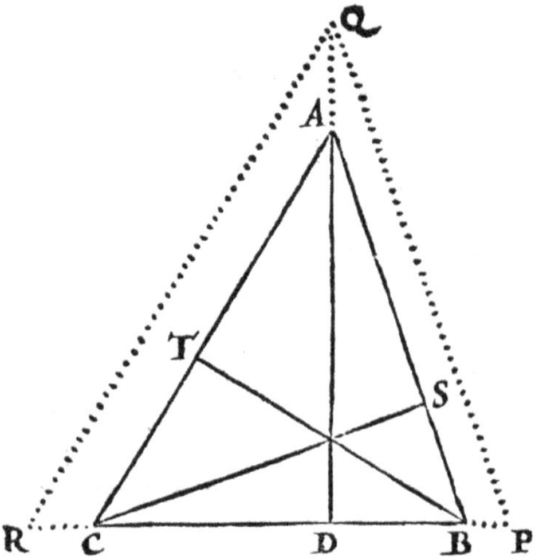

Figure 2.122: Question 96: the final triangle

Therefore, let us now construct (as in the 22nd proposition of the first book of Euclid) a triangle from [117] these three lines, being EK, EM and EO. This will be the triangle PQR here shown, whose perpendicular then is QD, and make AD equal to the largest perpendicular AB as QD is also the largest perpendicular of this triangle QPR, which falls on the shortest side PR which is equal to EK, as RQ is equal to EM and PQ equal to EO. At last, draw the line AC from point A parallel to QR and also AB parallel to QP so that they will cut PR at the points B and C, so that we get triangle ABC which then is the triangle one required, of which the perpendicular AD is equal to the given perpendicular AB and BT is equal to the given perpendicular BC and also CS equal to the given CD.

Question 97

Look at the following page.

Also, there are given four lines AB, BC, CD and DA, of which three are longer than one is by itself. Question: if one wants to make a quadrangle out of these four lines such that AB stands opposite to CD and BC to DA, and that this quadrangle is inscribed in a circle such that the four angles touch the circle, how much shall be the diameter of such a circle in which this quadrangle is inscribed ?

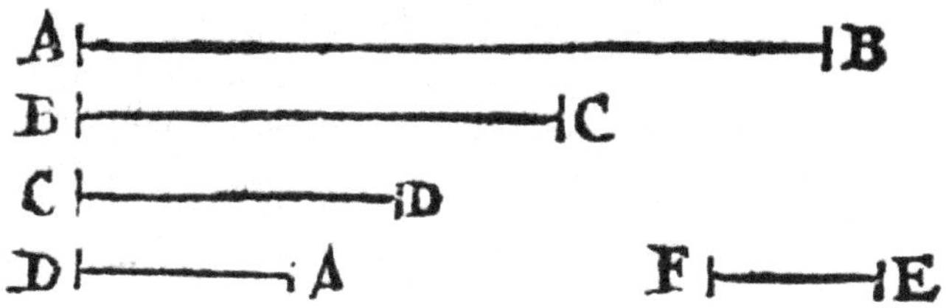

Figure 2.123: Question 97: the given sides

To do this, take any line as you desire, as here shown by FE. And say: the smallest given line DA gives me EF, what shall give me the given line BC? And one shall arrive at EG. Also say: the given line CD gives me EF and EG, what shall give me the largest given line AB? And one shall arrive at EH and EI.

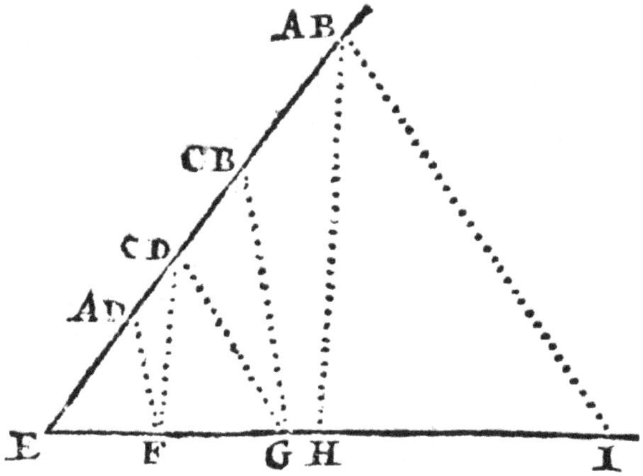

Figure 2.124: Question 97: the construction of the diagonals

Now, add EF to EI, which will become line FI. Also, add EG to EH which will become line HG. Construct a right-angled quadrangle with these two lines FI and HG, which will be of equal size to the quadrangle that can be inscribed in a circle, as the sides stand in the same proportion to each other as the sides of the required quadrangle, or the given lines AB, BC, CD and DA, and the diagonal lines would then be FI and HG. The right-angled quadrangle being constructed from the given lines FI and HG must now be transformed into a right-angled square with equal sides, which here will be the square $KLMN$.

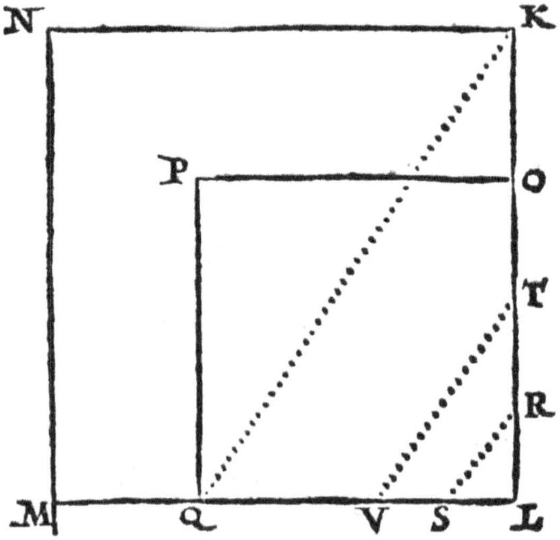

Figure 2.125: Question 97: the construction of the squares

Furthermore, also construct a right-angled quadrangle from the given lines AB and CD (which will stand opposite to each other), so that together with the right-angled quadrangle constructed from given lines BC and DA (which will stand opposite [119] to each other), these two quadrangles constructed by AB with CD and BC with DA, added together and brought to a right-angled square of equal sides, one shall arrive at the square which is here shown by $OPQL$. This square is as much as if one constructed a right-angled quadrangle with the diagonal lines of the the required quadrangle, which can be inscribed in a circle, with the given lines AB, BC, CD and DA, as we have been taught in our 53rd and 68th questions. Therefore, now as the side of the large square LK stands to the smaller square LQ, so shall the selected line FE stand to

DE, as [120] one required.

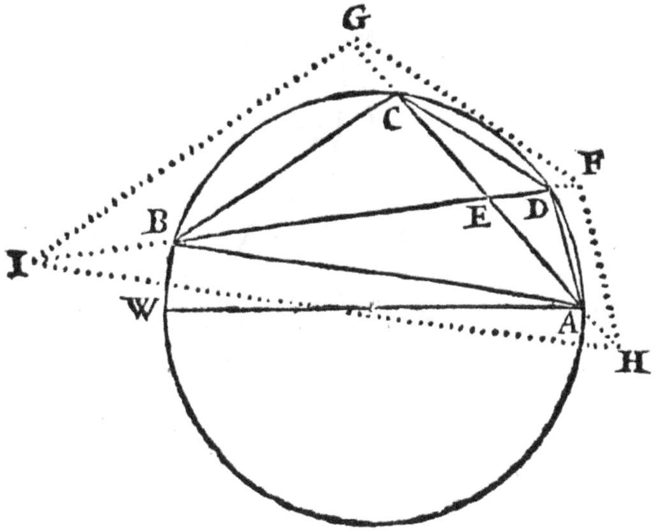

Figure 2.126: Question 97: the quadrangle inscribed in a circle

As the smaller part of the required diagonal line DB is cut by the other required diagonal line AC, as at point E in the required quadrangle $ABCD$, so we will make LR equal to the selected line FE, and draw RS parallel to KQ, which ends at point S, so will LS be equal to DE as required. Also, make LT equal to EG and draw TV parallel to KQ, ending at point V, so that LV is equal to EC as required. This, ED and EC being found make a triangle DEC with given line DC, and DEA, AEB and BEC then become known. Hence, also quadrangle $ABCD$ as one required. And by the 25th proposition of the third book of Euclid, the center of the circle and diameter AW will be known.

Question 98

There is proposed to me a pyramid $ABCD$ whose base is the triangle BCD of unequal sides, of which BC makes 13, CD 14 and DB 15 and from B to the top A is 14, CA 13, DA 15 feet. The question then is: how many cubic feet will be the volume of such pyramid?

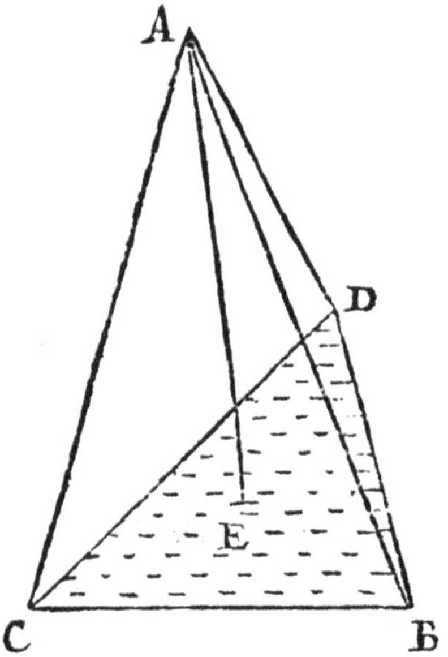

Figure 2.127: Question 98

First, look for the perpendicular AE as being taught in our 76th question, and one shall find $\sqrt{123\frac{3}{4}}$ for AE. From this take the $\frac{1}{3}$ which is $\sqrt{13\frac{3}{4}}$ and now look for the area of base BCD, which will be 84. This multiplied by the $\sqrt{13\frac{3}{4}}$ aforementioned, being the $\frac{1}{3}$ of the perpendicular AE, or the height [121] of the pyramid, one shall arrive at $\sqrt{97020}$ cubic feet for the volume, which is approximately $\sqrt{311\frac{1}{2}}$ [cubic] feet.

Question 99

There is another pyramid $ABCD$ or a stone, whose base is the triangle BCD of unequal sides, of which BC makes $8\frac{1}{2}$, CD 13 and DB $\sqrt{101\frac{1}{4}}$. The question is: how much will be the volume of this pyramid when from the top AC to C

is 25, from A to B $\sqrt{445\frac{1}{4}}$ and from A to D $\sqrt{596}$?

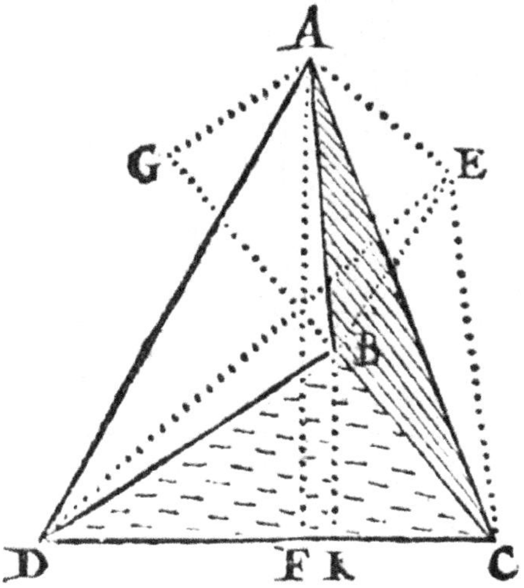

Figure 2.128: Question 99

First, look where the perpendicular will fall in triangle ACD, from A on line CD, which is FA and one shall arrive at FD $5\frac{5}{13}$ as in our first question. And for FC $7\frac{8}{13}$. Also look where the perpendicular in triangle ACD from A on line CBG, which is the perpendicular AG and this will fall outside the triangle ACB at G, and one shall arrive as in our 2nd question at BG $6\frac{11}{34}$ and for CG $14\frac{14}{17}$. And for CG $14\frac{14}{17}$. Furthermore, if one draws a line right-angled from F to the perpendicular of the pyramid AE falling to the floor, as in point E, such line as here FE will cut CG at point H. Let us therefore look for CH and to find such, first look as in our first question how much will be CK and then KB by the 47th proposition of the first book of Euclid, and one shall arrive at $5\frac{5}{13}$ for CK. And for BK $6\frac{15}{26}$. These two lines being known, so say: CK $5\frac{5}{13}$ gives me CB $8\frac{1}{2}$, what shall give me CF $7\frac{8}{13}$? One shall arrive at CH $12\frac{3}{140}$. This subtracted from CG $14\frac{14}{17}$, there remains [122] $2\frac{1909}{2380}$ for HG. Also KB $5\frac{5}{13}$ gives me CK $5\frac{5}{13}$, what shall give me HG $2\frac{1909}{2380}$? One shall arrive at $2\frac{5}{17}$ for EG.

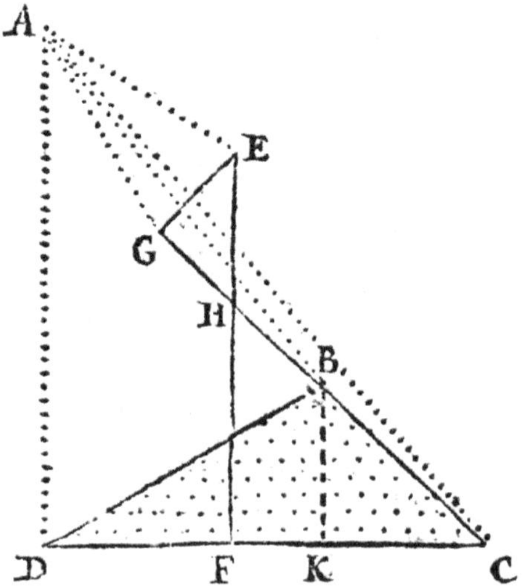

Figure 2.129: Question 99: solution

We already knew BG making $6\frac{11}{34}$ and as also AB is known, making $\sqrt{445\frac{1}{4}}$, so can AG be known by the 47th proposition of the first book of Euclid, and one shall arrive at $\sqrt{405\frac{76}{289}}$. Similarly, as EG is known to us, making $2\frac{5}{17}$, so can the perpendicular AE be known, and one shall arrive at AE 20. We now first look for the area of the base of the pyramid, or triangle BCD, which is $42\frac{3}{4}$. Then, as taught in our 98th question, multiply the area $42\frac{3}{4}$ with $\frac{1}{3}$ of the height AE of the pyramid, which is $6\frac{2}{3}$ and one shall arrive at 285 cubic feet for the volume of this reclining pyramid.

Question 100

Standing on a flat surface, straight away from a certain tower at the north, as here shown as C in the figure, I thus find that the top of the tower A is at an elevation of 30 degrees above the horizon, while at another place at D (200 feet away from the first place C) I then find at D, that the top A of the tower is at an elevation of 10 degrees above the horizon, being straight at north-east of the tower. The question is, how many feet is the tower above the horizon, and also, how many feet is there from the tower to the first place and second place, which is BC and BD?

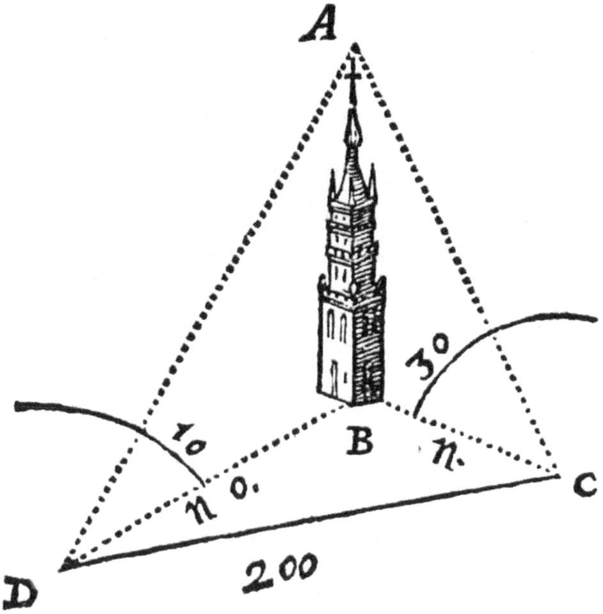

Figure 2.130: Question 100

First look for the tangent of angle BAC which is 60 degrees, and one shall arrive at 173205 for BC. Also, the tangent of angle BAD, being 80 degrees, arriving at 567128 for BD. So, one has a triangle DBC of which angle B is known, making 45 degrees, as north and north-east differ by 45 degrees, and the sides BD and BC are known in the smallest parts (as above), hence we must look for the small parts that make up side DC. To do this, look for the sine of 45 degrees, which is 70711 for EC. And say: 100000 BC, the whole of the sine gives me 173205, what shall give me 70711 EC? One shall arrive at 122475 for EB. This subtracted from 567128 DB, there remains 444653 for ED. The square of ED added to the square of EC, arrives at the square of DC. The $\sqrt{\mathfrak{z}}$ of this arrives at 461213 for DC. Now that also DC is known in its small parts, so say: [124] 461213 the small parts of DC give me 200 feet, what shall give me 100000, the small parts of AB? And one shall arrive at $43\frac{3}{8}$ feet for AB, being the height of the tower.

And to find BD and BC so say: AB 100000 small parts give me $43\frac{3}{8}$ feet, what shall give me 567128, the small parts of BD, and 173205, the small parts of BC. One shall arrive at 246 feet for BD and further 75 feet for BC.

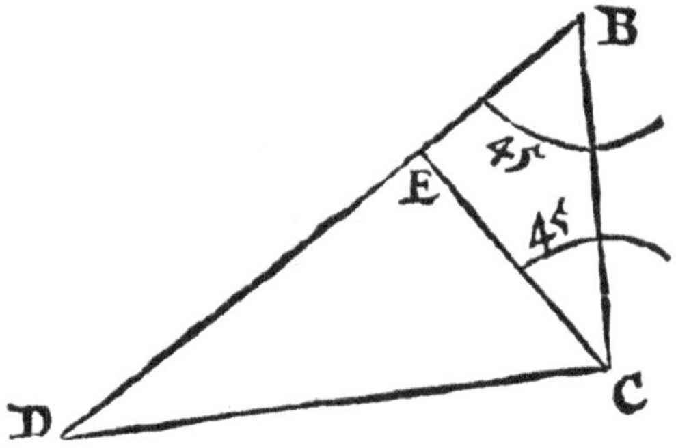

Figure 2.131: Question 100

Otherwise

Say: 50000 AB, the sine of 30 degrees gives me 17365 AB, the sine of 10 degrees, what shall give me 86603 BC, the sine of 60 degrees? One shall arrive at 30077 for BC, hence the proportion of AB to BC and BD is thus also known, so when AB makes 50000 small parts, so will BC 30077 and BD 98481, which is the sine of 80 degrees. Furthermore, say: 100000 the whole of the sine BC gives 30077 BC, what shall then be 70711 EC, the sine of 45 degrees? And one shall arrive at 21267 for EC. This subtracted from 98481 DB, there shall remain 77214 for ED. The square of this added to the square of EC together is the square of CD. Taking the $\sqrt{ʒ}$ of this arrives at 80080 for CD. So, CD is being known in its small parts, which will be equal to 200 feet. Therefore say: 80080 gives me 200 feet, what shall give me 17365? One shall arrive at $43\frac{3}{8}$ feet for AB, as we did before. Also, what shall give me 98481 and 30077? One shall arrive at 246 feet for BD and 75 feet for BC.

To conclude, now follows an arithmetical question, which I solved artfully without cossic numbers

[125] A merchant buys four kinds of cloth, knowing, of sort A as many pieces as each costs in Flemish pounds. Of B as many pieces as each piece cost B Flemish pounds. Of C as many as all the pieces of A and B together cost in Flemish pounds. And each piece of C costs as much as a piece of A and a piece of B together. Also, all the pieces of C amount to 715 pounds. At last, he bought as many pieces of D as the cost of all the pieces of B are

more in pounds than the amount of all the pieces of A. As each piece of D costs as many pounds as pieces of B cost more in pounds than these of A, this amounts together at 99 pounds for the pieces of D. Question: how many pieces he bought of every kind? Facit, 4 pieces of A, B 7, C 65 and 33 of D.

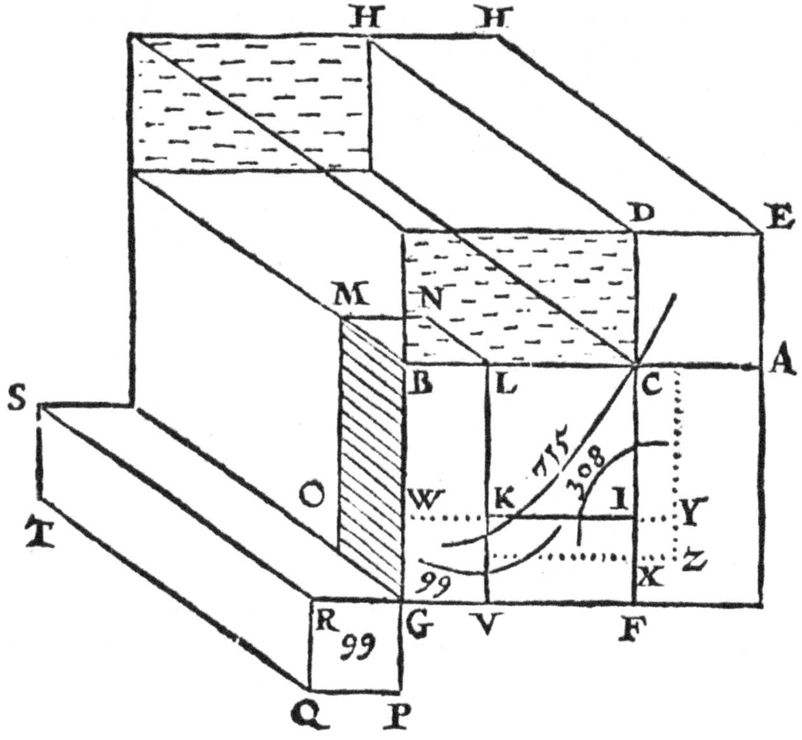

Figure 2.132: Question 101

To find this, I draw a line AB and divide this in two parts at point C. The part shown as AC I take [126] for how many pieces of cloth A, and CB as there are pieces of cloth B, the square $ACDE$ as much as the amount of Flemish pounds for the pieces of cloth A and the square $BCFG$ as much as the pounds amount for the cloth B. Also, these squares together for as many pieces of cloth there are of C. Furthermore, I construct two bodies, one with breadth and height of square $ACDE$ and with length EH or DH, which length is equal to AB. The other body with height and breadth as square $BCFG$, and also of the same length as EH or DH. The volume of these two bodies together then is 715, as much as all the pieces of cloth C amount in Flemish pounds, as the price of each piece of cloth is as much as a piece of A together with a piece of B, which is equal to line AB. Further, I construct the square $CIKL$ equal to $ACDE$, so that $IFGBLKI$ is as much as all the pieces of cloth D together, and hence LB or IF is as much as the cost of a piece of cloth D, which is as much as a piece of cloth B costs more in pounds as a piece of cloth A. Or here, as much as CB is more than AC. And if we construct a body from the plane $IFGBLKI$ and a length of LN or BM or GO, which are equal to LB, so that $LBMN$ becomes a square, this body will make 99. This is as much as the pieces of cloth D amount in Flemish pounds. And if I now construct a square $GPQR$ equal to $LBMN$ and make it into a body with length QT or RS (being of equal length as EH or AB), then I say that such a body also makes 99, because BG and FV together are as much as QT or RS, and the square $GPQR$ is equal to $LBMN$. And as the two bodies aforementioned are of equal length, I can remove the length of these bodies and take that the square $ACDE$ and the square $BCFG$ are together 715, and the square $GPQR$ then is 99. Hence, subtracting this 99 from 715 there remains 616, for as many as the four pieces [127] make together in a plane, knowing $ACDE$ and $CIKL$ with $IKVF$ and $LKWB$. Therefore, the half of this is 308 for the squares $CIKL$ and $IKVF$ together. Further, divide IF into 2 equal parts, which is at X, and as FI and GP are equal, so take the $\sqrt{\mathfrak{z}}$ of 99, arriving at $\sqrt{99}$ for IF. Its half is $\sqrt{24\frac{3}{4}}$ for IX, and now make IX equal to IY and multiply $\sqrt{24\frac{3}{4}}$ as a square, arriving at $24\frac{3}{4}$ for the square $IXZY$. This together with 308 is $332\frac{3}{4}$ for the area of the square that can be constructed from CX or YK. The $\sqrt{\mathfrak{z}}$ of this is $\sqrt{332\frac{3}{4}}$ for CX. From this IX [subtracted], being $\sqrt{24\frac{3}{4}}$, there remains $\sqrt{176}$ for CI, or CL or AC. Also add $\sqrt{99}$ (being [the length] of LB) to $\sqrt{176}$ AC, makes together for CB $\sqrt{539}$, this added to $\sqrt{176}$ CI, makes together $\sqrt{1331}$ for AB. The square of AB then contains 1331.

But as I left out the length of the bodies aforementioned, whose length were EH, equal to AB, so I now take them into account as before, so the body (with length EH, and height and breadth AB or EH) then is 1331. I therefore take the cubic $\sqrt{\ }$ and shall arrive at 11 for the true length of line AB. Multiply $\sqrt{176}$ AC as a square, arriving at 176 for the square $ACDE$, so the body constructed from this square and length EH 176 I therefore divide by 11, arriving at 16 for the area of the square $ACDE$, its $\sqrt{\mathfrak{z}}$ is 4 for AC, which is as much as the number of pieces bought of cloth A. This subtracted from

11, there remains 7 for CB, the pieces he bought of cloth B. Multiply 4 and 7 each as a square, arriving at 16 and 49, together is 65 for the pieces he bought of cloth C. At last, subtract 16 from 49, there remains 33 for the [number of] pieces of cloth D, and as such are very artfully found the number of pieces of cloth he bought of each kind, as one required. Also, you will have the prices of the cloth: a piece of cloth A is 4 pounds, B 7 pounds, C 11 pounds and D 3 pounds.

The End

Chapter 3

Mathematical commentary

3.1 Methods and techniques

3.1.1 Point multiplications

First, we consider the use of point multiplications ([Pijls 2004], part I), which is often used together with a variant of the Regula Falsi. The method amounts to a similarity transformation of a geometrical figure: every length is multiplied by a factor $\frac{1}{k}$ and as result, every area with $\frac{1}{k^2}$. This factor is fixed by giving a suitable chosen segment a length, usually 1 or 2, resulting in a relatively simple calculation. Every line segment in the original figure is multiplied by the factor $\frac{1}{k}$. The proportions however remain constant. As a consequence the given data such as the total area of total of the lengths remains unchanged. For example, instead of n given lengths the new problem has one length and $n-1$ proportions, as the length of one line segment is known and from the other original n given line segments the lengths have been changed while the proportions have remained the same. A well defined problem is transformed to a new equally well defined problem, though easier to solve because of the advantageously chosen line segment with length 1 or 2. Using algebra, or the 'Rule of Coss' as Cardinael refers to it, one would have given that suitable line segment the length of x. After solving the similar problem, the factor k can be calculated by the Regula Falsi by comparing some real given value from the original problem with the length found in the similar problem

$$k = \frac{real\ length\ of\ line\ \ segment\ in\ the\ original\ figure}{new\ length\ of\ line\ segment\ in\ the\ similar\ figure}$$

The real lengths belonging to the original problem can now be calculated by multiplying the newly found lengths with factor k, so by applying the inverse transformation of similitude. Instead of comparing an originally given length with the newly found length one can also compare the old and the new area if that is more convenient. In that case the multiplication factor is not k but k^2. By indicating the Questions where this technique has been applied, shown below, we have differentiated between metrical problems in which lengths are calculated and pure construction problems.

Point multiplications ([Heath 1956] 2, 231 – 232; [Pijls 2004], part I, 48 – 49)

Metrical problems: **Q16, 18 otherwise, 25, 30, 30 otherwise, 34, 38, 39 otherwise, 47, 53, 60, 61, 62, 68, 82, 83, 84.**
Construction problems: **Q95, 96, 97.**

3.1.2 Geometrical algebra

Another remarkable technique concerns the general geometrical algebra with the manipulation and transformation of rectangles and the application of area, as its most important parts. The Questions concerned can be distinguished by the characteristic use of shaded rectangles or squares in the accompanied figures. Most of the geometrical algebra consists of the multiple shifts, rotations and replacements of rectangles. The geometrical algebra is based on Elem.I: 35-47, Elem.II: 1-14 and Elem.VI: 27, 28, 29 and amounts roughly to

a. **Calculation with polygons**

   Every polygon can be transformed with the help of Elem.I-44 to an equal rectangle with one side given. In the same way polygons can be added by adding the rectangles.

b. **Calculation with squares**

   The Pythagorean theorem (Elem.I-47, §3.1.4 case 1 and §3.2.1: the titlepage): the square on the slanted side equals the sum of the squares on the right angled sides. Result: each square can be added to or subtracted from another square with a new square as sum or difference as result (with the help of Elem.I-47).

c. **Construction and calculation of the geometrical mean using geometrical algebra and without the use of proportions.**

   Every rectangle can with the help of geometrical algebra (elliptic application) be transformed to a square with an equal area (Elem.II-14,

[Dijksterhuis 1929-30] II, 19). This transformation is essentially the construction of the geometrical mean of two line segments.

d. Construction of the geometrical mean without geometrical algebra but with the use of proportions.

In Elem.VI-13, the construction of the geometrical mean of two given line segments is executed once more, but this time the proof is not based on geometrical application but instead uses similarity.

Assume that two line segments AD and DB are given, with

$$AD + DB = b$$

Construct a semi-circle (as in figure 3.1) with $AB = AD + DB$ the diameter and M the centre. Choose C on the circle so that CD is the perpendicular from C on AB, so $CD \leq \frac{1}{2}b$.

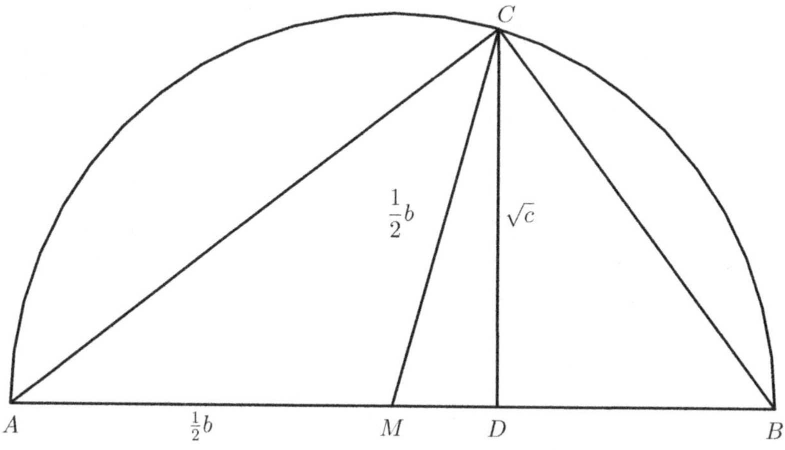

Figure 3.1: Construction of the mean proportional

According to Elem.III-31, angle C is a right angle. From the similarity of triangles BDC and CDA follows

$$DB : CD = CD : AD \tag{3.1}$$

CD, which is constructed in this way, is the geometrical mean between DB and AD. From 3.1 follows according to Elem.VI-16

$$CD^2 = AD \cdot DB \tag{3.2}$$

From 3.2 follows that the rectangle with sides AD and DB equals the square with sides CD. If we assume $CD = \sqrt{c}$, then AD, DB, c and b satisfy the system

$$\begin{cases} AD \cdot DB = c \ (= CD^2) \\ AD + DB = b \end{cases}$$

In the way as done in **c.** and **d.** we can construct the roots of a number by means of a line segment AD with that number as its length and line segment DB with length 1, as in that case $AD = CD^2$, from which follows that $CD = \sqrt{AD}$.
(**§3.2.1 subgroup 5; Q14, Q27 Otherwise 1, Q44, 45, 80**).

The other way around, the square of every number d can be constructed: construct a triangle CDB (see figure 3.1) with $CD = d$, $DB = 1$ and angle CDB a right angle. Then construct triangle ACB with angle ACB a right angle and A on the extension of BD. Then it follows from the above, that

$$d^2 = CD^2 = AD \cdot DB = AD$$

Furthermore, especially in §3.1.6, in the right-hand side of the geometrical equations we always find given squares with area c, p of q. Originally this could have been given polygons that first need to be reduced to rectangles with conservation of the areas in the standard way. After the transformation of these rectangles, in the way as described above, they are again transformed to squares with sides $\sqrt{c}$, $\sqrt{p}$ or $\sqrt{q}$ by means of the construction of the geometrical mean of the long and the short side (by construction and calculation!). The letter b always refers to a line segment with length b.

3.1.3 Application of area

A frequently used theme is the so called 'application of area'. This concerns an application of the geometrical identities Elem.II-5 and Elem.II-6 treated below in §3.1.5 case 8 and case 9 and as a special case, the identities of Elem.II-44,45 are treated in §3.1.4 case 2.

In our commentary on the Questions, 'application of area' denotes the geometrical operation of placing of a rectangle $ADEF$ with unknown lengths with AD alongside a given line segment AB, so that its area equals a given c and that the part $BDEG$ falling short to AB is a square $BDEG$ (figure 3.2) (see below §3.1.5 case 8 and §3.1.6 case 11) or that the part $BDEG$ exceeding AB (figure 3.3) is a square (§3.1.5 case 9 and §3.1.6 case 12).

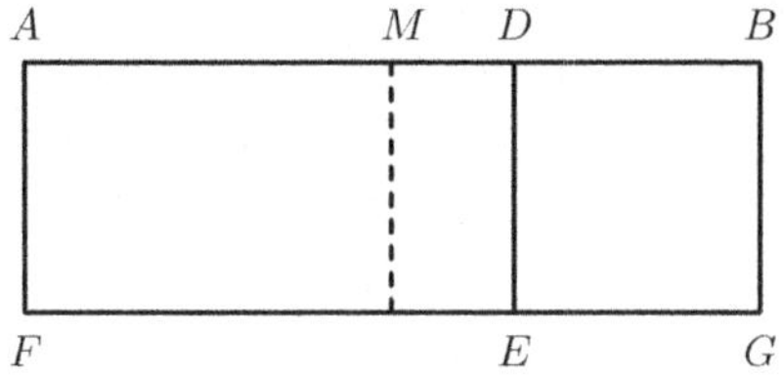

Figure 3.2: Application of area

In both cases, the unknown line segment AD can be constructed by pure geometrical means such as ruler and compass. If the square is falling short the application is called **elliptic** (figure 3.2) and if the square is exceeding it is called a **hyperbolic** application (figure 3.3).

The remaining case is called a **parabolic** application. Length AD must be equal to the given AB and the unknown DE follows from given $ADEF = c$.

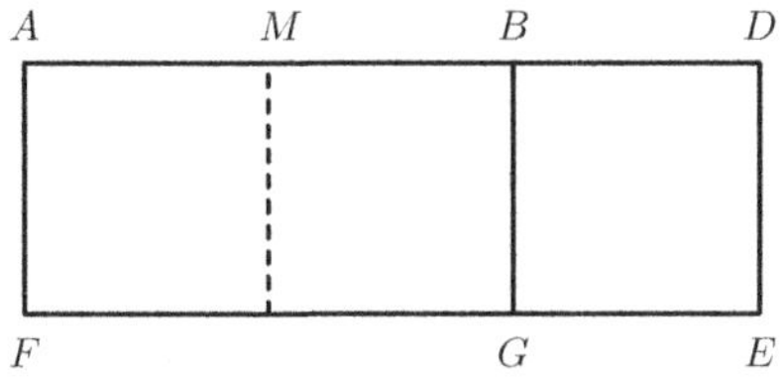

Figure 3.3: Parabolic application

The way Cardinael employs these techniques in the case of rectangles is based on the treatment of short falling or exceeding gnomons as treated by Euclid in Elem.II 5-6 (a third case, parabolic application, is treated by Euclid in Elem.I: 44-45). Generalizing to parallelograms in Elem.VI: 28-29, Euclid also uses the ideas of elliptic and hyperbolic application. In the case of application of area to rectangles the two propositions from Elem.II suffice. Cardinael always employs these in an implicit way. Nowhere in his HGQ one finds explicit reference to Euclid where the application of area is involved. We will use the qualifications 'parabolic', 'elliptic' and 'hyperbolic' only for the elucidation and analysis of Cardinael's solution methods, taking care its use is consistent with the meaning Euclid attributes to these concepts.

A geometrical quadratic equation with a line segment BD as unknown can for example be solved with 'geometrical algebra', or by 'hyperbolic or elliptic application of area'. For instance assume given the geometrical equation

$$AB \cdot BD + BD^2 = (\sqrt{c})^2 \tag{3.3}$$

with BD unknown, AB known and $(\sqrt{c})^2$ a square with side $\sqrt{c}$. The geometrical configuration of the left-hand side member, the rectangle

$$(AB + BD) \cdot BD$$

being the rectangle $ADEF$, exceeding by the square BD^2 outside AB (which is $BDEG$ in figure 3.3), corresponds exactly to the configuration described in Elem.II-6 (§3.1.5 case 9, §3.1.6 case 12), and based hereupon, the left-hand side of the equation can be replaced by the difference of the squares on sides DM and BM

$$DM^2 - BM^2 = (\sqrt{c})^2 \quad (\text{with } \ BM = \frac{1}{2}AB) \tag{3.4}$$

Because BM^2 and $\sqrt{c}$ are known, DM can be constructed (with ruler and compass) as well as calculated. Hence, also the unknown DB can be found, as $DB = DM - BM$ (§3.1.6 case 12 and also §3.1.7 case 15 type II).
If in the quadratic geometrical equation (3.3) the symbol x was used for the unknown DB, as is common practice today, there arises a regular quadratic equation from symbolic algebra of the form (with $AB = b$)

$$bx + x^2 = c$$

which is solved by completion of the left member to a complete square:

$$bx + x^2 + \left(\frac{1}{2}b\right)^2 - \left(\frac{1}{2}b\right)^2 = c \Leftrightarrow \left(\frac{1}{2}b + x\right)^2 - \left(\frac{1}{2}b\right)^2 = c$$

by which we have arrived at the translation into algebra of (3.4), from which the value of x can be easily obtained.

The following steps necessary to establish the algebraic solution of the quadratic equation (e.g. the formula for the roots) has a concrete geometrical meaning by its solution using application of area. There exists a 1-1 correspondence between the more or less abstract steps of the algebraic solution and the more concrete steps of the geometrical solution, which will treated below in greater detail. See §3.1.7 case 15 and **Q1, 40, 41, 42, 48**.
In the case of a linear equation

$$AB \cdot AL = b \cdot c$$

(b and c given and both ≥ 0) with AB given ($AB = a$) the goal is the construction and calculation of AL ($AL = x$). The solution is based on the construction of a rectangle $ABML$ with sides $AB = a$ and $AL = x$ so that $a \cdot x = b \cdot c$.

The construction (and calculation) of the unknown side AL is accomplished by the parabolic application of rectangle $ABML$ to AB (see §3.1.4 case 2 and Elem.I: 44-45).

The basis for these techniques of geometrical algebra by means of application of area is described by Euclid in Elem.I: 42-45 (parabolic application), Elem.II-5 (geometrical identity falling short by a square, see §3.1.5 case 8) and Elem.II-6 (geometrical identity exceeding by a square, see §3.1.5 case 9), ([Dijksterhuis 1929-30] II, $1-19$). The generalization of the theory of application of area to parallelograms is, as noticed above, treated in Elem.VI-27-29.

3.1.4 Two elementary examples of geometrical algebra and parabolic application of area

Case 1: The Pythagorean theorem (Elem.I-47 for another proof)

Figure 3.4 shows in an efficient way the validity of the Pythagorean theorem, convincingly demonstrating both the power and the simplicity of geometrical algebra.

4 triangles + inside square = outside square $\Leftrightarrow$
2 rectangles ab + inside square = outside square $\Leftrightarrow$
$2ab + c^2 = (a+b)^2 \Leftrightarrow$
$2ab + c^2 = a^2 + 2ab + b^2 \Leftrightarrow$
$c^2 = a^2 + b^2$

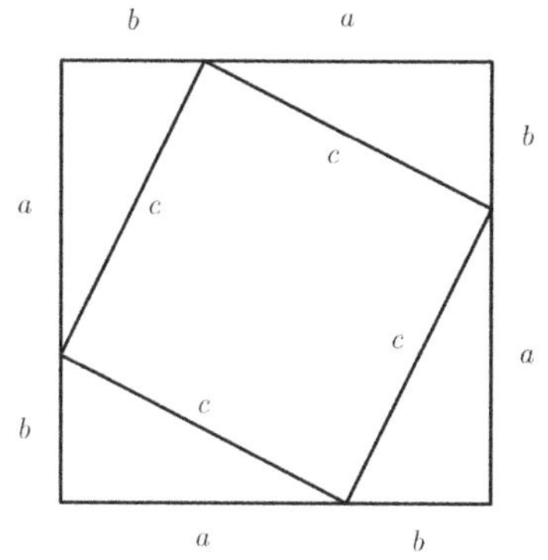

Figure 3.4: The Pythagorean theorem.

In the *Elements*, the lengths a, b, c or their squares a^2, b^2, c^2 are not mentioned but instead the sides BA, AC, BC and the squares on BA, AC and BC are used.

Case 2: Solving linear equations by parabolic application

Given the linear equation

$$ax = bc$$

by means of of algebra follows

$$x = \frac{bc}{a}$$

The same equation can also arise in a geometrical context when constructing a rectangle $ABML$ with $AB = a$ (a given and $a > 0$) and its area equal to a given rectangle R with area bc $(b, c > 0)$(see figure 3.5)

$$ABML = R$$

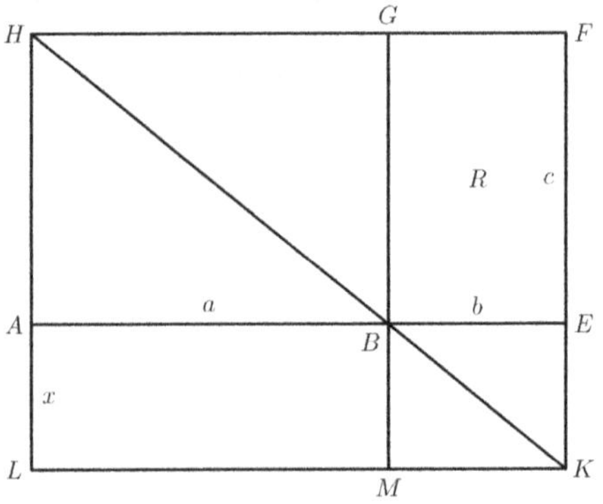

Figure 3.5: Solving linear equations geometrically

The solution depends on the parabolic application of a rectangle $ABML$ with area bc to a line segment AB with length a as described in §3.1.3 and as defined in Elem.I-44. In this case 2 the angle D is a right angle and triangle C is replaced by rectangle R with area bc ([Heath 1956] vol.1, p.342).

If we assume $AL = x, BE = b$ and $EF = c$ then the construction demonstrates that $ax = bc$ and the constructed AL is the solution of the equation $ax = bc.$[1].

[1] in Cardinael's era and certainly in his works a, b and c are always concrete numbers and not indeterminate values

3.1.5 Geometrical identities and the application of area

By means of introduction we list here a number of geometrical identities used by Cardinael that facilitate the use of geometrical algebra and in particular the special case of application of area (most of them are treated in Elem.II prop.1 to 14).

AB^2 denotes a square with side AB and $AB \cdot CD$ denotes a rectangle with sides AB and CD. The 'gnomon' is the part remaining when from a square a smaller square around the diagonal (at a vertex) is removed (Elem.II-definition 2, [Dijksterhuis 1929-30], I, 200; II, 2).

Often there are identities with at the left and the right member certain combinations of rectangles and squares (Elem.II-1 tot II-14). The validity can be established by adding suitable combinations of rectangles in combination with their translation and rotation.

Case 3: The Law of Cosine for the sharp and obtuse triangle

The two propositions on projection in Elem.II-13 (sharp-angled triangle) and Elem.II-12 (obtuse triangle), will henceforth be denoted as the Law of Cosine.

The Law of Cosine, corresponding with Elem.II-13, is applied in Chapter 2, **Q1**, Answer (Figure 2.3) with our commentary in Chapter 4, **Q1** Comment on Answer (Figure 4.2) in the form

$$AC^2 - AB^2 = DC^2 - DB^2$$

which is proven by Cardinael by means of elementary geometrical algebra.

The Law of Cosine, corresponding with Elem.II-12, is applied in Chapter 2, **Q2** (Figure 2.3) with our commentary in Chapter 4, **Q2**, Comment on Answer (Figure 4.4) in the form

$$AB^2 - AC^2 = DB^2 - DC^2$$

also proven by Cardinael using elementary geometrical algebra.

Our use of the term 'Law of Cosine' is justified by the derivation for the sharp-angled triangle ABC following from Chapter 2, **Q1**, Otherwise (Figure 2.2) with our commentary in Chapter 4, **Q1**, Comment on Otherwise (Figure 4.3), ammounting to

$$\begin{aligned}
AC^2 - AB^2 &= DC^2 - DB^2 \Leftrightarrow \\
AC^2 - AB^2 &= BC^2 + E + F \Leftrightarrow \\
AC^2 - AB^2 &= BC^2 + 2 \cdot DB \cdot BC \Leftrightarrow \\
AC^2 &= AB^2 + BC^2 + 2 \cdot DB \cdot BC
\end{aligned}$$

which is the classical Law of Cosine for a sharp-angled triangle ABC.

For the obtuse triangle ABC follows from Chapter 2, **Q2**, (Figure 2.4) with our commentary in Chapter 4, **Q2**, Comment on Otherwise (Figure 4.5)

$$\begin{aligned}
AC^2 - AB^2 &= DC^2 - DB^2 \Leftrightarrow \\
AC^2 - AB^2 &= BC^2 + E + F \Leftrightarrow \\
AC^2 - AB^2 &= BC^2 + 2DB \cdot BC \Leftrightarrow \\
AC^2 &= AB^2 + BC^2 + 2DB \cdot BC
\end{aligned}$$

which again is the classical Law of Cosine for the obtuse triangle ABC.

Case 4: Two identities that connect gnomon and rectangle

The identities have the form

$$a^2 - d^2 = (a + d) \cdot (a - d) \quad \text{with } a > d \tag{3.5}$$

or, written in another way

$$ad = \left(\frac{1}{2}a + \frac{1}{2}d\right)^2 - \left(\frac{1}{2}a - \frac{1}{2}d\right)^2 \tag{3.6}$$

expressing how $a^2 - d^2$, that is every difference of two squares, can be transformed to a rectangle and the other way around, a rectangle ad to a gnomon (3.5) and (3.6) are consequences of the cases 7 and 8 below.

Case 5: Elements, II-4

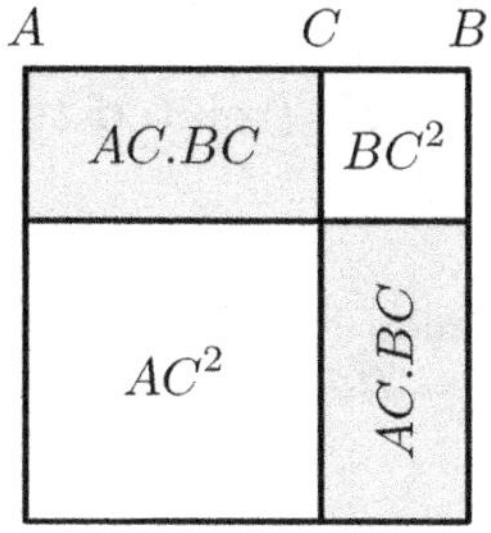

Figure 3.6: Elem.II-4

Point C lies on line segment AB by which the squares AB, AC and BC are drawn. From figure 3.6 follows that

$$(AC + BC)^2 = AC^2 + 2 \cdot AC \cdot BC + BC^2$$

Case 6: Elements, II-7

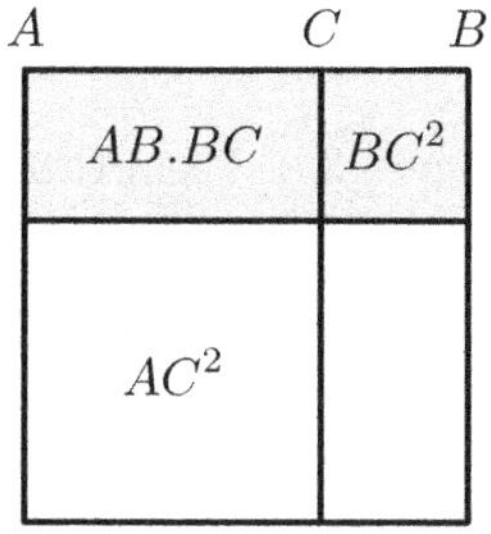

Figure 3.7: Elem.II-7

Next, figure 3.7 demonstrates that

$$AC^2 = AB^2 - 2AB \cdot BC + BC^2 \Leftrightarrow$$

$$(AB - BC)^2 = AB^2 - 2AB \cdot BC + BC^2$$

The following, case 7 is a general case, using (3.5) to transform a gnomon (left member) to a rectangle (right member). The rectangle has as sides the sum and the difference of the sides of the squares.

Case 7: Gnomon and rectangle (related to Elem.II: 5- 6)

Figure 3.8 shows that

$$\begin{aligned} AB^2 - BC^2 &= \text{gnomon } ACPQRSA = \text{ rectangle } UQRT = \\ &= (AB - BC) \cdot (AB + BC) \end{aligned}$$

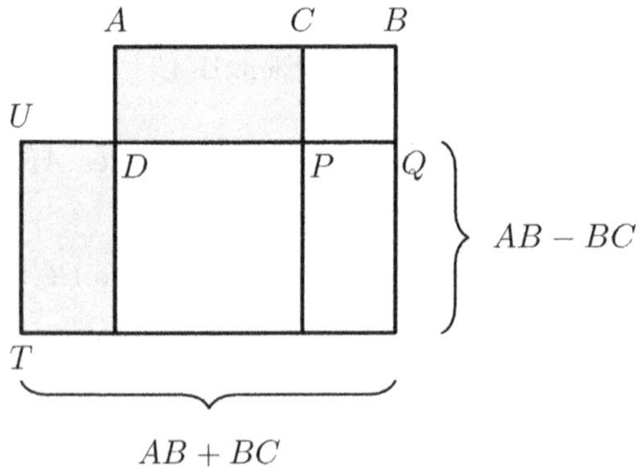

Figure 3.8: Elem.II 5-6

The geometrical interpretation of this identity is the rotation and then the translation of rectangle $DPCA$ to rectangle $TSDU$, so that the gnomon is transformed to a rectangle.

In the HGQ this is applied in

Q1: gnomon $DEFGHKD$ = rectangle $NHGM$,

Q2: gnomon $DCHMEF$ = rectangle $KMEL$,

Q31: gnomon → rectangle' is drawn twice in the figure.

Q56 otherwise: gnomon ($EG^2 - CG^2$)= rectangle $CE \cdot HG$.

Cases 8 and 9 concern a specific situation of the rectangle and the gnomon (see the figures below for cases 8 and 9).

The identities of the cases 8 and 9 are often used in the way of (3.6) to transform a rectangle to a gnomon.

We shall refer to the identity of case 8 as the elliptic identity. It corresponds to Elem.II-5 and is used in the elliptic application of area (see §3.1.3 and the application in case 11).

In case 9 we speak of the hyperbolic identity. This corresponds to Elem.II-6 and is used in hyperbolic application of area (see §3.1.3 and also the application in case 12).

Case 10, or the Golden Section, is a specific application of Case 9.

Case 8: The elliptic identity (anticipating §3.1.6 case 11)

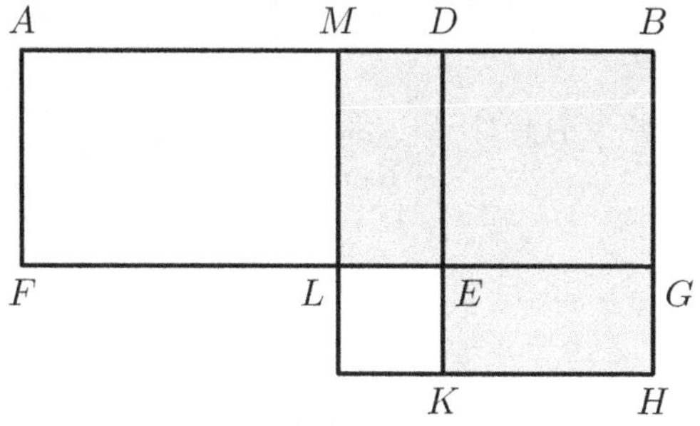

Figure 3.9: Gnomon and rectangle

AB is partitioned in two equal parts by M and into unequal parts by D . From the figure follows the equation (see figure 3.9)

$$\begin{aligned} BM^2 - DM^2 &= \text{gnomon } BHKELM \\ &= \text{rectangle } ADEF \\ &= DA \cdot DB \\ &= (BM + DM)(BM - DM) \end{aligned}$$

It does not matter if D is on the right-hand side or on the left-hand side of M as all equations remain valid (see Elem.II-5. Henceforth we shall interchange references to Elem.II-5 and case 8) .

Following Dijksterhuis ([Dijksterhuis 1929-30], 14) we shall refer to the geometrical configuration of case 8 by stating that the rectangle $ADEF$ is an elliptic application to AB falling short by the square $EGBD$.

If $ADEF = DA \cdot DB$ must have a prescribed size c and $AB = DA + DB$ a prescribed length b (see §3.1.6 case 11), the common terminology states that rectangle $ADEF$ has to be applied to the given line segment AB by falling short by the square DB^2 (AB is the sum of the long and the short side of the rectangle) and we refer to it as an elliptic application of the rectangle $ADEF$ to AB with the two unknowns DA and DB.

For an application, see §3.1.6 case 11: the construction and calculation of D and DB with

$$\begin{cases} DA \cdot DB = c \\ DA + DB = b \end{cases}$$

Case 9: The hyperbolic identity (anticipating §3.1.6 case 12)

AB is partitioned by M into two equal parts but now D is outside of AB. This leads to the equation

$$\begin{aligned} DM^2 - BM^2 &= \text{shaded gnomon} \\ &= \text{rectangle } ADEF \\ &= DA \cdot DB \\ &= (DM + BM)(DM - BM) \end{aligned}$$

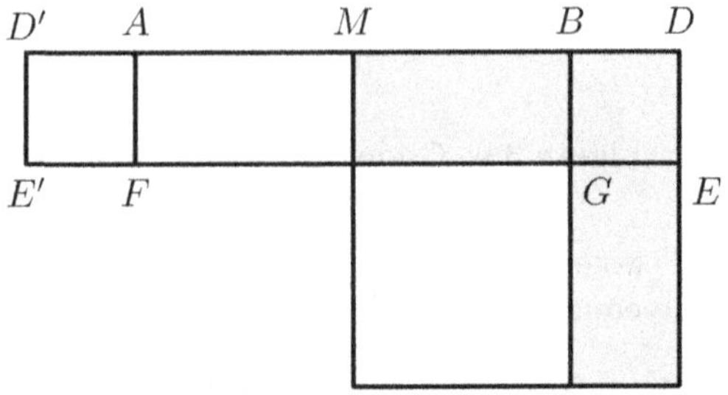

Figure 3.10: The hyperbolic identity

See Elem.II-6. Henceforth the references to Elem.II-6 or to case 9 are interchanged. Again following Dijksterhuis ([Dijksterhuis 1929-30], 14) we denote the geometrical configuration of case 9 by stating that $ADEF$ is a hyperbolic application of AB exceeding by the square $EGDB$.

The common interpretation is that the rectangle above on the left side is translated at A into the shaded rectangle below on the right-hand side at E, and the rectangle is transformed to a gnomon (or the other way around). In the case that $ADEF = DA \cdot DB$ must have a prescribed size c and $AB = DA - DB$ a prescribed length b (see §3.1.6 case 12), it is stated that the rectangle $ADEF$

is a hyperbolic application to the given AB (the segment which equals the difference of the long side and the short side of the rectangle) exceeding by the square DB^2, with the two unknowns DA and DB.

As in case 8 it does not matter for the application if D is on the extension of AB or on the extension of BA. All equations hold in the same way except that in the last case (D' on the left-hand side of A) rectangle $ADEF$ has to be replaced by rectangle $BD'E'G$. In this case rectangle $BD'E'G$ is applied to BA exceeding by the square $(AD')^2$.

For an application see §3.1.6 case 12: the construction and calculation of D when

$$\begin{cases} DA \cdot DB = c \\ DA - DB = b \end{cases}$$

with $c > 0$, $b > 0$, with D on the extension of AB;

Q4 (rectangle to gnomon),
Q81 (rectangle to gnomon).

Case 10: A special case of hyperbolic application: The Golden Section or partitioning in mean and extreme ratio

Definition (The Golden Section) If a segment AU is divided by a point F in segments AF and FU such that

$$FU : AF = AF : AU$$

this is denoted as the partitioning by F according to the Golden Section or partitioning in mean and extreme ratio.

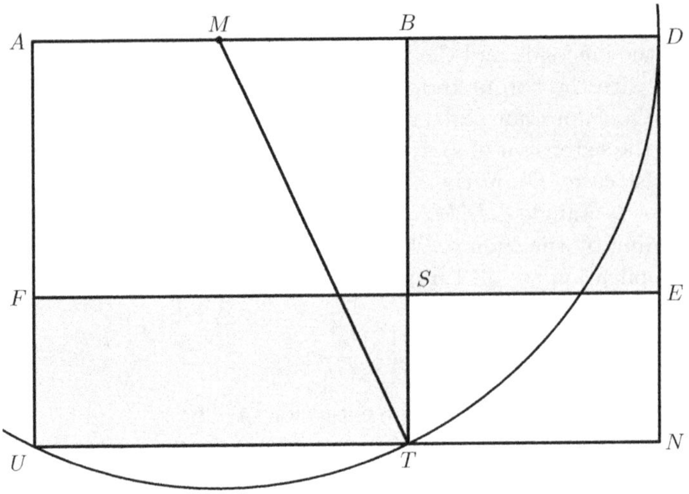

Figure 3.11: Constructing the golden section

Given a rectangle $ADEF$, which is a hyperbolic application to AB, exceeding by the square $BDES$.

Let M be the midpoint of AB and $ABTU$ the square on AB and N a vertex of the rectangle $ADNU$ (figure 3.11). From §3.1.5 case 9 follows

$$DM^2 - BM^2 = ADEF$$

In this case we have the following 5 equivalent properties

1. ... $ADEF = \square ABTU$
2. ... $FSTU = \square BDES$
3. ... AD is partitioned by B according to the Golden Section:

$$DB : AB = AB : DA$$

4. ... AU is partitioned by F according to the Golden Section:

$$FU : AF = AF : AU$$

5. ... $TM = DM$

Proof

(1) $\Leftrightarrow$ (2)

This is a direct consequence of the configuration (in figure 3.11) of $ABSE$, $BDEF$ and $FSTU$:

$$ABSF + \square BDES = ADEF$$

$$ABSF + FSTU = \square ABTU$$

(1) $\Leftrightarrow$ (3)

$$ADEF = \square ABTU \Leftrightarrow DA \cdot DB = AB^2 \Leftrightarrow DB : AB = AB : DA$$

(2) $\Leftrightarrow$ (4)

$$FSTU = \square BDES \Leftrightarrow SF \cdot ST = BS^2 \Leftrightarrow AU \cdot FU = AF^2 \Leftrightarrow$$

$$FU : AF = AF : AU$$

(1) $\Leftrightarrow$ (5)

$$\begin{aligned} ADEF = \square ABTU \quad &\Leftrightarrow DM^2 - BM^2 = AB^2 \\ &\Leftrightarrow DM^2 = BM^2 + AB^2 \\ &\Leftrightarrow DM^2 = BM^2 + BT^2 \\ &\Leftrightarrow DM^2 = TM^2 \\ &\Leftrightarrow DM = TM \end{aligned}$$

First consequence: Construction of point F dividing AU according to the Golden Section.

From (5) follows that in order to divide AU according to the Golden Section it suffices to construct the square $ABTU$ on AU, to choose the midpoint M of AB and to circle TM around M to DM with D on the extension of AB. If N is the vertex of the rectangle $ADNE$ and F, S, E are on AU, BT, DN such that $AF = BS = DE$, then it follows from the above that AU and BT are partitioned by F and S according to the Golden Section.

Second consequence: Calculation of AF and UF.

From (5) it follows that

$$\begin{aligned} AF \quad &= DB = DM - BM \\ &= TM - BM \\ &= \sqrt{AU^2 + (\tfrac{1}{2}AU)^2} - \tfrac{1}{2}AU \\ &= \sqrt{\tfrac{5}{4}AU^2} - \tfrac{1}{2}AU \\ &= \tfrac{1}{2}AU\sqrt{5} - \tfrac{1}{2}AU \\ &= \tfrac{1}{2}AU(\sqrt{5} - 1) \end{aligned}$$

and

$$UF = AU - AF = \frac{1}{2}AU(3 - \sqrt{5})$$

Remark: **Q35** is an application of the above, with
$AC = 10$ and $CD : AD = AD : AC$, so from (3) now follows:
$AD = 5(\sqrt{5} - 1)$ and $CD = 5(3 - \sqrt{5})$.

Compare with: Q25, 35, 69.

3.1.6 The solution of geometrical systems of two equations with two unknowns

As mentioned in §3.1.2 d, the letters c, p and q in the cases 11-14 always denote given squares with area c, p of q, and $\sqrt{c}$, $\sqrt{p}$ and $\sqrt{q}$ indicate the lengths of a side of such a square. With b is denoted the line segment with length b.

Cases 11 and 12, anticipated by §3.1.5 cases 8 and 9, are the geometrical version of the algebraic operation, known as completing the square.

In case 11 the equation $DA \cdot DB = c$ with the position of D on AB as the unknown is replaced by the (geometrical) identity $BM^2 - DM^2 = c$ where the unknown D only appears in the term DM^2. Hence, the equation can be solved in an easy way.

In case 12 something similar happens: $DA \cdot DB = c$ is replaced by

$$DM^2 - BM^2 = c$$

Case 11: system I (elliptic application of area with two unknowns)

Assume that the **system I**(DA, DB, c, b) is given with the line segments DA and DB as unknowns

$$\textbf{system I}(DA, DB, c, b) \left\{ \begin{array}{l} DA \cdot DB = c \\ DA + DB = b \end{array} \right.$$

with c a given square with side $\sqrt{c}, c > 0, b > 0, DA + DB = AB$ and M the midpoint of AB (see figure 3.12). What is asked is the calculation and construction of the pair of unknowns $\{DA, DB\}$.

Considering the square with sides $DA + DB = b$ (see §3.1.5 case 5 with C replaced by D) it can be proven by elementary geometrical algebra (as D moves between A and B, the maximum value of the area $DA \cdot DB$ is assumed if $DA = DB = \frac{1}{2}b$, so $0 < DA \cdot DB \leq (\frac{1}{2}b)^2$) that **system I** (DA, DB, c, b) implies that

$$0 < c \leq \left(\frac{1}{2}b\right)^2 \text{ with } c = \left(\frac{1}{2}b\right)^2 \Leftrightarrow DA = DB = \frac{1}{2}b$$

So the **system I** (DA, DB, c, b) is only solvable if this condition is satisfied. For that reason, in the case of **system I** (DA, DB, c, b) we shall always assume that the condition $0 < c \leq (\frac{1}{2}b)^2$ is fulfilled. Therefore from now on we shall always write

$$\textbf{system I}(DA, DB, c, b) \text{ with } 0 < c \leq \left(\frac{1}{2}b\right)^2$$

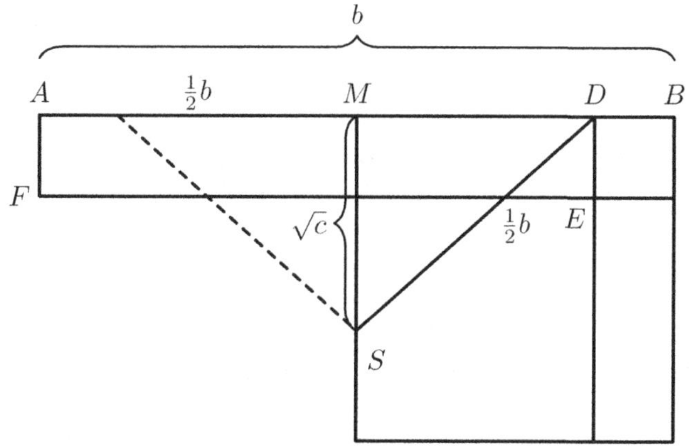

Figure 3.12: The system I(DA, DB, c, b)

According to §3.1.5 case 8 and the corresponding comment, case 11 concerns an elliptical application of area of rectangle $DAFE$ (figure 3.12) with $DE = DB$, to the line segment

$$AB = DA + DB$$

According to case 8, from this follows

$$DA \cdot DB = BM^2 - DM^2$$

so that

$$BM^2 - DM^2 = c \Leftrightarrow$$

$$DM^2 = \left(\frac{1}{2}b\right)^2 - c \Leftrightarrow$$

$$DM = \sqrt{(\frac{1}{2}b)^2 - c}$$

and hence follow directly the two solutions, mirrored in point M, for DA. The first solution ($DA \geq DB$):

$$DA = AM + DM \text{ and } DB = BM - DM \Leftrightarrow$$

$$DA = \frac{1}{2}b + \sqrt{\left(\frac{1}{2}b\right)^2 - c} \text{ and } DB = \frac{1}{2}b - \sqrt{\left(\frac{1}{2}b\right)^2 - c}$$

and the second solution ($DA \leq DB$) follows from the symmetry of the **system I** and arises from interchanging the lengths of the line segments DA and DB

$$DA = AM - DM \text{ and } DB = BM + DM \Leftrightarrow$$

$$DA = \frac{1}{2}b - \sqrt{\left(\frac{1}{2}b\right)^2 - c} \text{ and } DB = \frac{1}{2}b + \sqrt{\left(\frac{1}{2}b\right)^2 - c}$$

Hence, it also follows that there is exactly one solution for $c = (\frac{1}{2}b)^2$. The locus of D on AB (and so the length of DB to be determined) can also be constructed (figure 3.12). By locating S so that MS is perpendicular on AB and $MS = \sqrt{c}$ and by circling around S a segment with length $\frac{1}{2}b$ both positions of D on AB can be found, because DM is a right-angle side of the right-angled triangle with hypothenuse $\frac{1}{2}b$ and right-angle side $\sqrt{c}$.

Remark. From the above follows that solving **system I**(DA, DB, c, b) corresponds to calculating the sides DA and $DE = DB$ of rectangle $DAFE$ (figure 3.12) with area c and perimeter $2b$.

For applications see

§3.1.7 case 15 type I;

Q5 with

$$\begin{cases} DC \cdot DA = 576 \\ DC + DA = 50 \end{cases}$$

Q32, 44.

Case 12: system II (hyperbolic application of area with two unknowns)

Assume given the **system II**(DA, DB, c, b) with the line segments DA and DB as unknowns

$$\textbf{system II}(DA, DB, c, b) \begin{cases} DA \cdot DB = c \\ DA - DB = b \end{cases}$$

with $c > 0, b > 0$, c a given square with side $\sqrt{c}$ and with $DA - DB = AB$ and M the midpoint of AB. One asks the calculation and construction of the pair of unknowns $\{DA, DB\}$.

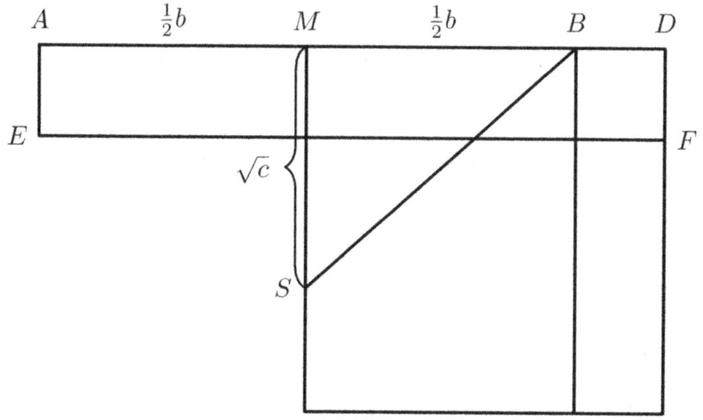

Figure 3.13: Hyperbolic application of area

According to §3.1.5 case 9 and the given comment thereupon case 12 concerns the geometrical identity belonging to the hyperbolic application of $DAFE$ to the line segment $AB = DA - DB$

$$DA \cdot DB = DM^2 - BM^2$$

so

$$DM^2 - BM^2 = c \Leftrightarrow$$

$$DM^2 = \left(\frac{1}{2}b\right)^2 + c$$

As DM^2 can be any positive number, contrary to case 11, no extra condition is needed for $c > 0$. For all values of $c \geq 0$ there is exactly one solution for the locus of point D. With $BM = \frac{1}{2}b$, this means for the locus of point D

$$DM^2 = \left(\frac{1}{2}b\right)^2 + c \Leftrightarrow$$

$$DM = \sqrt{\left(\frac{1}{2}b\right)^2 + c}$$

and so DA and DB are known

$$DA = AM + DM \text{ and } DB = -BM + DM \Leftrightarrow$$

$$DA = \frac{1}{2}b + \sqrt{\left(\frac{1}{2}b\right)^2 + c} \text{ and } DB = -\frac{1}{2}b + \sqrt{\left(\frac{1}{2}b\right)^2 + c}$$

To find the locus of D, a right-angled triangle BMS is constructed with $BM = \frac{1}{2}b$ and $MS = \sqrt{c}$ and a right angle in M. Then SB has the length of DM, so by circling around SB from point M the point D on the extension of AB can be found.

With D', on the extension of BA, so on the left side of A (see also case §3.1.5 case 9), the **system II** changes into

$$\textbf{system II}' \begin{cases} D'B \cdot D'A = c \\ D'B - D'A = b \end{cases}$$

This system arises by replacing in **system II** DA by DB and D by D'. In this way we get with the help of the hyperbolic application to BA

$$D'A = -\frac{1}{2}b + \sqrt{\left(\frac{1}{2}b\right)^2 + (\sqrt{c})^2} \text{ and}$$

$$D'B = \frac{1}{2}b + \sqrt{\left(\frac{1}{2}b\right)^2 + (\sqrt{c})^2}$$

For applications see

§3.1.7 case 15 type II: quadratic equations

Q40, 41, 48 , 50.

Case 13: system III

Assume, given the system

$$\textbf{system III}\begin{cases} AB \cdot BC = q \\ AB^2 + BC^2 = p \end{cases}$$

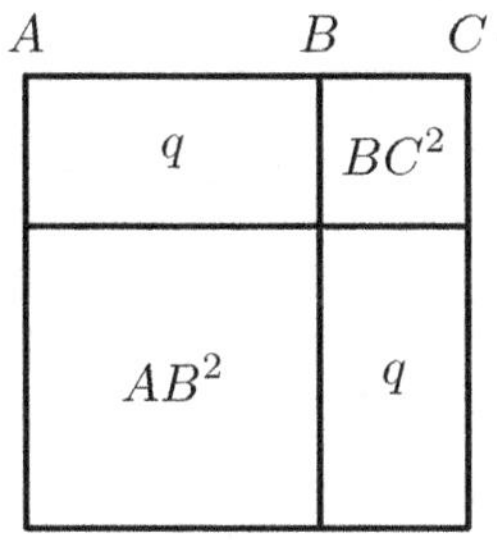

Figure 3.14: Quadratic equations, system III

From figure 3.14 follows geometrically

$$(AB + BC)^2 = p + 2q \Leftrightarrow$$

$$AB + BC = \sqrt{p + 2q}$$

and **system III** is in a geometrical way reduced to

$$\textbf{system I}(BA, BC, q, \sqrt{p+2q})\begin{cases} BA \cdot BC = q \\ BA + BC = \sqrt{p+2q} \end{cases}$$

which is according to case 11 only solvable if

$$0 \leq c \leq \left(\frac{1}{2}b\right)^2 \Leftrightarrow$$

$$0 \leq q \leq \frac{1}{4}(p + 2q) \Leftrightarrow$$

$$0 \leq q \leq \frac{1}{2}p$$

Case 14: system IV

Finally, let be given the system

$$\textbf{system IV} \begin{cases} AB + BC = p \\ AB^2 + BC^2 = q \end{cases}$$

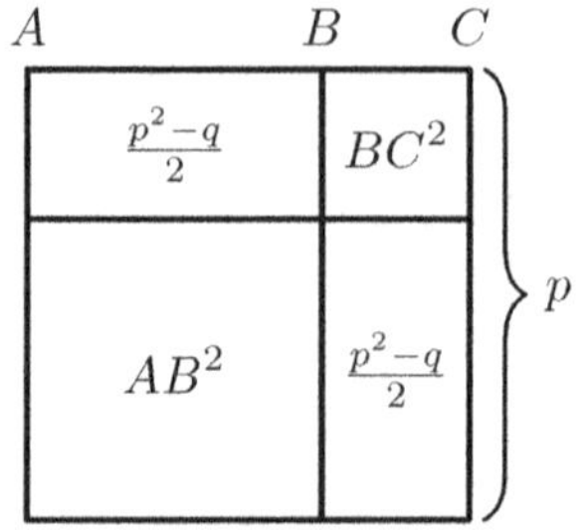

Figure 3.15: Quadratic equations, system IV

Figure 3.15 shows that (using §3.1.5 case 5)

$$(AB + BC)^2 = p^2 \Leftrightarrow$$

$$AB^2 + 2 \cdot AB \cdot BC + BC^2 = p^2 \Leftrightarrow$$

$$AB \cdot BC = \frac{p^2 - q}{2}$$

And so **system IV** is also transformed in a geometrical way to **system I** $(BA, BC, \frac{p^2-q}{2}, p)$

$$\begin{cases} BA \cdot BC = \frac{p^2-q}{2} \\ BA + BC = p \end{cases}$$

if according to case 11

$$\frac{1}{2}p^2 \leq q < p^2$$

is satisfied.

3.1.7 Quadratic equations

Case 15: Elliptic and hyperbolic applications of area

Every quadratic equation in which the coefficient of x^2 equals 1 has necessarily the form:

type I: $x^2 - bx + c = 0$

type II: $x^2 + bx - c = 0$

type III: $x^2 + bx + c = 0$

type IV: $x^2 - bx - c = 0$

with $b \geq 0$ and $c \geq 0$

Because type III only has negative solutions, equations of this form will not arise within the framework of the practical mathematics of Cardinael and his contemporaries. Negative solutions to quadratic equations were not accepted at the beginning of the seventeenth century and were therefore called 'false roots' ([Kinckhuysen 1661], 44) or 'ghedichte (imagined) roots' [Stampioen 1639]. That is why we limit our discussion to the types I, II and IV. For these three cases, modern algebraic solution methods will be connected with the geometrical way of solving as still practiced in Cardinael's era, he himself being a convinced adept of the geometrical method.

The structure of equations of type II and type IV point to a more direct correspondence between their solution methods: if x is a solution of type II than $-x$ is a solution of type IV and vice versa. So we could limit our discussion to type I and type II. However, it was common for Cardinael and his contemporaries to treat these three types separately.

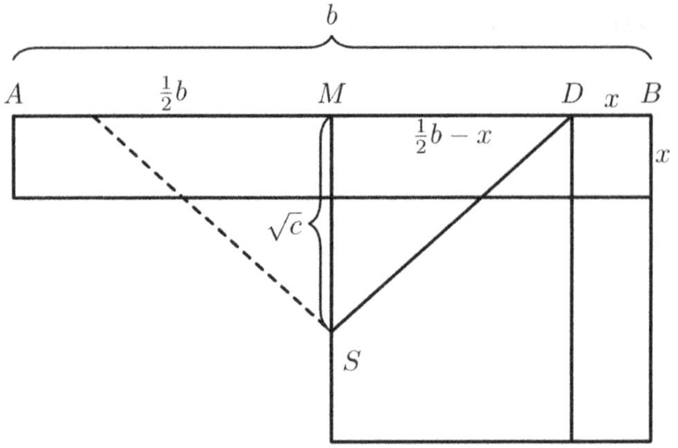

Figure 3.16: Quadratic equations, type I

type I

From the structure of this type of equation, it is known that

$$x(b-x) = c \text{ with } (b \geq 0, c \geq 0)$$

has only solutions if

$$0 \leq c \leq (\frac{1}{2}b)^2$$

In that case there are two positive solutions if $0 \leq c < (\frac{1}{2}b)^2$, namely

$$x = \frac{1}{2}b - \sqrt{\frac{1}{4}b^2 - c} \text{ or } x = \frac{1}{2}b + \sqrt{\frac{1}{4}b^2 - c}$$

and there is only the solution

$$x = \frac{1}{2}b \text{ if } c = \left(\frac{1}{2}b\right)^2$$

This equation (with $b > 0$)) also arises from the geometrical framework of an elliptic application of area as described in §3.1.6 case 11 if one assumes $AB = b$ and $DB = x$ with D between A and B.

$$\textbf{system I} \begin{cases} DA \cdot DB = c \\ DA + DB = b \end{cases}$$

is then replaced by

$$\text{type I:}(b-x)\cdot x=c$$

From the treatment of §3.1.6 case 11 it is known that the loci of the required point(s) D between A and B are fixed by the relations

$$DB=\frac{1}{2}b-\sqrt{\frac{1}{4}b^2-c}\text{ or }DB=\frac{1}{2}b+\sqrt{\frac{1}{4}b^2-c}$$

with $c\leq\frac{1}{4}b^2$. Substituting $DB=x$ leads to the algebraic solutions

$$x=\frac{1}{2}b-\sqrt{\frac{1}{4}b^2-c}\text{ or }x=\frac{1}{2}b+\sqrt{\frac{1}{4}b^2-c}\text{ with }c\leq\frac{1}{4}b^2$$

The geometrical method of solving is in this case fully equivalent to the algebraic way. As will be proven below, the equivalence of the geometrical and algebraic solution for equations of type II and IV will not apply in the same way because of the occurrence of negative solutions.

Remark: assuming that $DA=x$ instead of $DB=x$ gives the same algebraic equation, in contrast with the application of §3.1.6 case 12 in type II and type IV below.

type II
The equation

$$x(b+x)=c\text{ with }b\geq 0\text{ and }c\geq 0$$

similarly corresponds (though now partly) to §3.1.6 case 12 of hyperbolic application. The modern algebraic method leads to two solutions

$$x=-\frac{1}{2}b-\sqrt{\left(\frac{1}{2}b\right)^2+c}$$

and

$$x=-\frac{1}{2}b+\sqrt{\left(\frac{1}{2}b\right)^2+c}$$

By assuming in the geometrical context of §3.1.6 case 12 (D on extension of AB), $DB=x$ and $AB=b$ the

$$\textbf{system II}\begin{cases}DA\cdot DB=c\\ DA-DB=b\end{cases}\quad\text{with }c\geq 0\text{ and }b>0$$

is replaced by

$$x(b+x) = c \text{ with } c \geq 0 \text{ and } b > 0$$

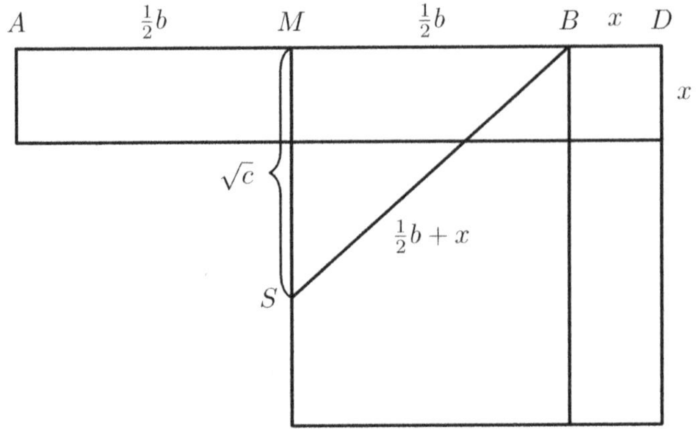

Figure 3.17: Quadratic equations, type II

For §3.1.6 case 12 with D on the extension of AB we have proven that for every $c \geq 0$ there is only one solution:

$$DB = -\frac{1}{2}b + \sqrt{\left(\frac{1}{2}b\right)^2 + \left(\sqrt{c}\right)^2}$$

The corresponding positive solution of type II is

$$x = -\frac{1}{2}b + \sqrt{\left(\frac{1}{2}b\right)^2 + c}$$

So, in this way, the negative solution does not occur!

type IV
The equation

$$\textbf{system II}: x(x-b) = c \text{ with } b \geq 0 \text{ and } c \geq 0$$

also arises from the hyperbolic application of §3.1.6 case 12 (again D on the extension of AB). In this case, algebra leads us to the two solutions

$$x = \frac{1}{2}b - \sqrt{\left(\frac{1}{2}b\right)^2 + c} \text{ and } x = \frac{1}{2}b + \sqrt{\left(\frac{1}{2}b\right)^2 + c}$$

By assuming now in the geometrical situation of §3.1.6 case 12 that $AD = x$ instead of $BD = x$ (as in the case of type II) and so $BD = x - b$

$$\textbf{system II}\begin{cases} DA \cdot DB = c \\ DA - DB = b \end{cases} \quad \text{with } c \geq 0 \text{ and } b \geq 0$$

is replaced by

$$\text{type IV} : x(x - b) = c$$

Again according to §3.1.6 case 12 there is for $c \geq 0$ only one solution

$$DB = -\frac{1}{2}b + \sqrt{\left(\frac{1}{2}b\right)^2 + (\sqrt{c})^2}$$

and so

$$DA = AB + DB = b + DB = \frac{1}{2}b + \sqrt{\left(\frac{1}{2}b\right)^2 + c}$$

The corresponding solution of type IV is

$$x = \frac{1}{2}b + \sqrt{\left(\frac{1}{2}b\right)^2 + c}$$

because it was assumed that $DA = x$. Also in the geometrical solution of this case, only the positive solution is found.

3.1.8 The equivalence of system I and the AH-system

In §3.1.6 case 11, with the help of an elliptic application, as solution of

$$\textbf{system I}\begin{cases} DA \cdot DB = c \\ DA + DB = b \end{cases} \tag{3.7}$$

with c being a given square with sides $\sqrt{c}$, $0 < c \leq (\frac{1}{2}b)^2$, where b is known and DA and DB and unknown, leads to the following solutions

$$DA = \frac{1}{2}b + \sqrt{\left(\frac{1}{2}b\right)^2 - c} \text{ and } DB = \frac{1}{2}b - \sqrt{\left(\frac{1}{2}b\right)^2 - c} \tag{3.8}$$

The second solution, arrising from the symmetry of the system by interchanging these values, can be represented by the locus of D on the left-hand side of M (see figure in §3.1.6 case 11)

$$DA = \frac{1}{2}b - \sqrt{\left(\frac{1}{2}b\right)^2 - c} \text{ and } DB = \frac{1}{2}b + \sqrt{\left(\frac{1}{2}b\right)^2 - c} \qquad (3.9)$$

In §3.1.6, case 11 also is treated with respect to the constructing of the two solutions.

A second way to solve such a system which is frequently applied by Cardinael, does not depend on the application of area. To explain this method we use the idea of an **AH-system** (see also **Q5**).

An

$$\textbf{AH-system } (CD, AB, \sqrt{c}, b) \qquad (3.10)$$

is defined as a right triangle ABC with right angle C and **A**ltitude $CD = \sqrt{c}$ from C on the **H**ypotenuse $AB = b$ with DA and DB unknown. The question asks for the calculation and construction of the line segments DA and DB.

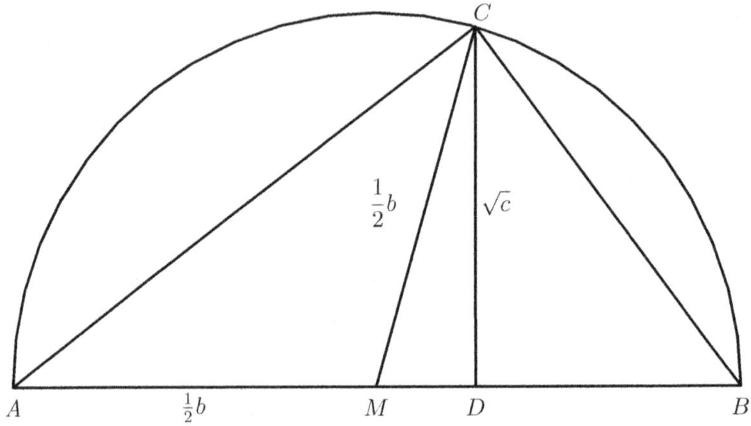

Figure 3.18: The AH-system

Case 16: Equivalence of AH-system (3.10) and system I (3.7)

This equivalence can be proven as follows (figure 3.18):

AH-system $\Rightarrow \triangle BDC$ is similar to $\triangle CDA \Rightarrow$
$DB : DC = DC : DA \Rightarrow$
$DA \cdot DB = DC^2 \Rightarrow$

$$\begin{cases} DA \cdot DB = c \\ DA + DB = b \end{cases} \tag{3.11}$$

From Elem.III-31 (Thales's theorem) follows that AB is the diameter of the circle with centre M, going through A, B and C. A direct consequence (with the help of $\triangle MDC$) is that

$$\sqrt{c} \leq \frac{1}{2}b \tag{3.12}$$

From (3.11) and (3.12) follows

$$\textbf{AH-system} \ \ (3.10) \Rightarrow \textbf{system I} \ (3.7) \tag{3.13}$$

Because of (3.12) we shall henceforth always assume, that in the case of an **AH-system** always $c \leq (\frac{1}{2}b)^2$.

And the other way around, we also have

System I (3.7)$\Rightarrow$ Choose on the semi-circle with diameter

$$AB = AD + DB = b$$

and centre M a point C so that for the point D of intersection of the altitude from C op AB with AB we have $DC = \sqrt{c}$. This is always possible because $\sqrt{c} \leq \frac{1}{2}b$. From Elem.III-31 follows, that if angle C is a right angle, then

$$\Longrightarrow \textbf{AH-system} \ \ (3.10) \tag{3.14}$$

From (3.13) and (3.14) follows

$$\textbf{AH-system} \ (3.10) \Longleftrightarrow \textbf{system I} \ (3.7) \tag{3.15}$$

or

$$\textbf{AH-system} \ (CD, AB, \sqrt{c}, b) \Longleftrightarrow \textbf{system I} \ (DA, DB, c, b) \tag{3.16}$$

In both cases the unknowns are DA and DB.

Solution of the AH-system $(CD, AB, \sqrt{c}, b)$
We now can come to a direct solution of the **AH-system** (3.10) without using application of area (figure 3.18). According to Thales's Theorem the points A, B and C are on the circle with diameter AB and centre M. In triangle MBC

$$DM^2 = CM^2 - DC^2 = \left(\frac{1}{2}b\right)^2 - c$$

so that

$$DM = \sqrt{\left(\frac{1}{2}b\right)^2 - c}$$

and the solutions for the **AH-system** $(CD, AB, \sqrt{c}, b)$ are

$$DA = \frac{1}{2}b + \sqrt{\left(\frac{1}{2}b\right)^2 - c} \text{ and } DB = \frac{1}{2}b - \sqrt{\left(\frac{1}{2}b\right)^2 - c} \tag{3.17}$$

The second solution for D left of M arises by interchanging the solutions of the first equation

$$DA = \frac{1}{2}b - \sqrt{\left(\frac{1}{2}b\right)^2 - c} \text{ and } DB = \frac{1}{2}b + \sqrt{\left(\frac{1}{2}b\right)^2 - c} \tag{3.18}$$

From the method of calculation follows the method of construction for the solutions. The above means that **system I** (DA, DB, c, b) and the equivalent **AH-system** $(CD, AB, \sqrt{c}, b)$ can each be solved in two ways.

Case 16a: system I is solved directly by means of application of area (according to §3.1.6 case 11 or §3.1.8 (3.8), (3.9).

Case 16b: system I is solved indirectly without using application of area (i.e. with the help of the equivalent **AH-system** which is solved according to (3.17), (3.18)).

Case 16c: AH-system is solved directly (according to (3.17), (3.18)).

Case 16d: AH-system is solved indirectly by means of application of area (i.e. the equivalent **system I** according to §3.1.6 case 11 or §3.1.8 (3.8), (3.9)).

Cardinael uses all variants (**16abcd**).

A third way to solve **system I** (DA, DB, c, b) uses hyperbolic application. The equivalence of this method with the two last methods shall be demonstrated in case 17.

3.1.9 Hyperbolic application instead of elliptic application

Case 17: The use of system II instead of system I

In figure 3.19 and 3.20 below, we always have for the sides of rectangle $ABCD$ that c is a given square with sides $\sqrt{c}$, $BA = a, BC = d, ad = c$ and $a + d = b, a - d = v$ or $d - a = w, M_1$ the midpoint of AC_1, M_2 midpoint of AC_2, $BC_1 = BC_2 = BC$.

In §3.1.6 case 11 it has been proven that the solution of

$$\textbf{system I}(BA, BC_1, c, b) \begin{cases} BA \cdot BC_1 = c \\ BA + BC_1 = b \end{cases} \tag{3.19}$$

by means of of elliptic application amounts to the calculation of the sides BA and $BC = BC_1$ of a rectangle $BADC$ (see figure 3.19 and figure 3.20) with area c and perimeter $2b$. The rectangle $BADC$ is then applied in an elliptic way to the line segment $AC_1 = BA + BC_1$ falling short by the square $(BC_1)^2$. For the two solutions $\{BA, BC_1\}$ with respectively $BA \geq BC_1$ and $BA \leq BC_1$ it was found in §3.1.6 case 11:

$BA \geq BC_1$ (figure 3.19)

$$BA = \frac{1}{2}b + \sqrt{\left(\frac{1}{2}b\right)^2 - c} \text{ and } BC_1 = \frac{1}{2}b - \sqrt{\left(\frac{1}{2}b\right)^2 - c}$$

$BA \leq BC_1$ (figure 3.20)

$$BA = \frac{1}{2}b - \sqrt{\left(\frac{1}{2}b\right)^2 - c} \text{ and } BC_1 = \frac{1}{2}b + \sqrt{\left(\frac{1}{2}b\right)^2 - c}$$

For the original rectangle $BADC$ this implies (for $BA \geq BC_1$)

$$BA = \frac{1}{2}b + \sqrt{\left(\frac{1}{2}b\right)^2 - c} \text{ and } BC = \frac{1}{2}b - \sqrt{\left(\frac{1}{2}b\right)^2 - c} \tag{3.20}$$

or (if $BA \leq BC_1$)

$$BA = \frac{1}{2}b - \sqrt{\left(\frac{1}{2}b\right)^2 - c} \text{ and } BC = \frac{1}{2}b + \sqrt{\left(\frac{1}{2}b\right)^2 - c} \tag{3.21}$$

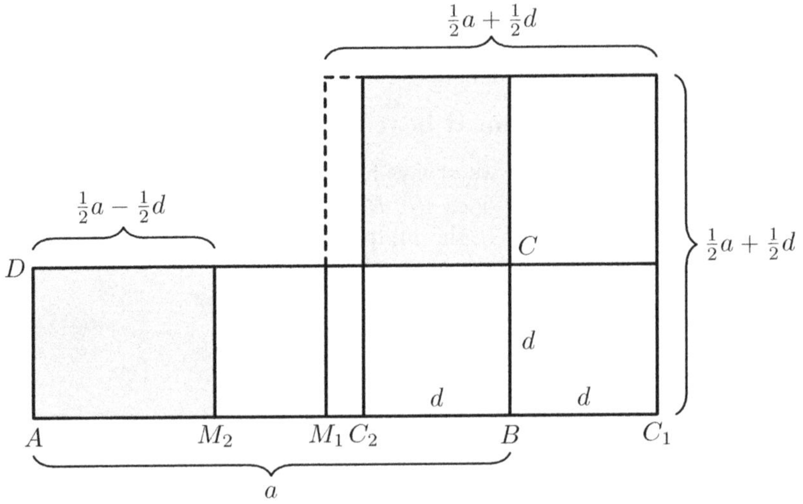

Figure 3.19: The case where $BA \geq BC$

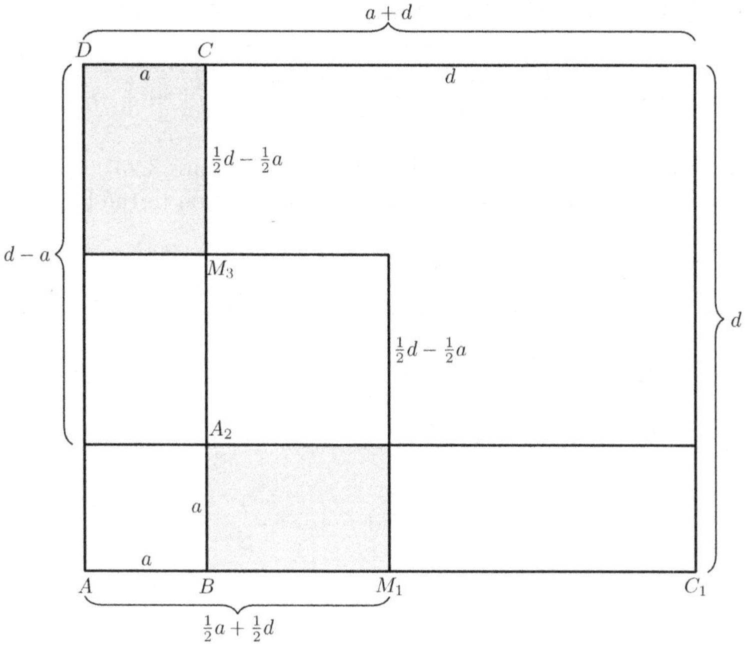

Figure 3.20: The case where $AB \leq BC$

However, it is also possible to find the solution(s) for the pair unknowns {BA,BC} of **system I** by interpreting a rectangle $BADC$ with a given area and perimeter as a hyperbolic application of area. In that case we differentiate a priori between $BA \geq BC$ and $BA \leq BC$

In the case where $BA \geq BC \Leftrightarrow BA \geq BC_2$ (figure 3.19), we interpret the rectangle $BADC$ as the hyperbolic application of area to the line segment

$$AC_2 = BA - BC_2 = v \text{ with } BC_2 = BC$$

exceeding with the square $(BC_2)^2$, that is described in §3.1.6 case 12 as

$$\textbf{system II } (BA, BC_2, c, v) \begin{cases} BA \cdot BC_2 = c \\ BA - BC_2 = v \end{cases} \quad v = a - d \tag{3.22}$$

The solution according to §3.1.6 case 12 can be expressed as

$$BA = \frac{1}{2}v + \sqrt{\left(\frac{1}{2}v\right)^2 + c} \text{ and } BC_2 = -\frac{1}{2}v + \sqrt{\left(\frac{1}{2}v\right)^2 + c}$$

For the original rectangle $BADC$ this implies (if $BA \geq BC_2$)

$$BA = \frac{1}{2}v + \sqrt{\left(\frac{1}{2}v\right)^2 + c} \text{ and } BC = -\frac{1}{2}v + \sqrt{\left(\frac{1}{2}v\right)^2 + c} \qquad (3.23)$$

In the case where $BA \leq BC \iff BC \geq BA_2$ (figure 3.20), the rectangle $BADC$ can be interpreted as the hyperbolic application to the line segment

$$CA_2 = BC - BA_2 = w \text{ with } BA_2 = BA$$

exceeding by the square $(BA_2)^2$, in §3.1.6 case 12 described as

$$\textbf{system II}(BC, BA_2, c, w) \begin{cases} BC \cdot BA_2 = c \\ BC - BA_2 = w \end{cases} \quad w = d - a \qquad (3.24)$$

The solution by means of §3.1.6 case 12 can be expressed as

$$BC = \frac{1}{2}w + \sqrt{\left(\frac{1}{2}w\right)^2 + c} \text{ and } BA_2 = -\frac{1}{2}w + \sqrt{\left(\frac{1}{2}w\right)^2 + c}$$

For the original rectangle $BADC$ this means (if $BC > BA_2$)

$$BA = -\frac{1}{2}w + \sqrt{\left(\frac{1}{2}w\right)^2 + c} \text{ and } BC = \frac{1}{2}w + \sqrt{\left(\frac{1}{2}w\right)^2 + c} \qquad (3.25)$$

With

$$c = ad, b = a + d \text{ and } v = a - d$$

it can easily be demonstrated that

$$(3.20) \iff (3.23)$$

and by means of of

$$c = ad, b = a + d \text{ and } w = d - a$$

it can be proven in the same way that

$$(3.21) \iff (3.25)$$

Cardinael, at several times uses the alternative solution method, a hyperbolic application, where an elliptic application is more evident. Doing so, he provides a more direct solution than we have done above. Therefore, the following calculations are discussed, more in style with Cardinael. Four cases can be distinguished i $BA > BC$, elliptic application

ii $BA > BC$, hyperbolic application

iii $BA < BC$, elliptic application

iv $BA < BC$, hyperbolic application

Each of these cases will now be discussed in detail.

(i): $BA > BC$ (so $a > d$), elliptic application (figure 3.19)

Choose C_1 on the extension of AB so that $BC_1 = BC$ and let the center point of AC_1 be M_1 so that $AM_1 = M_1C_1 = \frac{1}{2}b$. Rectangle $BADC$ can be interpreted as an elliptic application to AC_1 falling short by the square $(BC_1)^2$. According to §3.1.5 case 8, the following identities hold

$$\begin{aligned}
BA \cdot BC_1 &= (C_1M_1)^2 - (BM_1)^2 \iff \\
BA \cdot BC_1 &= (\tfrac{1}{2}BA + \tfrac{1}{2}BC_1)^2 - (C_1M_1 - BC_1)^2 \iff \\
BA \cdot BC_1 &= (\tfrac{1}{2}BA + \tfrac{1}{2}BC_1)^2 - (\tfrac{1}{2}BA + \tfrac{1}{2}BC_1 - BC_1)^2 \iff \\
BA \cdot BC_1 &= (\tfrac{1}{2}BA + \tfrac{1}{2}BC_1)^2 - (\tfrac{1}{2}BA - \tfrac{1}{2}BC_1)^2
\end{aligned} \tag{3.26}$$

system I is therefore replaced by

$$\begin{cases} (\tfrac{1}{2}BA + \tfrac{1}{2}BC_1)^2 - (\tfrac{1}{2}BA - \tfrac{1}{2}BC_1)^2 = c \\ BA + BC_1 = b \end{cases} \iff$$

$$\begin{cases} \tfrac{1}{2}BA - \tfrac{1}{2}BC_1 = \sqrt{(\tfrac{1}{2}BA + \tfrac{1}{2}BC_1)^2 - c} \\ BA + BC_1 = b \end{cases} \iff$$

$$BA = \frac{1}{2}b + \sqrt{\left(\frac{1}{2}b\right)^2 - c} \text{ and } BC = \frac{1}{2}b - \sqrt{\left(\frac{1}{2}b\right)^2 - c} \tag{3.27}$$

For applications see
Q14, 42

(ii): From $BA > BC$ (so $a > d$), hyperbolic application (figure 3.19)

C_2 is situated between A and B such that $AC_2 = BA - BC_2$. Point M_2 is the center point of the line segment AC_2 so $AM_2 = M_2C_2 = \frac{1}{2}BA - \frac{1}{2}BC_2$. The rectangle $BADC$ can be interpreted as a hyperbolic application to AC_2 exceeding by the square $(C_2B)^2$. According to §3.1.5. case 9, the following identities hold

$$\begin{aligned}&BA\cdot BC_2=(BM_2)^2-(C_2M_2)^2\iff\\&BA\cdot BC_2=(BA-AM_2)^2-(AC_2-AM_2)^2\iff\\&BA\cdot BC_2=(BA-(\tfrac{1}{2}BA-\tfrac{1}{2}BC_2))^2-(BA-BC_2-(\tfrac{1}{2}BA-\tfrac{1}{2}BC_2))^2\\&\iff BA\cdot BC_2=(\tfrac{1}{2}BA+\tfrac{1}{2}BC_2)^2-(\tfrac{1}{2}BA-\tfrac{1}{2}BC_2)^2\end{aligned}\tag{3.28}$$

This equation is the same as the one of (3.26), with BC_1 replaced by BC_2 and with $BA+BC_1=b$ replaced by $BA+BC_2=b$ so also here we arrive (3.27) at the solution

$$BA=\frac{1}{2}b+\sqrt{\left(\frac{1}{2}b\right)^2-c}\ \text{ and }\ BC=\frac{1}{2}b-\sqrt{\left(\frac{1}{2}b\right)^2-c}\tag{3.29}$$

For applications see: **Q14, 15, 42, 51.**

(iii): $BA<BC$ (so $a<d$), elliptic application (figure 3.20)

C_1 and M_1 again as in (i). The rectangle $BADC$ can be interpreted also in this case as an application to AC_1 falling short by the square $(BC_1)^2$. According to §3.1.5 case 8, the following identities hold

$$\begin{aligned}&BA\cdot BC_1=(C_1M_1)^2-(BM_1)^2\iff\\&BA\cdot BC_1=(\tfrac{1}{2}BA+\tfrac{1}{2}BC_1)^2-(M_1A-BA)^2\iff\\&BA\cdot BC_1=(\tfrac{1}{2}BA+\tfrac{1}{2}BC_1)^2-(\tfrac{1}{2}BA+\tfrac{1}{2}BC_1-BA)^2\iff\\&BA\cdot BC_1=(\tfrac{1}{2}BA+\tfrac{1}{2}BC_1)^2-(\tfrac{1}{2}BC_1-\tfrac{1}{2}BA)^2\end{aligned}\tag{3.30}$$

By means of the last equation of 3.30 **system I** is replaced by

$$\begin{cases}(\tfrac{1}{2}BA+\tfrac{1}{2}BC_1)^2-(\tfrac{1}{2}BC_1-\tfrac{1}{2}BA)^2=c\\BA+BC_1=b\end{cases}\iff$$

$$\begin{cases}\tfrac{1}{2}BC_1-\tfrac{1}{2}BA=\sqrt{(\tfrac{1}{2}BA+\tfrac{1}{2}BC_1)^2-c}\\BA+BC_1=b\end{cases}\iff$$

$$BA=\frac{1}{2}b-\sqrt{\left(\frac{1}{2}b\right)^2-c}\ \text{ and }\ BC=\frac{1}{2}b+\sqrt{\left(\frac{1}{2}b\right)^2-c}\tag{3.31}$$

(iiii): $BA < BC$ hyperbolic application (figure 3.20)

A_2 is situated on BC such that $BA = BA_2$, so $CA_2 = BC - BA_2$. Point M_3 is the center point of the line segment CA_2. Rectangle $BADC$ is a hyperbolic application to CA_2 exceeding by the square $(BA_2)^2$. According to §3.1.5 case 9, the following identities hold

$$\begin{aligned}
&BA_2 \cdot BC = (BM_3)^2 - (A_2M_3)^2 \Longleftrightarrow \\
&BA_2 \cdot BC = (BC - CM_3)^2 - (CA_2 - CM_3)^2 \Longleftrightarrow \\
&BA_2 \cdot BC = (BC - (\tfrac{1}{2}BC - \tfrac{1}{2}BA_2))^2 - (BC - BA_2 - (\tfrac{1}{2}BC - \tfrac{1}{2}BA_2))^2 \Leftrightarrow \\
&BA_2 \cdot BC = (\tfrac{1}{2}BC + \tfrac{1}{2}BA_2)^2 - (\tfrac{1}{2}BC - \tfrac{1}{2}BA_2)^2
\end{aligned} \tag{3.32}$$

As this equation is identical to last equation of (3.30) with BA replaced by BA_2 and C_1 by C we find here the same solution of **system I** as in (3.31)

$$BA = \frac{1}{2}b - \sqrt{\left(\frac{1}{2}b\right)^2 - c} \quad \text{and } BC = \frac{1}{2}b + \sqrt{\left(\frac{1}{2}b\right)^2 - c} \tag{3.33}$$

For applications see

Q15, 43

From the first equation of (3.26) and the first equation of (3.30) it can be observed that the two solutions by means of an elliptic application are based on the same elliptic identity.

The two solutions using hyperbolic applications are based upon slightly different hyperbolic identities as the first equation of (3.28) and the first equation of (3.32) belong to different geometrical configurations.

From the last equation of (3.26), the last equation of (3.28), the last equation of (3.30) and the last equation of (3.32) it becomes clear that the use of elliptic or hyperbolic applications for solving the **system I** always amount to the same as an application of the geometrical identity from §3.1.5 case 4 (3.6)

$$ad = \left(\frac{1}{2}a + \frac{1}{2}d\right)^2 - \left(\frac{1}{2}a - \frac{1}{2}d\right)^2 \tag{3.34}$$

expressing the equality between a rectangle (right member) and a gnomon (left member). The rectangle is always replaced by a gnomon, so the identity (3.34) is used from the left to the right:

$$BA \cdot BC_1 = \left(\frac{1}{2}BA + \frac{1}{2}BC_1\right)^2 - \left(\frac{1}{2}BA - \frac{1}{2}BC_1\right)^2 \quad \text{last equation of (3.26)}$$

$$BA \cdot BC_2 = \left(\frac{1}{2}BA + \frac{1}{2}BC_2\right)^2 - \left(\frac{1}{2}BA - \frac{1}{2}BC_2\right)^2 \quad \text{last equation of (3.28)}$$

$$BA \cdot BC_1 = \left(\frac{1}{2}BA + \frac{1}{2}BC_1\right)^2 - \left(\frac{1}{2}BC_1 - \frac{1}{2}BA\right)^2 \quad \text{last equation of (3.30)}$$

$$BA_2 \cdot BC = \left(\frac{1}{2}BC + \frac{1}{2}BA_2\right)^2 - \left(\frac{1}{2}BC - \frac{1}{2}BA_2\right)^2 \quad \text{last equation of (3.32)}$$

From (3.27), (3.29), (3.31) and (3.33) follows that for **system I** both with the elliptic as with the hyperbolic application the same two pairs of solutions are found, (3.27) and (3.31) with elliptic application, (3.29) and (3.33) with hyperbolic application. Because of the symmetry of **system I**, the solutions of (3.27) and (3.31) or (3.29) and (3.33) are interchangeable by swapping the values of AB and BC. Hence, in practice it suffices to calculate only one solution. This is the likely reason why Cardinael never gives more than one solution, with exception of the two questions **Q42** and **43**.

Furthermore, it is remarkable that Cardinael mostly uses the hyperbolic application, while this method is more laborious as there are two different geometrical configurations to consider: in case (ii) with $AB > BC$ the center point M_2 of segment $AC_2 (= AB - BC)$ is used, and in case (iiii) with $AB < BC$ is used the center point M_3 of segment $CA_2 (= BC - AB)$. Hence, a complete solution is required with the use of two different drawings as one can see by comparing the questions **Q42** and **Q43**.

Cardinael applies this all in the questions **Q14 otherwise, Q15, 42, 43** and **Q51** and is explicated by us in **Q14 otherwise, Q42, 43** and **Q51**.

3.1.10 Inscribed and circumscribed square

The construction of a constant boundary extension is defined as follows. Given a convex quadrilateral $BCDA$, it is required to calculate and to construct an extension of constant width and given area, along two connected sides. Our explanation is based on **Q81** (see figure 3.21), by means of the inscribed quadrilateral $BALK$ and the circumscribed quadrilateral $BMOI$, which in this simple case are both squares. The circumscribed quadrilateral is always characterized by having a boundary extension with the same outer length and width as the boundary extension of the original quadrilateral.

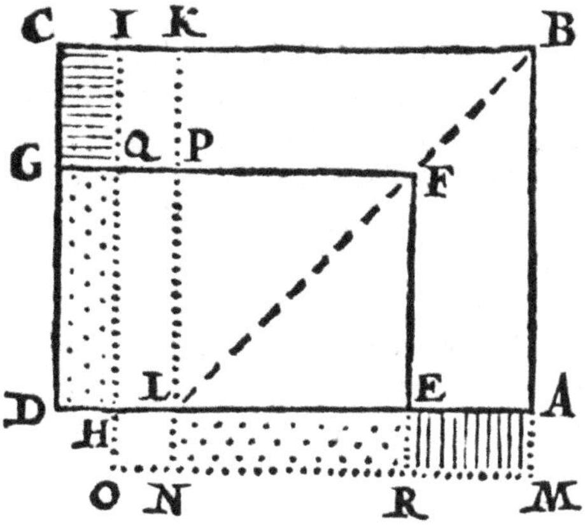

Figure 3.21: Circumscribed and inscribed square

Case 18: using Q81 for the construction of a constant boundary extension (§3.2.2 and §5.3)

Point K is chosen on BC such that $BK = BA = 12$. The drawing illustrates how a rectangle $BCDA$ with a constant boundary extension with constant width $ABCGFE$ along the two connected sides CB and BA can be transformed firstly, into an inscribed square $BALK$ and secondly, into a circumscribed square $BMOI$ with its boundary extension of the same outer length as the boundary extension of the original square $BCDA$.

This construction , which will be generalized in subgroup 11, requires several steps. Firstly, the boundary extension bisector of angle B (the middle vertex of the boundary extension) is drawn intersecting DA in L. Side BA is extended by AM such that $AM = IC = \frac{1}{2} \cdot KC$. Because $AB + BC = MB + BI$, the extension of $BCDA$ can be moved over the length CI so that this extension is just along MB and BI within the square $BMOI$.

Square $BMOI$ now has, along its boundary extension, the same outer length $MB + BI$ as the outer length $AB + BC$ of the boundary extension of $DCDA$. However, there is an important difference between the rectangle

BCDA and the circumscribed square *BMOI*. When the extension is cut off from *BCDA* the result is a rectangle *FEDG* not similar to the original rectangle. In the case of square *BMOI* though, after cutting off the boundary extension arises a square *FROQ* which is similar to the original square *BMOI*.

Both (not similar) rectangles *BCDA* and *FEDG* with as difference the constant extension *BCGFEA* are replaced by the two (similar) squares *BMOI* and *FROQ* with as difference the constant extension *BIQFRM* with the same outer length, width and area as the extension *BCGFEA*. This idea of 'similarity after cutting off a constant extension' will play a key role in subgroup 11 in the generalization by Cardinael of the problem of constant boundary extensions for any convex quadrilateral.

Another recurring theme connected with this construction is the coincidence of the diagonal and boundary extension bisector , the bisector of the angle at the middle vertex of the boundary extension. For the occurrence of 'similarity by cutting off or adding a constant extension' it is necessary that the boundary extension bisector also lies the diagonal of the inscribed square *BALK* and of the circumscribed square *BMOI*. It does not matter if the extension is cut off or added. Cardinael only generalizes in the case of cutting off the extension (**Q81-84**). He then uses the circumscribed quadrilateral (see **Q84**), a generalization of the circumscribed square (see **Q81**).

3.1.11 Cardinael's methods and techniques in their historical context

Cardinael's approach in early seventeenth-century geometry is unique in the sense that his solutions do not depend on the use of cossic symbols, avoiding algebraic equations which became more common in the geometry of that time. Instead, he places the unknown line segment in a geometrical correspondence (connected in a natural way to the given question) which often has the character of an elliptic or hyperbolic application of area and then he uses the methods as treated in §3.1.6 case 11 or 12 (only in those questions that seem to be quadratic, so questions that translated into algebra lead to a quadratic equation). Typical examples are **Q40 t/m 44** and **Q48**.

Another aspect which places Cardinael outside of the practices of many of his contemporaries is his treatment of the problem of elliptic application. He distinguishes only one of the two possible positive solutions §3.1.6 (case 11). As a result, he fails to see **Q42** and **Q43** as a single problem with two different solutions and he therefore misses out on the second solution in the questions **Q14, 27,** and **Q32**.

An interesting situation arises in Cardinael's solution of **Q32**. In this question, it is rather difficult to determine within this geometrical configuration the required geometrical connection where the unknown line segment EF can be found applying an elliptic application. The geometrical condition for the required line segment $GF(=EF)$, can be described by the equation

$$EF \cdot (67\frac{3}{5} - EF) = 313$$

In solving this equation, he constructs outside the configuration of the original drawing an isolated elliptic application problem from which EF can be derived in the style of §3.1.6 case 11 (**Q32**).

Well considered, Cardinael's geometrical construction is very similar to an algebraic solution for this problem. By posing for EF the cossic symbol for the unknown, one finds the same equation that Stevin or Petri would have found by posing for EF this cossic symbol from the start.

Cardinael's principal objection against the Rule of Coss is probably not so much the use of algebraic expressions, but rather the algorithmic methods of solving such equations, devoid from any geometrical interpretations and having no real control on the geometrical meaning of such derivations.

From the second half of the sixteenth century, algebraic solutions depended on a blind confidence in the correctness of the rules for solving the different cases for quadratic equations. Mathematicians like Cardano and Bombelli strongly believed in the epistemic validity of these rules which facilitated the acceptance of anomalies such as negative solutions and roots of negative numbers (see [Heeffer 2010]). However, Cardinael and some of his contemporaries lacked such confidence and therefore relied on geometrical intuitions and techniques for solving problems.

Another motivation for avoiding the use of algebra may have been Cardinael's concern for his audience. As explained in the introduction, the book was mostly bounded and sold together with the work on surveying by Sems and Dou, originally from 1600. The intended audience was therefore mostly lay culture, surveyors, engineers, craftsmen and reckoning masters whose daily practices depended on a basic knowledge of practical geometry and arithmetic. Cardinael's work goes well beyond the basic needs of this mathematical culture. However, those practitioners with an interest to expand their knowledge of mathematics would benefit more from Cardinael's approach which connected with their intuitions and practices than from an abstract translation of geometrical problems into algebraic expressions.

3.1.12 Some final remarks

The parabolic application (not exceeding or short-falling, based on Elem.I-44, see application in §3.1.4 case 2), the elliptic application (short-falling by a square, based on Elem.II-5, see §3.1.5 case 8) and the hyperbolic application (exceeding by a square), Elem.II-6, see §3.1.5 case 9) are important achievements of Ancient Greek mathematics as witnessed by their frequent appearance in Euclid's *Elements*.

Cardinael uses these construction constantly, often implicit, in geometrical problems of a linear or quadratic type and many of his contemporaries too use it in the last resort by justifying geometrically the algebraic solutions of linear and quadratic equations.

A generalization, with the falling short or exceeding square replaced by a parallelogram, is treated in Elem.VI-25, 27, 28, 29. An impressing application of this is the theory of conic sections as developed by Apollonius of Perga (262-190 v.C.), one of Euclid's successors. The words parabola, ellipse and hyperbola are used since then to characterize the three kinds of curves in which the conic sections can be divided on the base of the type of application used in their definition ([Kline 1972] I, 91).

For reference, we provide here a summary of the Questions where the geometrical application of areas is employed. We distinguish the three most important variations:

1. General (without application of area)

   Q1, 2, 14, 27, 28, 29, 31, 32, 33, 36, 37, 49, 52, 54, 55, 56, 56 otherwise, 57, 66, 69, 67, 68, 78, 79, 80, 81, 82, 83, 84, 85, 86, 87, 88, 89, 90, 91, 92, 93, 94, 95, 97.

2. Parabolic application (mostly linear problems)

   construction of the solution of the equation $ax = bx$ with the help of Elem.I- 43, 44, 45 ([Heath 1956], 346, 347)

   Q27, 96.

3. Elliptic or hyperbolic application of area (mostly quadratic problems)

   application of area with the help of §3.1.5 cases 8 and 9, §3.1.6 cases 11 and 12, §3.1.8 cases 16 and 17 and based on Elem.II-5, II-6 or II-11 ([Heath 1956], p.382-388 and 402)

   Q4, 5, 14, 15, 25, 32, 35, 40, 41, 42, 43, 44, 48, 50, 51, 69, 81.

3.2 Structure and some special themes of the HGQ

3.2.1 Classification in groups and subgroups

Though Cardinael does not provide any clues for a possible systematic division the following distinction into five groups and fourteen subgroups is an obvious choice.

I triangles (**Q0-10, 12-24, 26-29, 31-34**)

subgroup 0

the title page of the HGQ

An elegant and original proof of the Pythagorean theorem by Cardinael.

Q0

subgroup 1

the transversal in a triangle

Q1, 2, 4, 5, 7, 13, 19-21, 31-34

Application of Stewart's theorem

subgroup 2

stand alone Questions

Q3, 6, 8-10, 12, 14-18, 22-24, 26-29

Many practical exercises of surveyor geometry. In **Q18** a first application of Elem.VI-31, the generalisation of Elem.I-47.

II rectangles (**Q36-53**)

subgroup 3

squares, rectangles and soldiers in order of battle

Q36-39, 49-53

With exercises about the triple and quadruple ratio, about adding or cutting of an extension and in **Q53** the second application of Elem.VI-31

subgroup 4

elliptic and hyperbolic application

Q40-44, 48

System I and application of §3.1.8 case 17 with hyperbolic application.

In **Q48** adding an extension with given area and unknown constant width along two succeeding sides of a rectangle

subgroup 5

similarity and geometrical mean

Q45-47

The geometrical mean of the two succeeding rectangles around a diagonal in a rectangle (**Q45**)

III circles (**Q25, 35, 54-66, 68, 69**)

subgroup 6

circles

Q54-65

In which the radius of the inscribed and circumferential circle of a triangle is calculated, with given sides, application of area and multiplication transformations.

subgroup 7

the quadrilateral inscribed in a circle

Q66, 68

Calculation of the basis of a triangle with given slanted sides and the diameter of the circumscribed circle. And also, as an application of Ptolemy's theorem, the calculation of the circumferential circle and the diagonals of a quadrilateral inscribable in a circle with the sides given.

subgroup 8

the Golden Section, pentagon and decagon

Q25, 35, and **69**

Calculating the segments of the Golden Section with the geometrical mean known (**Q25**) and its application to the decagon and pentagon (**Q35** and **Q**69).

IV polygons, no rectangles (**Q11, 30, 70-75, 77-79**)

subgroup 9

'suitable' and 'unsuitable' squares

Q70-75

Calculations in quadrilaterals, with five data.

subgroup 10

transformation of form with equal area

Q11, 80, 87-89

Transformation of a triangle to a trapezium and vice versa without changing area (construction s). The square with maximal area inside a triangle. In **Q88** the third application of Elem.VI-31.

subgroup 11

partitioning convex polygons, the problem of extensions

Q77-79, 81-86, 90-95 (Q81 concerns a rectangle)

Partitioning a triangle, trapezium, quadrilateral and hexagon according to prescribed proportions (construction s). The Questions **Q81-84, 95** concern truncating or adding of a constant boundary extension parallel to two succeeding sides of a convex quadrilateral. Required is the width of the extension with prescribed area. In **Q92** and **Q93** we find the fourth (finest) and fifth application of Elem.VI-31, used for the division into equal parts.

subgroup 12

calculation and construction of a triangle and a quadrilateral inscribable in a circle

Q30, 96 and **Q97**

Construction of a triangle with (geometrical) given altitudes and the construction of a quadrilateral inscribable in a circle with (geometrical) given sides.

V spacial geometry (**Q67, 76, 98-101**)

subgroup 13

tetrahedrons

Q67 (parrot), **76** (parrot), **98-100**

Reduction to problems from plane geometry. Calculation of the capacity of a tetrahedron if the lengths of the edges are given.

Establishing the height of a tower with the help of trigonometry.

subgroup 14

arithmetical Questions

Q101

An arithmetical problem as a bonus question is geometrically translated and the solution is obtained by means of of geometrical algebra and without the application of the Rule of Coss.

For an index to the questions by subgroup, see Table 3.1.

Table 3.1: Index to the questions by subgroup

Q	subgroup	Q	subgroup	Q	subgroup	Q	subgroup
1	1	26	2	51	3	76	13
2	1	27	2	52	3	77	11
3	2	28	2	53	3	78	11
4	1	29	2	54	6	79	11
5	1	30	12	55	6	80	10
6	2	31	1	56	6	81	11
7	1	32	1	57	6	82	11
8	2	33	1	58	6	83	11
9	2	34	1	59	6	84	11
10	2	35	8	60	6	85	11
11	10	36	3	61	6	86	11
12	2	37	3	62	6	87	10
13	1	38	3	63	6	88	10
14	2	39	3	64	6	89	10
15	2	40	4	65	6	90	11
16	2	41	4	66	7	91	11
17	2	42	4	67	13	92	11
18	2	43	4	68	7	93	11
19	1	44	4	69	8	94	11
20	1	45	5	70	9	95	11
21	1	46	5	71	9	96	12
22	2	47	5	72	9	97	12
23	2	48	4	73	9	98	13
24	2	49	3	74	9	99	13
25	8	50	3	75	9	100	13

3.2.2 Some special themes from the HGQ and their distribution over the subgroups

Before proceeding to the mathematical commentary of each of the hundred geometrical questions we conclude this section with a survey of some special themes such as Stewart's theorem, The Golden ratio, the problem of extensions and partitioning a triangle or a quadrilateral, which are specific to the respective subgroups 1, 8 and 11.

Stewart's theorem

The idea of this proposition is important for solving questions **Q1, 2, 4, 5, 7, 13, 19–21, 31–34** all appearing in subgroup 1.

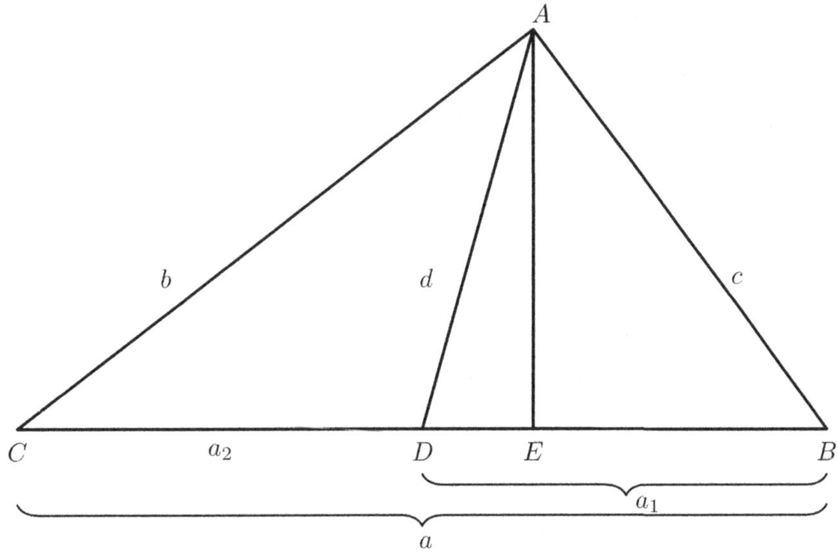

Figure 3.22: Stewart's theorem

In each question there is a connection between a transversal from a vertex of a triangle and the parts of the opposite side divided by the transversal . In 1746, Matthew Stewart ([Stewart 1746], 2 – 5), formulated a general relation for this connection since then known as Stewart's theorem

$$d^2a = c^2a_2 + b^2a_1 - aa_1a_2$$

if the intersection point D is between C and B and

$$d^2a = c^2a_2 - b^2a_1 + aa_1a_2$$

if point D is on the extension of line segment CD, and

$$d^2a = -c^2a_2 + b^2a_1 + aa_1a_2$$

if D is on the extension of line segment BC (with $BC = a$ in the figure). Because

$$a_1 + a_2 = a \text{ or } a_2 - a_1 = a \text{ or } -a_2 + a_1 = a$$

all three cases give rise to two equations with six unknowns. For a determinate solution, four more equations are necessary. In each question of subgroup I, the problem is to find the complete solution by means of four supplementary relations between the unknowns a, a_1, a_2, d, b and c. Cardinael achieves this by his typical geometrical approach.

For completeness sake, we outline a (modern) proof of Stewart's theorem (see figure 3.22), so with D between C and B and with AE the perpendicular from A on BC. Assume $BE = p$, then in triangle ABD

$$AD^2 = c^2 + a_1^2 - 2a_1p$$

And in triangle ABC

$$b^2 = c^2 + a^2 - 2ap$$

Eliminating p from both equations leads us to

$$AD^2 = a_2c^2 + a_1b^2 - a_1a_2a$$

In Stewart's book ([Stewart 1746], p.2-5) the theorem appears as Proposition II in a form which is unrecognizable for the modern reader. The reason is that Stewart still formulates his proposition, more than hundred years after Descartes, in Cardinael's geometrical style. In Holland, Stewart's theorem was introduced in textbooks between 1902 and 1924, as Versluys [Versluys 1902] does not yet mention it but in [Molenbroek 1924] it is referred to by this name.

The Golden ratio

In **Q25, 35** and **69**, all belonging to subgroup 8, the division in mean and extreme reason is always mentioned (coined since 1835 as the Golden ratio by Georg Simon Ohm, see [Dijksterhuis 1929-30], II, 17, n.5). Cardinael does not use a special name for this special ratio but refers to Elem.II-11 and Elem.VI-30. Though in all cases the line segments are calculated, they can also be constructed based on Elem.II-11 or Elem.VI-30 (see §3.1.5 case 10). **Q35** can be considered as a preliminary study for **Q69**, though Cardinael hardly uses it.

Problems of extensions of convex quadrilaterals

Problems of extensions (mostly subgroups 3, 11) concerns the truncation or addition of a side area of constant width along two succeeding sides of a given square, rectangle or, more general, a convex quadrilateral in which the area of the extension or the ratio to the total area is given. The problem is to calculate the width of the extension. In his treatment of such problem Cardinael builds a kind of mini theory, based on the questions **Q36, 37, 48, 49, 50, 51, 81-84** and ending with **Q95**.

A method applied by Cardinael (see appendix §5.3: method 1) consists of constructing the extension in one line and the width is then calculated with the help of application of area (**Q48**).

A second method (see appendix §5.3: method 2) is based on the simple case of adding or truncating an extension of a quadrilateral. In the case of a square, the calculation is simple because the resulting area is again a square, similar to the original square. In case of a rectangle (**Q81**) or of any convex quadrilateral (**Q82-84**) the new shape is not any longer similar to the original quadrilateral.

In all these cases, Cardinael translates the original quadrilateral in an ingenious way to another quadrilateral with a boundary extension of the same width, area and outer length, which we name the circumscribed quadrilateral. In the case of a rectangle this is a circumscribed square (see §3.1.10 case 18) with the additional feature that adding or truncating the extension leads to a similar new quadrilateral. Then the situation arises which corresponds to the problem of the extension of a square. The width of the original extension then can be calculated in an easy way (**Q82-84, 95**).

Partitioning the triangle and the convex quadrilateral

The problem concerns the division of a polygon (subgroup 11), in particular the triangle, into two parts according to a prescribed ratio by means of a line drawn from a point inside or outside that polygon. At the beginning of the seventeenth century this was a favorite subject in the geometry curriculum, undoubtedly because of its relevance to prevailing land surveying techniques.

Cardinael provides in **Q92, 93** a solution for the division of a triangle or a quadrilateral into two equal parts by means of a line through a point within the triangle or quadrilateral. He employs geometrical algebra ingenuously combined with Elem.VI-31, a generalization of the Pythagorean theorem (Elem.I-47).

Cardinael solution has the form of a construction , that can easily be translated into a calculation procedure. In §5.4 some historical particularities in connection with the solution of this kind of problems are discussed, starting with Euclid's work. For more technical details and a modern solution relying on the theory of affine transformations and the theory about the envelope of a family of one parameter of hyperboles, see [Sitters 2008].

Chapter 4

Comments per question

Question 0

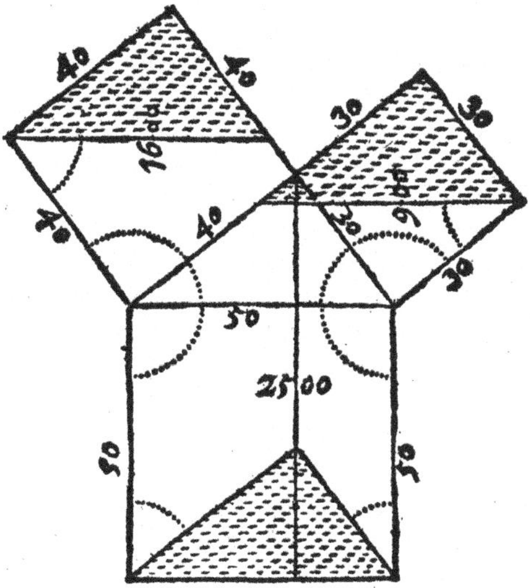

Figure 4.1: Question 0: subgroup 0, the Pythagorean theorem

The figure on the title page can already be seen as a first exercise, which we will denote as Question 0. This problem concerns an original and elegant proof of the Pythagorean theorem, which Cardinael consistently refers to as "propos. 47 of Euclid". The proof by Cardinael as implied by the picture is missing from known surveys on different proofs of Euclid's Elem. 1-47. Only the booklet by Versluys ([Versluys 1914], p. 56, proof 43 (adapted from [Hauff 1803]) with ninety six proofs of the Pythagorean theorem, includes the same picture, however the proof is different from what is suggested by Cardinael's picture as a proof by transformation and rotation. Cardinael's proof is also unmentioned in Bruno Ernst's [Ernst 2002] collection of the most interesting proofs of the Pythagorean theorem.

The picture (Figure 4.1) illustrates how the two rectangles below are together equal to the square on the hypotenuse. These two rectangles are transformed to two parallelograms under conservation of area and then rotated over a right angle. The shaded areas show that these parallelograms are equal to the squares on the sides of the rectangles. From this follows that these two squares have to be equal to the square on the hypotenuse.

Theory used

Transformation, rotation and translation of rectangles.

Compare with

[Ernst 2002], [Hauff 1803]

Question 1

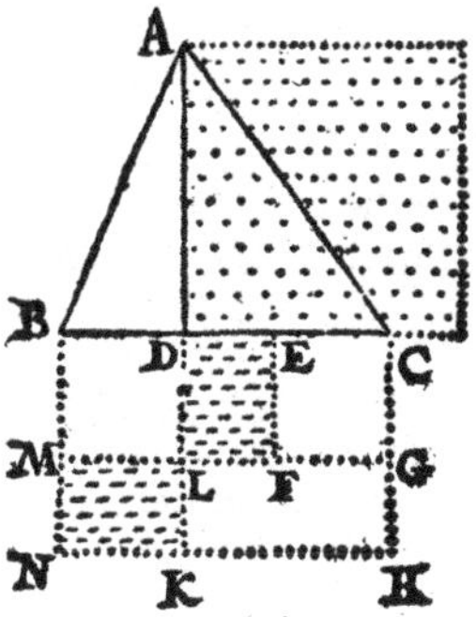

Figure 4.2: Question 1: subgroup 1, altitude in acute triangle

Comment on Answer

Take E, M, L, F, G, N, K and H so that $BD = EC$ and $KHCD$ and $FGCE$ are squares. The difference of the squares on AC and AB equals the difference of the squares of DC and BD, which is also equal to the difference of the squares of DC and EC, being 56

$$\begin{aligned} AC^2 - AB^2 = DC^2 - DB^2 &\Leftrightarrow \\ 56 = DC^2 - DB^2 &\Leftrightarrow \\ 56 = DC^2 - EC^2 & \end{aligned}$$

From this follows that the gnomon $DEFGHK$ has area 56 and that the area of rectangle $NHGM$ with side $NH = 14$ is known to be 56

$$\begin{aligned} 56 &= \text{gnomon} DEFGHK \Leftrightarrow \\ 56 &= NHGM \Leftrightarrow \\ 56 &= (DC + EC)(DC - EC) \Leftrightarrow \\ 56 &= 14 \cdot DE \Leftrightarrow \\ DE &= 4 \end{aligned}$$

from which follows

$$2DB + 4 = 14 \Leftrightarrow DB = 5 \quad \text{and} \quad DC = 9$$

and with the help of Elem.I-47 we get

$$AD = \sqrt{AB^2 - DB^2} = \sqrt{169 - 25} = 12$$

The calculation is characteristic for geometrical algebra: the difference of two squares is always a gnomon, here the figure $DEFGHK$. The equality of gnomon $DEFGHK$ and rectangle $NHGM$ is nothing else than the application of the geometrical identity (see §3.1.5 case 7)

$$DC^2 - EC^2 = (DC + EC) \cdot (DC - EC)$$

A solution by means of algebra could have been as follows. Suppose $DE = x$ then it follows from $DB = EC$

$$x + 2DB = 14 \tag{4.1}$$

from $DC^2 - DB^2 = 56$ we have

$$(x + DB)^2 - DB^2 = 56 \tag{4.2}$$

from (4.2) follows

$$(x + 2DB) \cdot x = 56 \tag{4.3}$$

with the help of (4.1)

$$14 \cdot x = 56 \Leftrightarrow \tag{4.4}$$

$$x = 4 \tag{4.5}$$

and again with the help of (4.1):

$$2 \cdot y = 10 \tag{4.6}$$

$$y = 5 \tag{4.7}$$

and so finally

$$DB = 5 \quad \text{and} \quad DC = 9$$

By interpreting every step of this this solution geometrically, Cardinael's calculation by means of geometrical algebra becomes clearly recognizable

4.1 corresponds to: $DE + 2 \cdot DB = BC$ and $BC = 14$,
4.2 corresponds to: $DC^2 - DB^2 = 56$,
4.3 corresponds to: $MGHN = 56$,
4.4 corresponds to: $14 \cdot DE = 56$,
4.5 corresponds to: $DE = 4$,
4.6 corresponds to: $2 \cdot DB = 14 - 4$,
4.7 corresponds to: $DB = 5$.

Comment on Otherwise

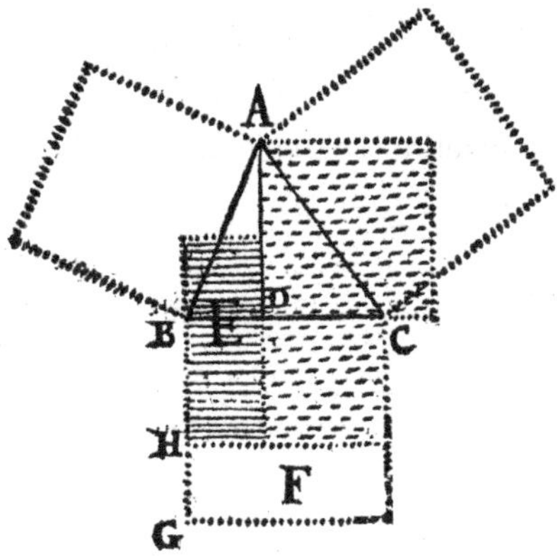

Figure 4.3: Question 1, alternative solution

This calculation corresponds to a proof of Elem.II-13. Cardinael here provides an alternative proof for the law of cosines for acute triangles:

$$AC^2 = AB^2 + BC^2 - 2DB \cdot BC$$

Slightly streamlined, Cardinael's proof proceeds as follows. From figure 4.3 we may conclude that

$$AB^2 + BC^2 - AC^2 = E + F \Leftrightarrow$$
$$AB^2 + BC^2 - AC^2 = 2DB \cdot BC$$
$$AC^2 = AB^2 + BC^2 - 2BC \cdot DB$$

With AC, AB and BC known, DB can now simply be calculated leading to the final solution.

General comment

Cardinael provides in **Otherwise** a proof of the Law of cosines for acute triangles that is clearly much simpler than the proof of Elem.II-13 in [Heath 1956].

Theory used

Answer (geometrical algebra, Pythagoras)

§3.1.5 case 4 (3.5)(3.5) and §3.1.5 case 7: $a^2 - d^2 = (a + d) \cdot (a - d)$, Elem.I-47.

Otherwise (geometrical algebra, Pythagoras)

Elem.I-47

Compare with

[Brasser 1663], p.287

[Petri 1583], question 11

[Sitters 2008], 5.2.2.2, ex. 9

Q2

Elem.II-13

Question 2

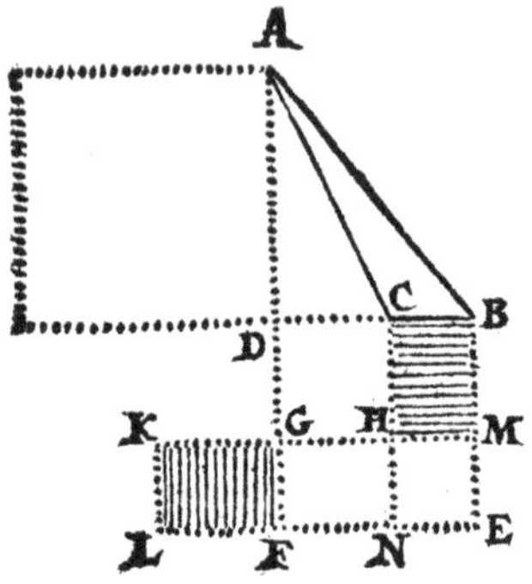

Figure 4.4: Question 2: subgroup 1, altitude in obtuse triangle

Comment on Answer

The difference of the squares on DB and DC is calculated in the same way as in **Q1**:

$$\begin{aligned} AB^2 - AC^2 &= DB^2 - DC^2 \Leftrightarrow \\ 56 &= DB^2 - DC^2 \end{aligned}$$

This is equal to the gnomon $CBEFGH$ and as figure 4.4 shows, this gnomon equals the rectangle $KMEL$, so

$$\begin{aligned} 56 &= \text{gnomon } CBEFGH \Leftrightarrow \\ 56 &= KMEL \Leftrightarrow \\ 56 &= (DB + DC)(DB - DC) \Leftrightarrow \\ 56 &= (DB + DC) \cdot 4 \Leftrightarrow \\ DB + DC &= 14 \Leftrightarrow \\ 2DC + 4 &= 14 \Leftrightarrow \\ DC &= 5 \end{aligned}$$

With the help of Elem.I-47 follows

$$AD = \sqrt{AC^2 - DC^2} = \sqrt{169 - 25} = 12$$

Comment on Otherwise

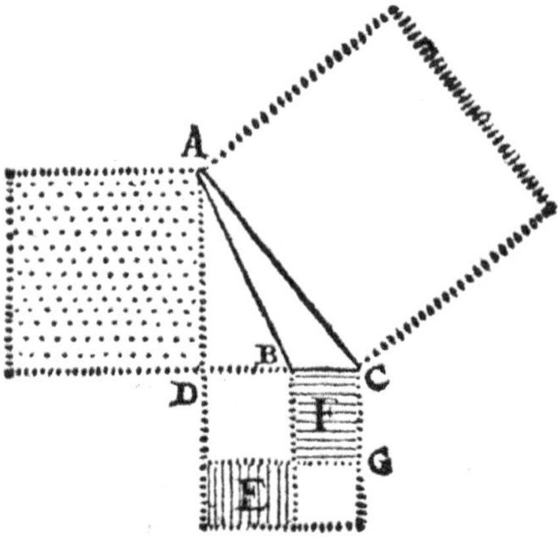

Figure 4.5: Question 2, alternative solution (letters B and C should be interchanged to correspond with the previous figure)

This calculation by Cardinael corresponds to a proof of Elem.II-12. He provides an alternative proof for the law of cosines for obtuse triangles (see figure 4.5), corresponding with Chapter 2 (Figure 2.4. Cardinael's proof can be streamlined as follows

$$\begin{aligned} AC^2 - AB^2 &= DC^2 - DB^2 \Leftrightarrow \\ AC^2 - AB^2 &= E + F + BC^2 \Leftrightarrow \\ AC^2 &= AB^2 + BC^2 + 2 \cdot BC \cdot DB \Leftrightarrow \\ AC^2 &= AB^2 + BC^2 + 2 \cdot BC \cdot DB \end{aligned}$$

which is the classical Law of Cosine. With AC, AB and BC known, DB can be calculated as

$$DB = 5$$

from Elem.I-47 it follows that

$$AD = \sqrt{AB^2 - DB^2} = \sqrt{169 - 25} = 12$$

General comment

Also in this alternative solution, Cardinael gives a proof of the law of cosines for obtuse triangles which is much simpler than the proof in Elem.II-12 as shown by [Heath 1956].

Theory used

Answer (geometrical algebra, Pythagoras)

§3.1.5 case 4 (3.5) 3.5 and §3.1.5 case 7: $a^2 - b^2 = (a + b) \cdot (a - b)$; Elem.I-47.

Otherwise (geometrical algebra, Pythagoras)

Simple geometrical algebra and Elem.1-47

Compare with

[Sitters 2008], 5.2.2.2 ex. 8

Q1

Question 3

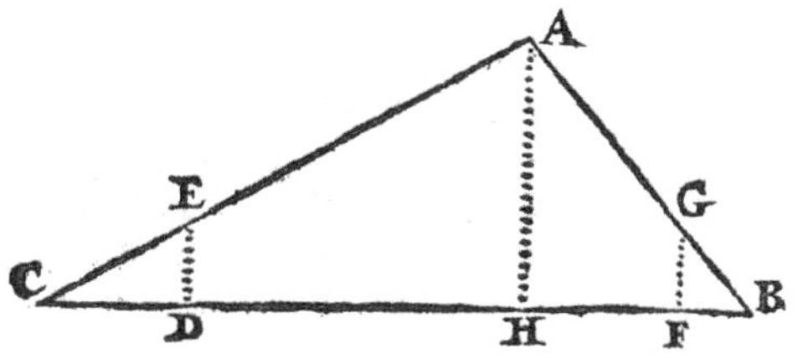

Figure 4.6: Question 3: subgroup 2, triangle with two lines known parallel to an altitude and how these lines cut the base

Comment on answer

By use of the proportions and the similarity of triangles the sides of the triangle ABC can be calculated as

$$AB = 130, \quad AC = 200 \quad \text{and} \quad BC = 210$$

General comment

A typical surveyors exercise just like **Q6, 8, 9, 10** and **12**.

Theory used

Answer (proportions)

Similarity and the rule of three

Question 4

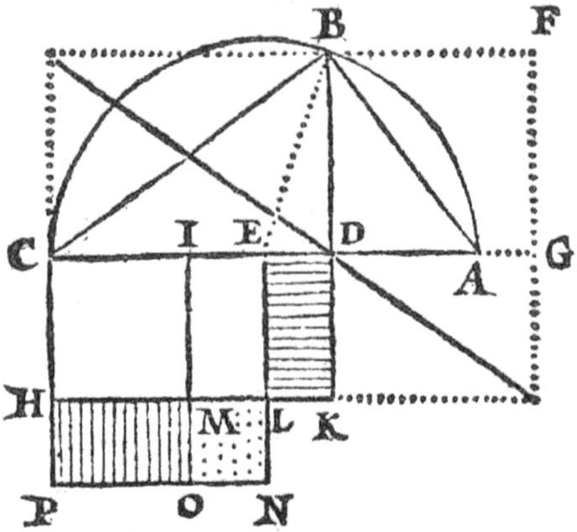

Figure 4.7: Question 4: subgroup 1, altitude of a right-angled triangle unknown but the segments of the base divided by this altitude known

Comment on answer

Here Cardinael uses a combination of Elem.IV-5 and III-31: the circle through A, B and C exists and has AC as diameter and E as centre, so

$$BE = EC = AE = 25$$

and from Elem.I-47 follows

$$BD = \sqrt{BE^2 - DE^2}$$

Comment on Otherwise 1

This solution depends on the similarity of ADB and BCD (Elem.VI-8), so that

$$AD : DB = DB : DC$$

and from Elem.VI-17

$$DB^2 = AD \cdot DC \text{ and } DB = \sqrt{576} = 24$$

Comment on Otherwise 2

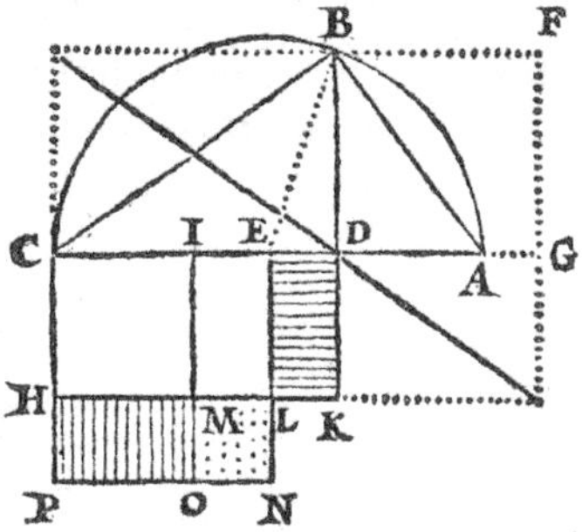

Figure 4.8: Question 4: With K such that $DA = DK$ and $CI = AD$

Cardinael attempts to prove the identity $BD^2 = AD \cdot DC$ without using proportions. From §3.1.5 case 9 it follows

$$CE^2 - IE^2 = CD \cdot CI = DC \cdot AD \tag{4.8}$$

so that

$$\begin{aligned} BD^2 &= EB^2 - ED^2 \\ &= CE^2 - IE^2 \quad \text{(uses } EB = CE) \\ &= DC \cdot AD \end{aligned}$$

This is an example of the hyperbolic identity from §3.1.5 case 9, because $DKHC$ is applied to DI with $ICHM$ as the exceeding square ($DKHC$ can also be seen as an elliptic application to AC falling short by the square on DA (§3.1.5 case 8)

$$AE^2 - DE^2 = DC \cdot DA = DC \cdot AD$$

in agreement with (4.8).

Comment on Otherwise 3

This third alternative solution differs slightly from **Otherwise 1**. Multiplication by a factor $\frac{1}{18}$ makes the handling of the proportionality more simple. To arrive at the real values the result has to be multiplied with 18.

General comment

Cardinael takes great effort to avoid the use of proportions in **Otherwise 2** just as he has done in **Answer**. However, the method he proposes in **Otherwise 2** is not a real alternative because after all he refers to the equality $EC = EB$. The method of **Answer** is also based on this equality, though more direct and neither uses proportions.

Theory used

Answer (Thales , Pythagoras)

Elem.IV-5 and III-31

Otherwise 1 (proportions)

Similarity in right-angled triangle: Elem.VI-8

Elem.VI-17

Otherwise 2 (Pythagoras , Thales , geometrical algebra, application of area)

§3.1.5 case 9 (hyperbolic identity);

Elem.IV-5 and III-31

Otherwise 3 (Thales , Pythagoras , proportions)

Similarity

Elem.VI-17

Multiplication factor without Regula Falsi

Compare with

[Sitters 2008], 5.2.2.2 ex. 17

Q4 is opposite of **Q5**, because

- **Q4**: $AD, DC \rightarrow BD, AC$
- **Q5**: $AC, BD \rightarrow AD, DC$

Question 5

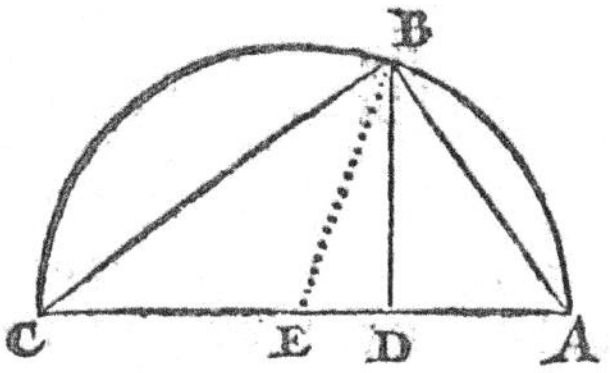

Figure 4.9: Question 5, subgroup 1: right-angled triangle with base and altitude known

Comment on answer

From Elem.IV-5, III-31 it follows that AC is the diameter of the circle through A, B and C and also that

$$BE = CE = 25$$

Applying Elem.I-47 on triangle BED gives

$$ED = \sqrt{25^2 - 24^2} = 7$$

so that $CD = 25$ and $AD = 18$

Considering figure 4.9 it becomes clear that there are two solutions for AD and DC. Generally, Cardinael pays no attention to this, probably because in most cases the results of one solution arise from the exchange with the results of the other solution. In §3.1.8 (3.17, 3.18), (3.17) (3.18) the method of solution used in **Answer** was extensively treated as the straightforward solution of the **AH-system** $(BD, CA, 24, 50)$ (see also §3.1.8 case 16c).

Comment on Otherwise

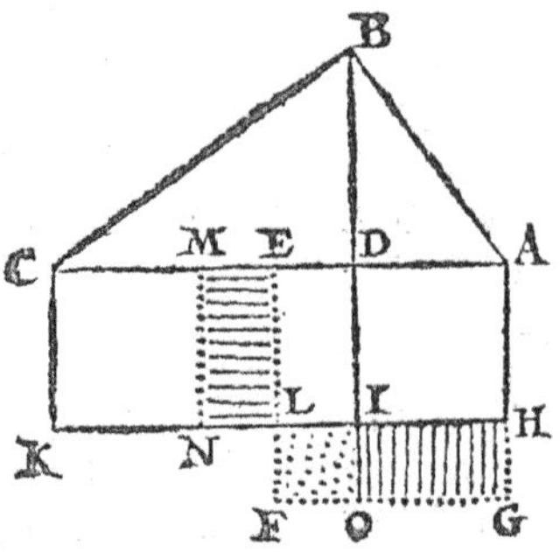

Figure 4.10: Question 5, $AHID$, $CMNK$ and $IOFL$ are squares, E is the midpoint of AC and of MD

From the similarity of ADB and BDC (Elem.VI-8) follows

$$AD \cdot DC = BD^2 = 576 \quad \text{(Elem.VI-17)}$$

so that $DIKC = 576$ and $AEFG - DIKC = 625 - 576 = 49$.

M is chosen so that

$$ME = HG$$

and

$$LEMN = HIOG$$

from which follows

$$AGOILE = AHNM$$

but as

$$AHNM = DIKC$$

so is

$$DIKC = AGOILE$$

and

$$AEFG - DIKC = AEFG - AGOILE = ILFO = 49$$

From $ILFO = ED^2$ follows

$$ED = 7$$

and

$$AD = 25 - 7 = 18 \quad \text{and} \quad CD = 32$$

Cardinael's argumentation in **Otherwise** is somewhat redundant. It is based on **system I**

$$\textbf{system I}(DC, DA, 24^2, 50)\begin{cases} DC \cdot DA = 576 \\ DC + DA = 50 \end{cases}$$

His solution is based on the idea of the elliptic application of area described in §3.1.6 case 11 (and §3.1.5 case 8), which is based on Elem.II-5. It seems that in the first part of the *HGQ*, Cardinael is not yet willing to simply make use of Elem.II-5, but prefers to provide the proof first by himself. Directly using Elem.II-5 would imply that rectangle $DIKC$ is an elliptical application to CA falling short by the square DA^2. According to Elem.II-5 (§3.1.5 case 8 or §3.1.6 case 11)

$$\begin{aligned} EA^2 - ED^2 &= DIKC \Leftrightarrow \\ 625 - ED^2 &= 576 \Leftrightarrow \\ ED^2 &= 49 \Leftrightarrow \\ ED &= 7 \end{aligned}$$

from which follows that $DC = 32$ and $DA = 18$.

The second solution $DC = 18$ and $DA = 32$, as calculated in §3.1.6 case 11, is not mentioned any more. As argued above, Cardinael probably did not consider this as fundamental different solutions as only the values of DC and DA are changed.

General comment

Q5 is an example of an **AH-system** $(BD, CA, 24, 50)$ with DA and DC unknown. In §3.1.8 case 16c,d the two ways in which this can be solved are already explained. In this case, Cardinael uses both possibilities, the direct way of case 16c without application of area (the **Answer**) and the indirect manner of case 16d with the help of **system I**$(DC, DA, 24^2, 50)$ (see **Otherwise**).

Theory used

Answer (Thales , Pythagoras)

Elem.IV-5, III-31 and I-47.

Otherwise (proportions, geometrical algebra, application of area)

Similarity in right-angled triangle: Elem.VI-17 $BD^2 = AD \cdot DC$;

§3.1.6 case 11: elliptic application, Elem.II-5.

Compare with

§3.1.8 case 16;

[Petri 1583], fol. 163 example 23;

Q5 is the opposite of **Q4**

- **Q5**: $BD, AC \to AD, DC$
- **Q4**: $AD, DC \to BD, AC$

Question 6

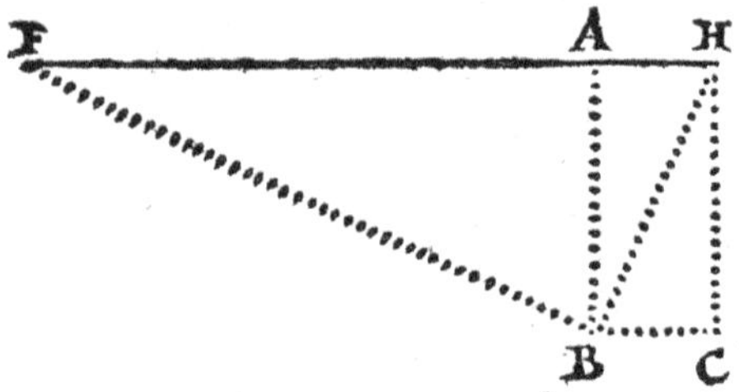

Figure 4.11: Question 6: subgroup 2, right-angled triangle with perpendicular and one segment of the base known

Comment on Answer

From the similarity of the triangles HAB and ABF (Elem.VI-8) follows

$$AB^2 = AH \cdot AF \quad \text{(Elem.VI-17)}$$

so that $AF = 1250$ and $HF = 1450$.

Theory used

Answer(proportions)

Similarity in right-angled triangle: Elem.VI-8

Question 7

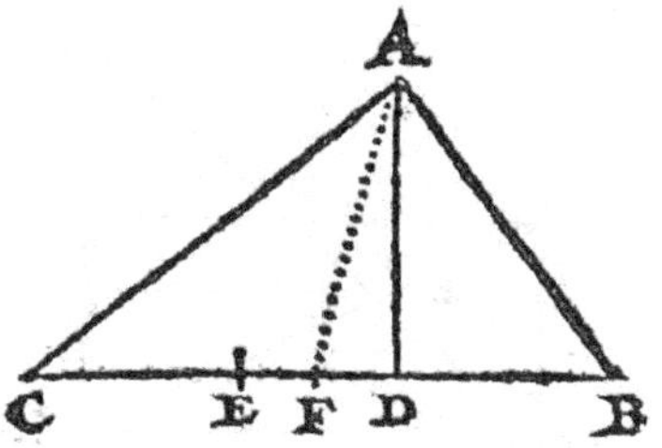

Figure 4.12: Question 7: subgroup 1, altitude AD in right-angled triangle and the difference of the segments DC and DB of the base known

Comment on Answer

By posing E such that $CE = BD$ it follows that $ED = 14$. If F is the midpoint of ED then $EF = FD = 7$. AF can then be calculated from the triangle ADF by means of Elem.I-47: $AF = 25$. It then follows from Elem.IV-5 and III-3 that CF is also known. Thus $CF = 25$ and so is $CD = 32$ and $BD = 18$, from which follows that $BC = 50$. AC and AB are calculated by applying Elem.I-47 on triangles ADC and ADB.

Theory used

Answer (Pythagoras , Thales)

Elem.IV-5 and III-31

Question 8

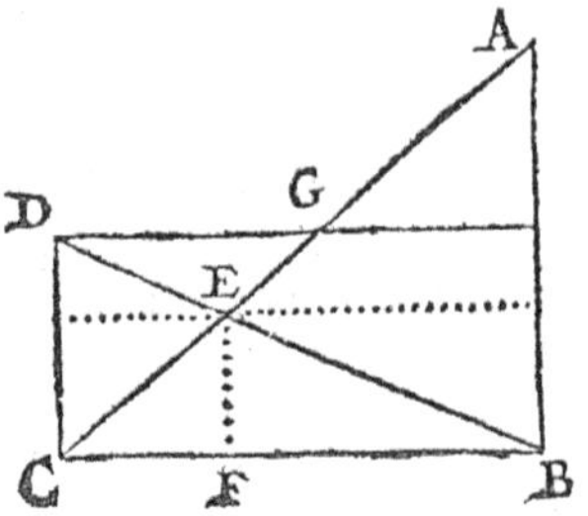

Figure 4.13: Question 8: subgroup 2, right-angled triangle with the sides and the direction of a transversal known

Comment on Answer

From the similarity of EDC and EBA follows

$$\begin{aligned} DC : AB &= FC : BF \Leftrightarrow \\ (DC + AB) : (FC + BF) &= DC : FC \Leftrightarrow \\ 15 : 10 &= 6 : FC \Leftrightarrow \\ FC &= 4 \end{aligned}$$

So

$$EF = 3\frac{3}{5} \quad \text{and} \quad EC = \sqrt{28\frac{24}{25}}$$

with the help of the similarity of CFE and CBA.
In the same way

$$EB = \sqrt{48\frac{24}{25}}, \ ED = \sqrt{21\frac{19}{25}}, \ EA = \sqrt{65\frac{4}{25}}$$

and the similarity of ECB and EGD finally yields

$$GD = \sqrt{44\frac{4}{9}}$$

Theory used

Answer (proportions)

Similarity

Elem.V-18: rule of proportions

Question 9

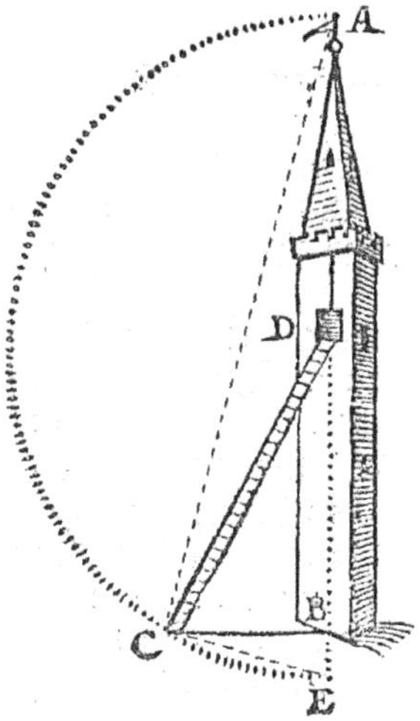

Figure 4.14: Question 9: subgroup 2, tower with known height of which the length of the ladder to the window equals the distance from the window to the top

Comment on Answer

From $DA = DC = DE$ follows that AE is the diameter of a semi-circle on which C. So ECA is a right angle and BC the geometrical mean of EB and AB. So $BE = 3$ can be calculated

$$BC^2 = BE \cdot AB \Leftrightarrow$$
$$225 = BE \cdot 75 \Leftrightarrow$$
$$BE = 3$$

From this follows $DA = DC = 39$ and $DB = 36$.

Theory used

Answer (proportions)

Elem.III-31: angle ECA is a right angle

Elem.VI-8: similarity in right-angled triangle

ABC and CBA are similar triangles

Elem.VI-17: calculation geometrical mean

Question 10

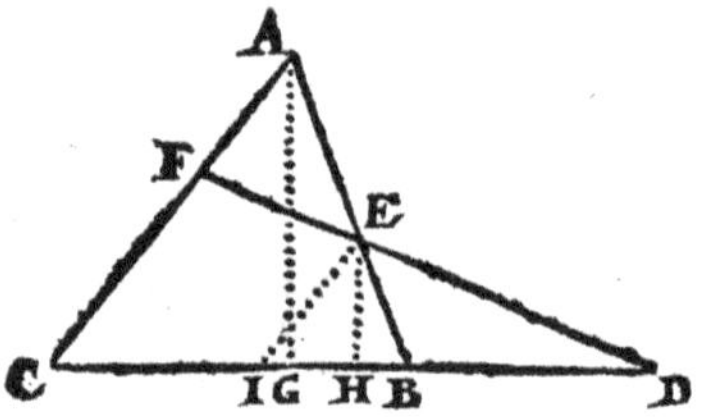

Figure 4.15: Question 10: subgroup 2, triangle with sides known and one side is extended with a known segment. Point I is chosen so that EI is parallel to FC.

Comment on Answer 1

First BI is calculated (similarity of BEI and BAC): $BI = 5\frac{5}{13}$. Then EI (again similarity of BEI and BAC): $EI = 5\frac{10}{13}$. With EI and DI known FC can be calculated (similarity of DEI and DFC) and then AF also is known: $FC = 9$ and $AF = 6$.

General comment 2

The auxiliary lines AG and EH are also needed. In [Sitters 2008] this exercise is also treated. As additionally also the lengths of DE and EF are required, AG and EH are needed for their calculation. The length of AG (and GB) is already calculated in **Q1**.

Theory used

Answer (proportions)

Only similarity of triangles

Compare with

Q1: the same triangle mirrored in the line AG;

[Sitters 2008], §5.2.2.2 ex. 21

Question 11

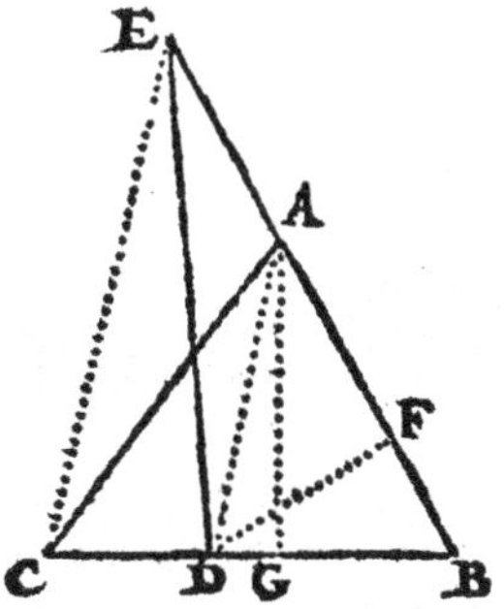

Figure 4.16: Question 11: Transforming a triangle without changing its area by removal of two vertices along their line of connection with the third vertex

Comment on Answer

Firstly, note that AD is parallel with EC, because ABC and EBD are equal (the triangles ADC and ADE are equal and according to Elem.I-39 then EC is parallel with AD). So the triangles ABD and EBC are similar and

$$BD : BA = BC : BE \Leftrightarrow$$

$$18 : 26 = 30 : BE \Leftrightarrow$$

$$BE = 43\frac{1}{3}$$

According to **Q1** (G is the base point of the perpendicular from A on BC) we have

$$BG = 13\frac{1}{5}$$

and therefore

$$AB : BG = BD : BF \Leftrightarrow$$
$$26 : 13\frac{1}{5} = 18 : BF \Leftrightarrow$$

$$BF = 9\frac{9}{65}(F \text{ is the base of the perpendicular from } D \text{ on } AB)$$

So

$$FE = BE - BF = 43\frac{1}{3} - 9\frac{9}{65} = 34\frac{38}{195}$$

and

$$FD^2 = BD^2 - BF^2 = 240\frac{2064}{4225}$$

such that finally

$$DE = \sqrt{FE^2 + FD^2} = \sqrt{(34\frac{38}{195})^2 + 240\frac{24064}{4225}}$$

$$DE = \sqrt{1409\frac{1183}{1522}}$$

Theory used

Answer (proportions)

Elem.I-39: similar triangles on the same base have the same height

Q1

Question 12

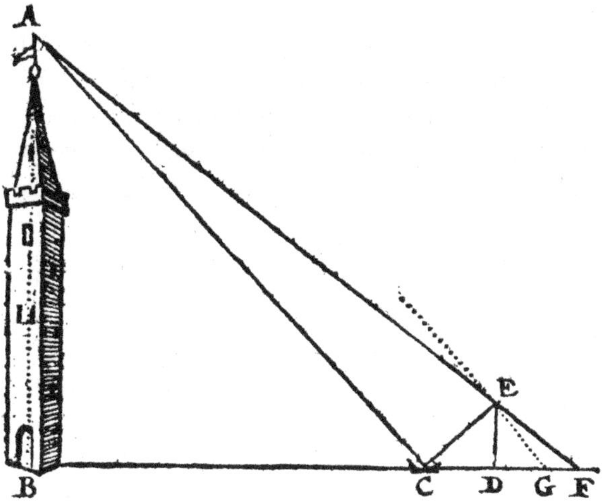

Figure 4.17: Question 12: subgroup 2, a tower at an unknown distance and of an unknown height

Comment on Answer

This ingenious solution is based on a classic technique from the surveyor tradition, as described in the fifteenth century by Leon Battista Alberti [Alberti 2010, 87]. The stick and mirror need to be positioned so that the top A of the tower can just be seen from the top E of the stick through the mirror C. As EG is parallel to AC, $ABCF$ is similar to $EDGF$ and AC corresponds to EG and CF to GF. The triangle EDF is a minute copy of ABF within ABF. The triangle $ABCF$ is not known except CF which can be measured via CD and DF:

$$CF = CD + DF = 8 + 9 = 17$$

$EDGF$ can be fully measured and so this figure is entirely known.

$$ED = 8, DF = 9 \text{ so } GF = 1$$

From $CF = 17$ and $GF = 1$ follows that between the triangles EDF and ABF there exists a multiplication factor 17. So we can conclude that

$$AB = 17 \cdot ED = 17 \cdot 6 = 102$$

and

$$BC = 17 \cdot DG = 17 \cdot 8 = 136$$

General comment

Particular to this solution is that only one stick is needed, that no angle had to be measured and both the distance to the tower and the height of the tower were unknown. However angles are rather important since the calculation is possible by the formula 'angle of incidence = angle of reflection'.

Theory used

Answer (proportions)

Only similarity of triangles

Compare with

Q17

Question 13

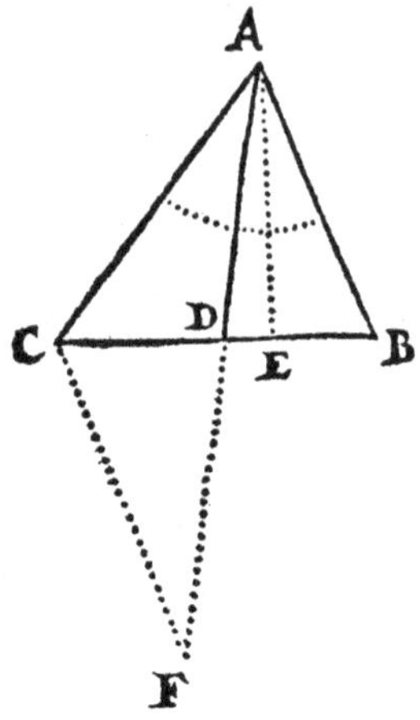

Figure 4.18: Subgroup 1, a bisector bisecting the opposite side

Comment on Answer

F is the intersection of the extension of AD and of a line through C parallel with AB). The similarity of DFC and DAB together with the similarity of the angles CFD and CAD lead to the proportionality $CD : BD = AC : AB$, as $FC = AB$, and so CD and BD can be calculated: $BD = 6\frac{1}{2}$ and $CD = 7\frac{1}{2}$. Using **Q1**, the perpendicular AE and the line segment BE can be calculated. From this follows the calculation of DE via $DE = BD - BE$. And AD is calculated by application of Elem.I-47 in triangle $ADE : AD = \sqrt{146\frac{1}{4}}$.

General comment

Cardinael's proof is similar to Elem.VI-3.

Theory used

Answer (proportions, Pythagoras)

Similarity;

Q1.

Compare with

Elem.VI-3

Question 14

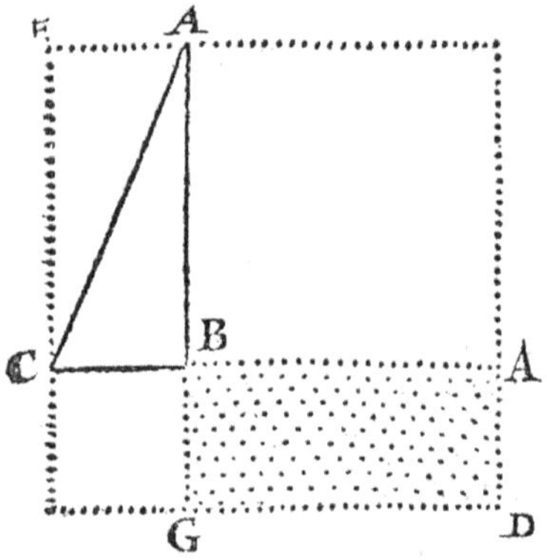

Figure 4.19: Subgroup 2, right-angled triangle with hypotenuse and the sum of the sides of the right angle know

Comment on Answer

Letter E is missing in figure 2.2 and G should be C. In figure 4.20 is intended a triangle with a right angle at the top. We shall indicate that top with D, so CAD is a right-angled triangle with altitude DB.

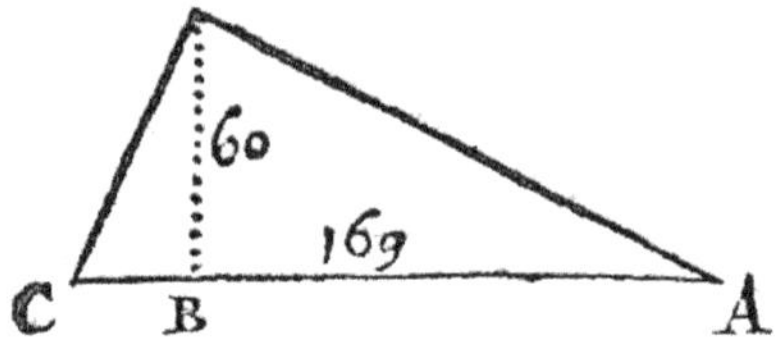

Figure 4.20: Question 14, explanation

Cardinael assumes in figure 2.2 that $AB > BC$. The square with sides $AB + BC = 17$ in figure 2.2 is known and so its area is 289. Both squares on the diagonal are also known and together their area is

$$AB^2 + BC^2 = AC^2 = 169$$

For the two complementary (equal) rectangles remains 120 and for that the rectangles $ABCD$ and $ABCE$ each are 60 (vertex E of the rectangle $ABCE$ is not indicated in figure 2.2, and G must be replaced by C). So

$$AB \cdot BC = 60$$

From figure 2.2 it appears that $AB \cdot BC$ is the geometric mean of BC^2 and AB^2 because

$$\begin{aligned}
&(\text{small square on } BC) : ABCD = ABCE : \ (\text{big square on AB}) \Leftrightarrow \\
&BC^2 : ABCD = ABCE : AB^2 \Leftrightarrow \ (\text{with } ABCE = AB \cdot BC) \\
&BC^2 : AB \cdot BC = AB \cdot BC : AB^2 \Leftrightarrow \\
&BC^2 : 60 = 60 : AB^2 \Leftrightarrow \\
&BC^2 \cdot AB^2 = 60^2
\end{aligned}$$

We now have a **system I**

$$\textbf{system I } (AB^2, BC^2, 60^2, 169) \begin{cases} AB^2 \cdot BC^2 = 60^2 \ \ (= c) \\ AB^2 + BC^2 = 169 \ (= b) \end{cases}$$

Cardinael wants to calculate AB^2 and BC^2 with the help of a right-angled triangle (figure 4.20) with right angle at D (not indicated in figure 4.20), altitude $DB = 60$ and hypotenuse AC with length 169. Figure 4.20 is rather misleading. Cardinael must have intended that the hypotenuse with length 169 is divided by the altitude DB with length 60 in the segments BC^2 and AB^2). This corresponds (see §3.1.8 (3.16)) to the (with **system I** equivalent) **AH-system**

$$\textbf{AH-system } (DB, AC, 60(=\sqrt{c}), 169(=b))$$

with unknowns BA (variable for AB^2) and BC (variable for BC^2).
From §3.1.8 (3.17), (3.18) we have the following two solutions

$$BA = \frac{1}{2}b + \sqrt{(\frac{1}{2}b)^2 - c}$$

and

$$BC = \frac{1}{2}b - \sqrt{(\frac{1}{2}b)^2 - c} \Leftrightarrow$$

$$AB^2 = \frac{169}{2} + \frac{119}{2}$$

and

$$BC^2 = \frac{169}{2} - \frac{119}{2} \Leftrightarrow$$

$$AB = 12 \text{ and } BC = 5$$

or

$$AB = 5 \text{ and } BC = 12$$

Apparently Cardinael opted for $AB > BC$, so $AB = 12$ and $BC = 5$ while the other solution is omitted. For the solution of the **AH-system** $(DB, AC, 60, 169)$ (see figure 4.20) with unknowns BA and BC, no application of area was involved. The **AH-system** is solved in the direct way of §3.1.8 case 16 (3.17), (3.18) with the help of the Thales's theorem and Pythagoras .

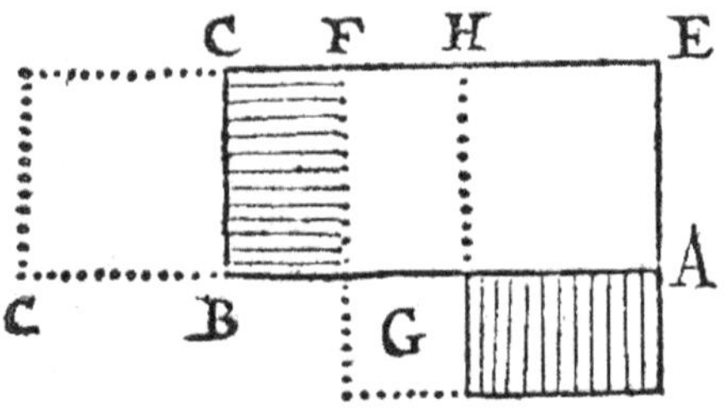

Figure 4.21: Question 14, alternative solution

Comment on Otherwise

Again Cardinael assumes in figure 4.21 that $AB > BC$. He now employs the application of area. Of the rectangle $ABCE$ in figure 4.21, the area is known (see **Answer**)

$$AB \cdot BC = 60$$

In figure 4.21 point C is given on the extension of AB such that the line segment $A(B)C = 17$, F is chosen on EC such that

$$EF = \frac{1}{2} \cdot A(B)C = \frac{1}{2}(AB + BC) = \frac{17}{2}$$

From this follows that

$$EF^2 = (\frac{17}{2})^2 = 72\frac{1}{4}$$

Furthermore, point H is chosen on EH such that

$$EH = EA$$

Now we can interpret the rectangle $ABCE$ in figure 4.19 as an elliptic application of area to $A(B)C$ falling short by the square on BC. From §3.1.5 case 8 follows with the help of the elliptic identity, that

$$\begin{aligned} BC \cdot BA = EF^2 - CF^2 &\Leftrightarrow \\ CF^2 = EF^2 - BC \cdot BA &\Leftrightarrow \text{ (with } EF = \frac{1}{2}A(B)C = \frac{17}{2}) \\ CF^2 = 72\frac{1}{4} - 60 &\Leftrightarrow \\ CF = 3\frac{1}{2} & \end{aligned}$$

So

$$BA = EF + CF = 12 \text{ and } BC = EF - CF = 5$$

This solution corresponds to §3.1.9 case 17 (i). Cardinael omits the possibility §3.1.9 (iii) as a consequence of his assumption $AB > BC$.
It is possible that Cardinael had in mind the alternative solution by means of a hyperbolic application of $ABCE$ to the segment CH (see §3.1.9 case 17 (ii)) with exceeding square HA. Figure 4.21 is not unambiguous on his choice.
With $EC > EA$, according to the hyperbolic identity of §3.1.5 case 9 (again with $EF = \frac{1}{2}AC = \frac{17}{2}$) we arrive at

$$\begin{aligned} EC \cdot EH = EF^2 - HF^2 &\Leftrightarrow \\ HF^2 = EF^2 - EC \cdot EH &\Leftrightarrow \\ HF^2 = 72\frac{1}{4} - 60 &\Leftrightarrow \\ HF = 3\frac{1}{2} & \end{aligned}$$

So

$$AB = EC = EF + HF = 12 \text{ and } BC = EH = EF - HF = 5$$

in full agreement with the preceding solution of the elliptic application.

General comment

The most simple and general solution of **Q14** would be the initial calculation $AB \cdot BC = 60$ (see **Answer**) which gives the following **system I**

$$\textbf{system I } (BA, BC, 60, 17) = \begin{cases} BA \cdot BC = 60 \ \ (= c) \\ BA + BC = 17 \ (= b) \end{cases}$$

and then solving this **system I** in the manner of §3.1.8 case 16a or §3.1.8 case 16b. Here is a brief sketch of both solutions:

§3.1.8 case 16a (direct solution of **system I** with application of area).
system I $(BA, BC, 60(= c), 17(= b))$ has according to §3.1.6 the following two solutions

$$BA = \frac{1}{2}b + \sqrt{(\frac{1}{2}b)^2 - c} \text{ and } BC = \frac{1}{2}b - \sqrt{(\frac{1}{2}b)^2 - c} \Leftrightarrow$$

$$BA = \frac{17}{2} + \sqrt{(\frac{17}{2})^2 - 60} = 12 \text{ and } BC = \frac{17}{2} - \sqrt{(\frac{17}{2})^2 - 60} = 5$$

or

$BA = 5$ and $BC = 12$.

§3.1.8 case 16b (indirect solution via the equivalent **AH-system** without application of area).

The equivalent **AH-system** of the **system I** $(BA, BC, 60(= c), 17(= b))$ has the form (see 3.16 (3.16))

$$\textbf{AH-system } DB, AC, \sqrt{60}(= \sqrt{c}), 17(= b)$$

(with unknowns BA and BC).

According to (3.17 3.17) (3.18 3.18) it has the following two solutions:

$$BA = \frac{1}{2}b + \sqrt{(\frac{1}{2}b)^2 - c} \text{ and } BC = \frac{1}{2}b - \sqrt{(\frac{1}{2}b)^2 - c} \Leftrightarrow$$

$$BA = \frac{17}{2} + \sqrt{(\frac{17}{2})^2 - 60} = 12 \text{ and } BC = \frac{17}{2} - \sqrt{(\frac{17}{2})^2 - 60} = 5$$

or

$$BA = 5 \text{ and } BC = 12$$

It remains unclear why Cardinael did not rely on this relatively simple solution method.

Theory used

Answer (proportions, Thales , Pythagoras)

Geometrical algebra §3.1.2 c: geometric mean, **Q5**

Otherwise (application of area)

§3.1.5 case 8 or §3.1.9 case 17 (*i*) and §3.1.5 case 9 or §3.1.9 case 17 (*ii*)

Compare with

Q29 and **Q52** for a generalisation of **Q14**

Q42: there too, given the figure used and the calculation, obscurity about the choice between elliptic or hyperbolic application;

Q44: in spite of apparent an other formula a solution fully analog to that of **Q14**, except that **Answer** and **Otherwise** are interchanged

Q50 and **Q51**

Question 15

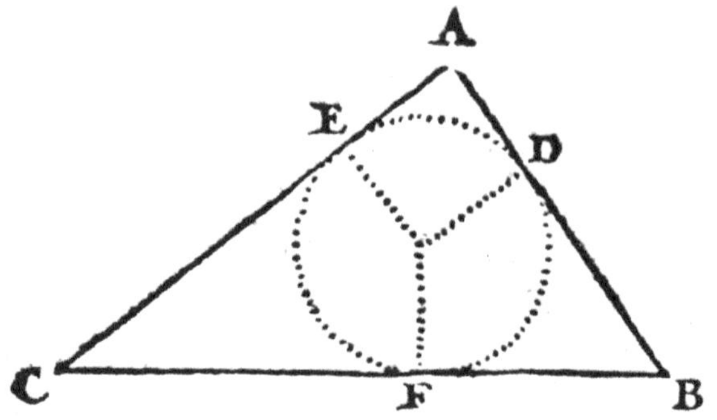

Figure 4.22: Subgroup 2, right-angled triangle with sum of the sides and the area known

Comment on Answer

E, D and F are the points of contact of the inscribed circle in triangle ABC with sides AC, AB and BC. With $r = \frac{O}{s} = \frac{600}{60} = 10$ we have

$$AD = AE = r = 10$$

so

$$DA + AE = 20$$

Then

$$2CF + 2BF = 100$$

so

$$BC = 50$$

and

$$AB + AC = 70$$

This configuration corresponds with **Q14 Otherwise**, because now we have

$$AB, AC = 1200 \text{ and } AB + AC = 70$$

and in the way explained above (as in **Q4 Comment on Otherwise**), AB and AC can be calculated with the help of elliptic or hyperbolic application:

$$AB = 30 \text{ and } AC = 40$$

General comment

Here, Cardinael implicitly relies on the known association between the radius r of the inscribed circle, the area O and the circumference s

$$r = \frac{O}{s}$$

and finds

$$AD = AE = r = 10$$

Just as remarked in the general comment on **Q14**, a more simple and general solution method would be to interpret the data

$$CA \cdot BA = 1200 \text{ and } CA + BA = 70$$

as a **system I**

$$\textbf{system I } (CA, BA, 1200, 70) = \begin{cases} CA \cdot BA = 1200 \\ CA + BA = 70 \end{cases}$$

which fulfills the condition $c \leq (\frac{1}{2}b)^2$ because $1200 \leq 35^2$
So according to §3.1.8 (16a) or §3.1.8 (16b) this leads to the solution

$$CA = \frac{1}{2}b + \sqrt{(\frac{1}{2}b)^2 - c} \text{ and } BA = \frac{1}{2}b - \sqrt{(\frac{1}{2}b)^2 - c} \Leftrightarrow$$

$$CA = \frac{35}{2} + \sqrt{(\frac{35}{2})^2 - 1200} \text{ and } BA = \frac{35}{2} - \sqrt{(\frac{35}{2})^2 - 1200} \Leftrightarrow$$

$$CA = 40 \text{ and } BA = 30$$

Theory used

Answer (application of area)

$r = \frac{O}{s}$ (r radius of the inscribed circle);

§3.1.5 case 8 or §3.1.5 case 9

Compare with

Q14 Otherwise

Q65

Question 16

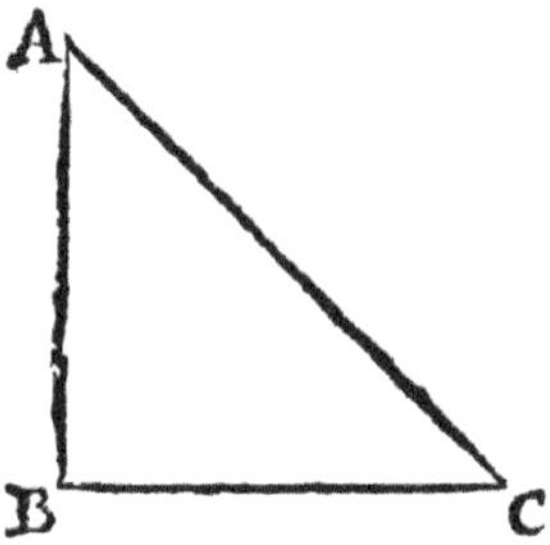

Figure 4.23: Subgroup 2, right-angled isosceles triangle with the sum of the sides known

Comment on Answer

By supposing $AB = BC = 1$, Cardinael calculates AC and the sum of the sides. From the real value of this sum, the multiplication factor can be determined and and hence real values of AB and BC

$$AB = 2, BC = 2, AC = \sqrt{8}$$

Theory used

Answer (Pythagoras , proportions)

Multiplication factor and Regula Falsi

Question 17

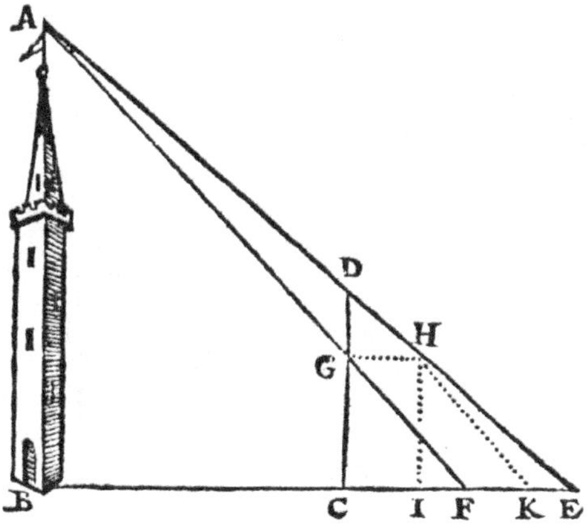

Figure 4.24: Subgroup 2, measuring the height of a tower by means of two sticks

Comment on Answer

Another classic technique from the surveyor tradition, a variation of problem 6 discussed by Alberti [Alberti 2010, 90]. From the similarity of the triangles ECD and HGD follows that $GH = 5$ and for that $KE = 8 - 5 = 3$.

Also the triangles EIH and EBA are similar and thus the corresponding line segments HK and AF are parallel. So $EK : EF = HI : AB$ and (with $EK = 3, EF = 8$ and $HI = 16$) $AB = 42\frac{2}{3}$.

General comment

The reasoning is the same as that of the similarity of the figures $EKIH$ and $EFBA$ and the corresponding line segments HK and AF of **Q12**. Now two sticks are needed instead of one stick and a mirror.

Theory used

Answer (proportions)

Similarity

Compare with

Q12;

Sitters 2008 §5.2.2.2 vb 20.

Question 18

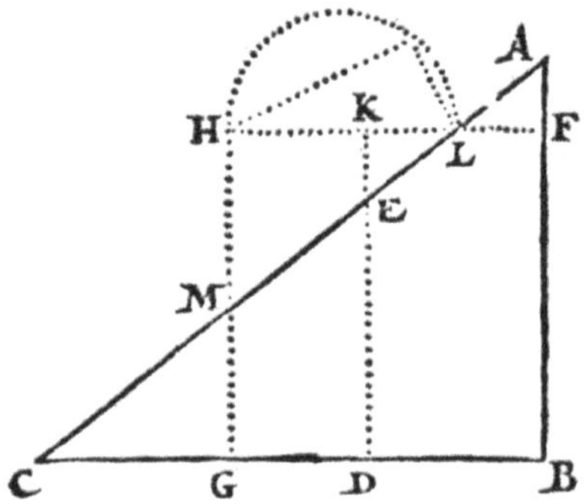

Figure 4.25: Subgroup 2, calculating the sides of a right-angled triangle given the area of two parts

Comment on Answer

Trapezium $ABDE$ is replaced by a rectangle $BDKF$ with equal area with $BD = 4$ and $BF = 7\frac{1}{2}$. Also triangle EDC is replaced by a rectangle $KDGH$ (with $DG = 3\frac{1}{5}$) connected with $BDKF$. Now, as rectangle $BFHG$ is equal to triangle ABC, the triangles AFL and MGC together are equal to triangle LHM. Furthermore, these triangles are similar and so (according to Elem.VI-31) there is a right-angled triangle with hypotenuse LH and sides of the right angle FL and GC. Because $FL = LK = 2$ and $LH = 5\frac{1}{5}$ so $GC = 4\frac{4}{5}$, $BC = 12, AB = 9$ and $AC = 15$.

While Cardinael does not mention it, he here employs for the first time Elem.VI-31, which is a generalization of Elem.I-47. The figure indicates such application by means of the semicircle with diameter LH and right-angle sides, apparently equal to FL and GC. Further on, this generalisation of the Pythagorean theorem is applied frequently (see Elem.VI-31).

Comment on Otherwise

The total area of triangle ABC is 30+24=54. That is $2\frac{1}{4}$ times the area of triangle EDC. So the sides of triangle ABC are $\sqrt{2\frac{1}{4}}$ times the corresponding sides of triangle EDC. Supposing that $CD = 1$, then $BC = \frac{3}{2}$ and so $BD = \frac{1}{2}$. But given is that $BD = 4$, therefore $DC = 8$.

Cardinael here opts for a more arithmetical approach by using point multiplication in combination with a version of Regula Falsi (see §3.1.1).

Theory used

Answer (proportions)

Elem.VI-31: generalization of the Pythagorean theorem

Similarity

Otherwise

Point multiplication and Regula Falsi

Question 19

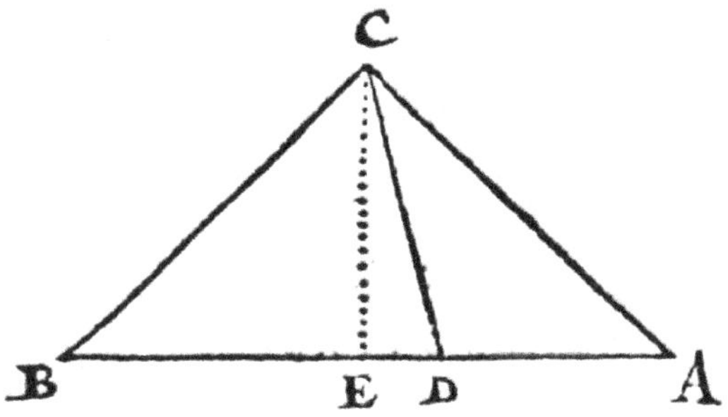

Figure 4.26: Subgroup 1, isosceles triangles

Comment on Answer

From given values follows that $AB = 22$. Hence, with E as the center of AB then $EB = 11$ and so $DE = 12 - 11 = 1$. From this follows

$$EC^2 = 105 \Leftrightarrow EC = \sqrt{105}$$

Hence, the sides AC and BC are also known.

Theory used

Answer (Pythagoras)

Elem.I-47

Question 20

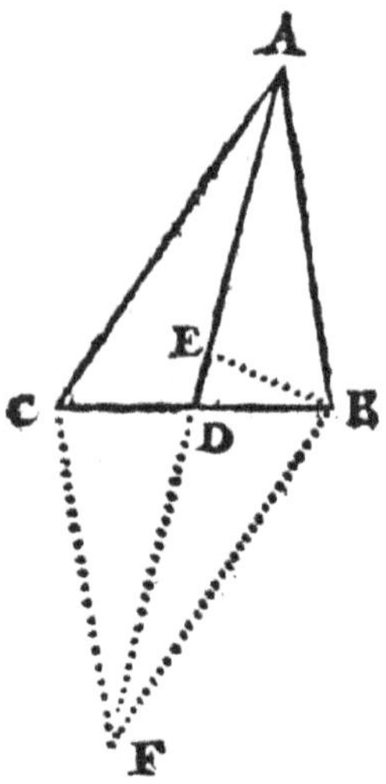

Figure 4.27: Subgroup 1, the median of a triangle

Comment on Answer

Construct F on the extension of AD so that $AD = DF$, then $AF = \sqrt{424}$. From the congruence of triangles BDF and CDA follows that $BF = AC = 12$. All three sides of triangle AFB are thus known

$$AF = \sqrt{424}, AB = 10 \text{ and } BF = 12$$

If E is the projection of B on AF, then using **Q1** the segments AE, ED and EB can be calculated

$$AE = \sqrt{85\frac{15}{106}},\ EB = \sqrt{14\frac{91}{106}}$$

So

$$ED = AD - EA = \sqrt{106} - \sqrt{85\frac{15}{106}} = \sqrt{1\frac{15}{106}}$$

For triangle BDE follows, by means of the Pythagorean theorem

$$BD = \sqrt{1\frac{15}{106} + 14\frac{91}{106}} = 4 \text{ and so } BC = 8$$

General comment

In parallelogram $ABFC$, with $AB = FC = 10$ and $AC = FB = 12$, hinging around A, B, C and F it is clear that based on continuity, the result of transformation of AD will go through every value between 11 (when $AF = 22$ and $BC = 2$) and 1 (when AF=2 and $BC = 22$) and at some point will necessarily also take the value of $\sqrt{106}$.

Theory used

Answer (Pythagoras)

Q1.

Compare with

Q57, Q70

Question 21

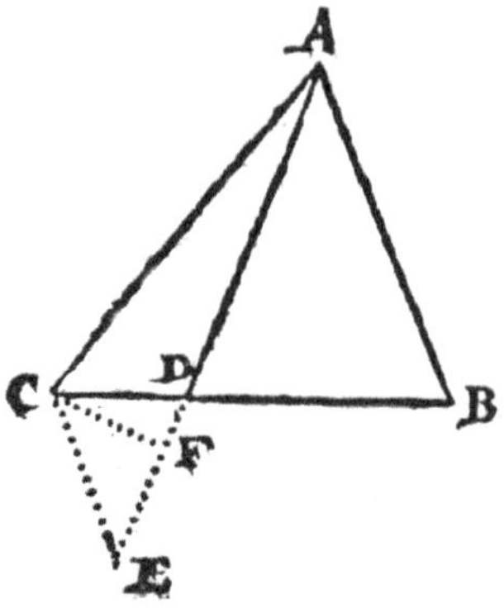

Figure 4.28: Subgroup 1, a line from the top of a triangle divides the base in segments in proportion of 1 to 2

Comment on Answer

Construct on the extension of AD a point E so that CE is parallel with AB. Triangle DCE is similar to triangle DBA. From $BD : DC = 2 : 1$ follows $AB : CE = 2 : 1$ and $AB : DE = 2 : 1$. From $AB = 13$ now follows $CE = 6\frac{1}{2}$. In the same way $DE = \sqrt{40\frac{25}{36}}$ and so

$$AE = AD + DE = \sqrt{366\frac{1}{4}}$$

Again from **Q1** follows that $FE = \sqrt{22\frac{5769}{5860}}$, hence

$$CF = \sqrt{19\frac{389}{1465}}$$

with F being the projection of C on AE. For the triangle EDC holds

$$FD = ED - EF = \sqrt{2\frac{6754}{13185}}$$

so

$$DC = \sqrt{FD^2 + FC^2} = 4\frac{2}{3} \text{ and } BD = 2 \cdot 4\frac{2}{3} = 9\frac{1}{3} \text{ and } BC = 14$$

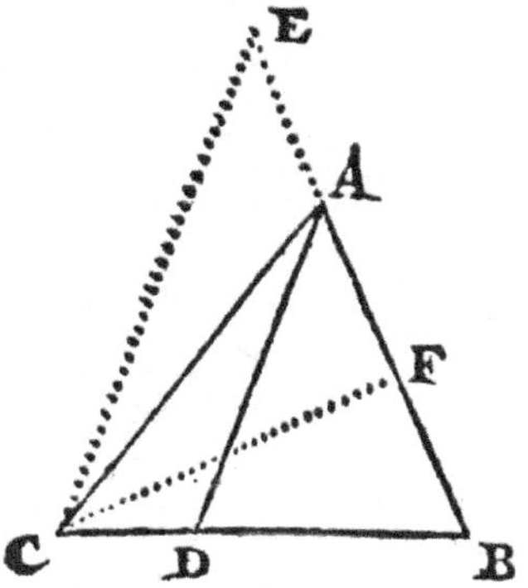

Figure 4.29: Question 21, alternative solution

Comment on Otherwise

Construct on the extension BA a point E so that CE is parallel with DA as in figure 4.29. Then triangles BCE and BAD are similar. From $AB = 13$ follows that $BE = 19\frac{1}{2}$ or $AE = 6\frac{1}{2}$ as $DC : BD = 1 : 2$. Now all sides of triangle EAC are known because $EC = \frac{3}{2} \cdot AD$, and following **Q2**, the segments AF and FC can be determined, with F being the projection of C on BE.

$$AF = 7\frac{2}{13} \text{ and } FB = 5\frac{5}{13}$$

so

$$BC = \sqrt{FB^2 + FC^2} = \sqrt{28\frac{168}{169} + 167\frac{1}{169}} = 14$$

General comment

This exercise is a generalization of **Q20**.

Theory used

Answer (proportions, geometrical algebra, Pythagoras)

Q1

Otherwise (Proportions, Geometrical algebra, Pythagoras)

Q2

Question 22

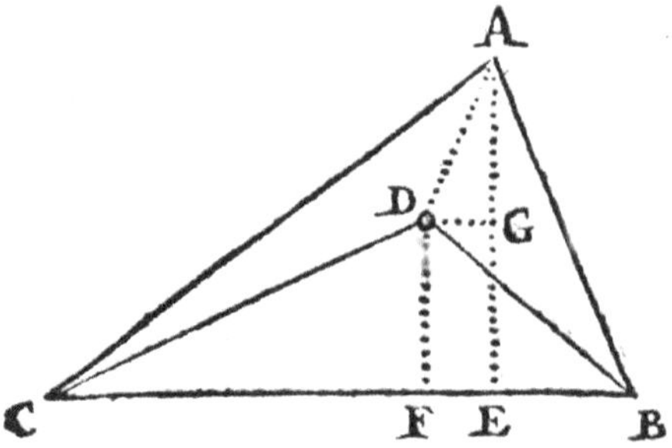

Figure 4.30: Subgroup 2, calculation of the distance of a point inside a triangle to the third vertex with the distances to both other vertices known

Comment on Answer

DF and AE in figure 4.30 are the perpendiculars from D and A on BC, DG is the perpendicular from D on AE. Following **Q1**, BE and AE can be found

$$BE = 10 \text{ and } AE = 24$$

In the same way BF and DF can be found

$$BF = 16 \text{ and } DF = 16$$

So

$$GD = BF - BE = 6$$

Also

$$GA = EA - FD = 8$$

so

$$AD = \sqrt{36 + 64} = 10$$

Theory used

Answer (geometrical algebra, Pythagoras)

Q1

Question 23

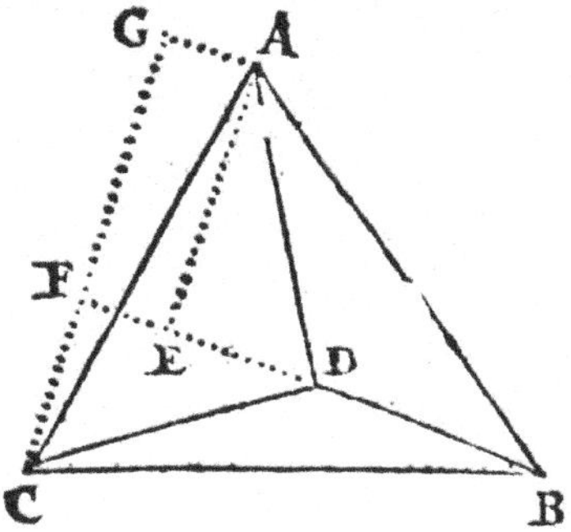

Figure 4.31: Subgroup 2, determining the third side of a triangle when the distance of a point within the triangle to the vertices and two sides are known

Comment on Answer

In figure 4.31, AE and CF are perpendiculars from A and C on the extension of BD, $AEFG$ is a rectangle. According to **Q2** the line segments BE and AE can be determined

$$BE = \sqrt{108\frac{52}{181}} \text{ and } AE = \sqrt{87\frac{129}{181}}$$

In the same way

$$BF = \sqrt{216\frac{108}{181}} \text{ and } CF = \sqrt{8\frac{73}{181}}$$

Adding FG, being equal to AE, with FC gives

$$GC = \sqrt{150\frac{75}{181}}$$

And further

$$AG = EF = BF - BE = \sqrt{216\frac{108}{181}} - \sqrt{108\frac{52}{181}} = \sqrt{18\frac{106}{181}}$$

In triangle ACG then is, according to Pythagoras

$$AC = \sqrt{150\frac{75}{181} + 18\frac{106}{181}} = 13$$

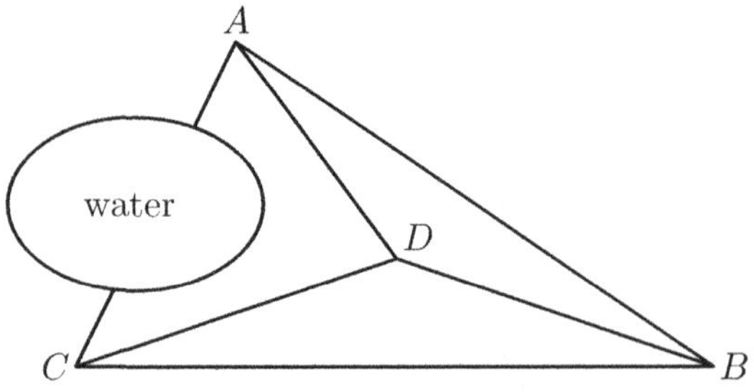

Figure 4.32: Measuring a distance blocked by water

General comment

This Question has a direct application in surveying for measuring the distance of two points A and C divided by water or another obstacle (as shown in figure 4.32). Measuring AB, BC, AD, BD and CD makes it possible to calculate in an easy way the distance AC.

A smart surveyor would of course select here point D on AB or on BC. It can be noticed that the roles of AD and AC in **Q22** and **Q23** are interchanged.

Theory used

Answer (geometrical algebra, Pythagoras)

Q2.

Question 24

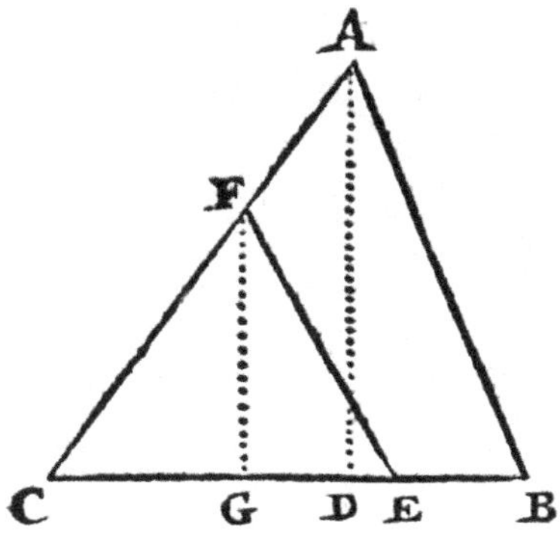

Figure 4.33: Subgroup 2, calculating the distance of two points on two different sides of a given triangle

Comment on Answer

In figure 4.33, AD and FG are perpendiculars from A and F on BC. Following **Q1**, the line segments AD and DC can be determined

$$AD = 12 \text{ and } DC = 9$$

The triangles ADC and FGC are similar, so $GF = 8$. From the same similarity follows that $GC = 6$ and so

$$EG = 14 - GC - EB = 14 - 6 - 4 = 4$$

From this follows directly that

$$EF = \sqrt{GF^2 + EG^2} = \sqrt{80}$$

Theory used

Answer (geometrical algebra, Pythagoras , proportions)

Q1;

Similarity

Question 25

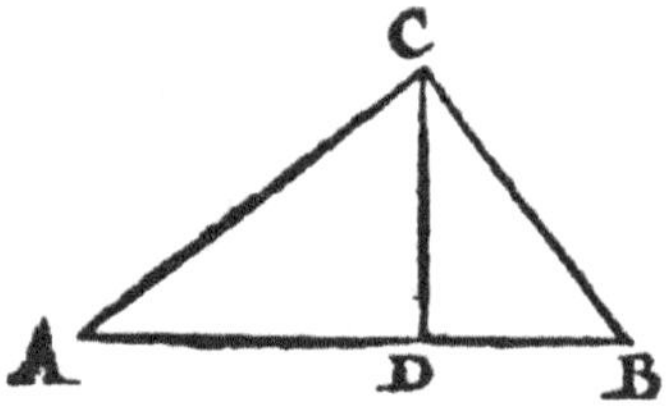

Figure 4.34: Subgroup 8, a right-angled triangle with a right angle side known and the base divided by the perpendicular according to the Golden Section

Comment on Answer

In **Answer** Cardinael starts from the configuration in figure 4.35 where $ABCD$ is a square, $DMOC$ a rectangle and $OBLK$ a square such that

$$DMOC = OBKL$$

E and F are the center points of AB and DC. From the equality of $DMOC$ and $OBKL$ follows

$$OC \cdot OM = OL \cdot OB \Leftrightarrow OC \cdot BC = OB^2$$

and therefore

$$OC : OB = OB : BC$$

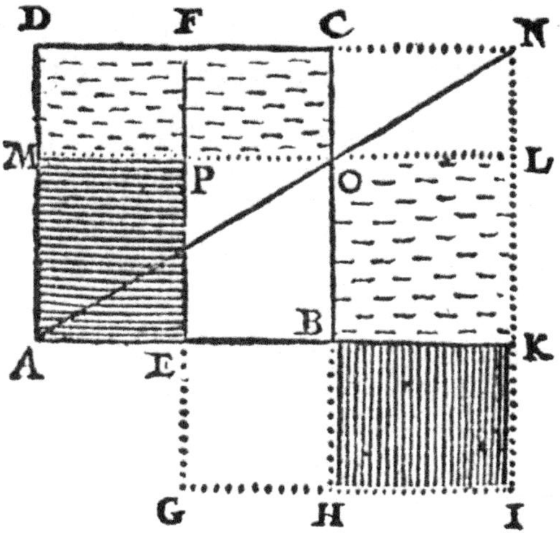

Figure 4.35: Question 25: solution

In this way it is shown that BC becomes divided by O in an extreme and mean ratio as AB by D is divided in figure 4.34. As a direct consequence

$$AKLM = ABCD$$

and that the extensions of KL and DC intersect in point N, also on the extension of AO as

$$NL : LO = OB : BA$$

Cardinael is not quite right in referring to Elem.II-11 and VI-30 because there the construction and demonstration is done the other way around, dividing a line segment in extreme and mean ratio, though he uses the same constellation of rectangles and squares as in figure 4.35. The calculation of OB and OC commences with $AB = 2$ and depends on the special case of hyperbolic application (§3.1.5 case 10): $AKLM$ is applied to AB exceeding by the square $BKLO$ such that

$$\text{rectangle } AKLM = \text{ square } ABCD$$

According to (§3.1.5 case 9) this goes together with the hyperbolic application

$$EK^2 - EB^2 = AKLM \Leftrightarrow EK^2 - EB^2 = ABCD \Leftrightarrow EK^2 - 1 = 4 \Leftrightarrow$$

$$EK^2 = 5 \Leftrightarrow EK = \sqrt{5} \Leftrightarrow BK = \sqrt{5} - 1$$

So

$$OB = \sqrt{5} - 1 \text{ and } OC = 3 - \sqrt{5}$$

Next, Cardinael relies on the unproven identity

$$AD = BC \text{ (in figure 4.34)}$$

and from that concludes

$$AD = 10 - \sqrt{20}$$

and that we must also have (in figure 4.35)

$$OB = 10 - \sqrt{20}$$

while above it was also determined that

$$OB = \sqrt{5} - 1$$

Which means that the real value of AB (in figure 4.34) is not 2 but follows from

$$AB = \frac{10 - \sqrt{2}}{\sqrt{5} - 1} \cdot 2 = \sqrt{80}$$

Only at the end of the calculation, Cardinael demonstrates that $AD = BC$ (figure 4.34) using Elem.VI-8. from the similarity of triangles BDC and BCA follows

$$BD : BC = BC : AB$$

and because

$$BD : AD = AD : AB$$

follows

$$BC = AD$$

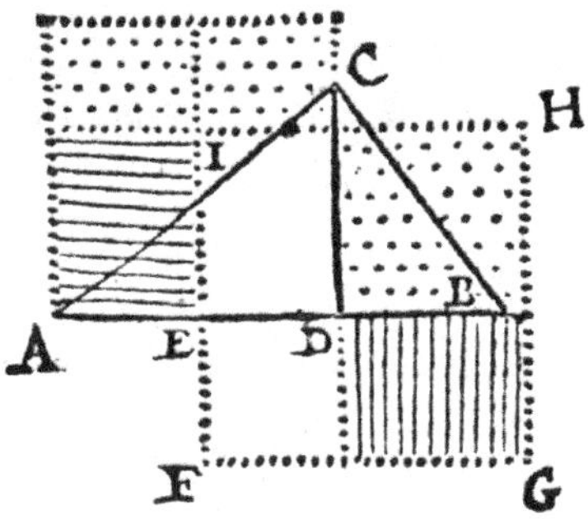

Figure 4.36: Question 25: alternative solution

Comment on Otherwise

In figure 4.36, H is chosen so that $BH = BD$ and I so that $EI = BH$ and F so that $ED = EF$. In this alternative solution Cardinael proceeds more directly and systematically. For triangle ABC is firstly demonstrated by means of Elem.VI-8, that

$$AD = BC = 10 - \sqrt{20}$$

As AB is divided in an extreme and mean ratio by D, we have

$$BH \cdot AB = AD^2 \text{ (Elem.VI-17)}$$

Then follows a special case of the hyperbolic application, in this case directly in the given triangle ABC (figure 4.36). E is the center point of AD and rectangle AH is applied to the line segment AD exceeding by the square DH, so from the hyperbolic identity of §3.1.5 case 9 follows

$$EB^2 - ED^2 = BD \cdot AB \Leftrightarrow$$
$$EB^2 - ED^2 = BH \cdot AB \text{ (see above)} \Leftrightarrow$$
$$EB^2 - ED^2 = AD^2$$

and therefore

$$EB^2 = ED^2 + AD^2 \Leftrightarrow$$
$$EB^2 = (5 - \frac{1}{2}\sqrt{20})^2 + (10 - \sqrt{20})^2 \Leftrightarrow$$
$$EB^2 = \sqrt{125} - 5$$

from which finally follows

$$DB = 6\sqrt{5} - 10 \text{ and } AB = \sqrt{80}$$

General comment

The calculation method in **Otherwise** is in Cardinael's point of view definitely superior to the method in **Answer** since the concealed use of algebra by posing $AB = 2$, becomes unnecessary.

Theory used

Answer (geometric algebra, proportions, application of area)

A multiplication factor, therefore the Rule of Coss

§3.1.5 case 10 and §3.1.5 case 9: hyperbolic application (special case: Golden Section)

Elem.VI-8: similarity in right-angled triangle

Otherwise (proportions, geometrical algebra, application of area)

Elem.VI-8: similarity in right-angled triangle

§3.1.5 cases 9 and 10: hyperbolic application (special case: Golden Section)

Compare with

Q35 and **Q69**;

[Petri 1583], ex. 29 and 30;

[Brasser 1663], questien 32 and 33;

[Sitters 2008], 5.2.2.2 ex. 5

Question 26

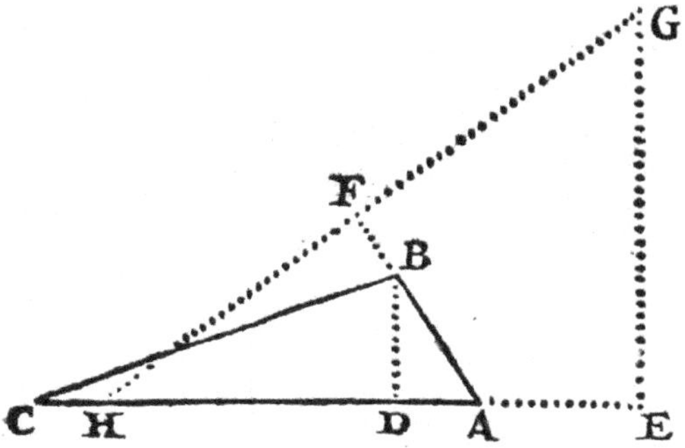

Figure 4.37: Subgroup 2, the length of the two intersecting perpendiculars on the extension of two of the sides of a triangle

Comment on Answer

Following **Q1** it can be determined that $AD = 6$ and $BD = 8$. If GF is extended to the intersection on AC in H, this leads to three similar triangles AFH, GEH and ADB. From their similarity follows that $AH = 25$, and $EH = 36$. Also $FH = 20, EG = 27$ and $GH = 45$. So $GF = 45 - 20 = 25$ and $EG = 27$.

Theory used

Answer (geometrical algebra, Pythagoras , proportions)

Q1

Similarity

Question 27

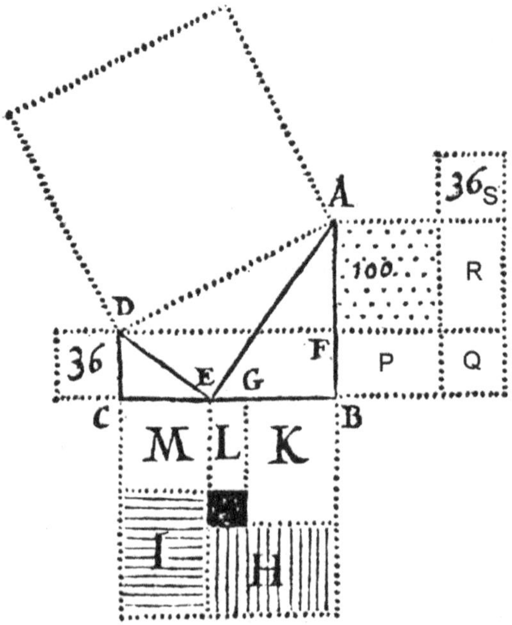

Figure 4.38: Subgroup 2, two ladders at the same point leaning against opposite walls

Comment on Answer

In figure 4.38, H and I are the equal rectangles $BE \cdot EC$ and since $BG = CE$ also the rectangles K and LM are equal to $BE \cdot EC$. The letters P, Q, R and S are added to the original drawing.
From

$$AB^2 + BE^2 + EC^2 + CD^2 = AD^2 \text{ (Elem.I-47)}$$

and

$$AF^2 + BC^2 = AD^2$$

follows

$$AF^2 + BC^2 = AB^2 + BE^2 + EC^2 + CD^2$$

$$AB^2 - AF^2 + BE^2 + EC^2 + CD^2 = BC^2$$
$$AB^2 - AF^2 + CD^2 = BC^2 - (BE^2 + EC^2)$$

or

$$AB^2 - AF^2 + CD^2 = H + I$$

The four equal segments H, I, K and LM are contained in BC^2. The area of these segments can be calculated as follows

$$H + I = P + Q + R + CD^2$$

in which the equal P and R and the square ($Q = 36$) inside the square on AB are the complement of the dotted area AF^2. From this follows

$$H + I = P + CD^2 + R + CD^2 \Leftrightarrow$$

$$H + I = 2(P + CD^2) = 2 \cdot (P + Q) = 2 \cdot 16 \cdot 6 = 192$$

so

$$H + I + K + LM = 2(H + I) = 4H = 384$$

From

$$BC^2 = 400$$

we then have

$$EG^2 = BC^2 - 4H = 400 - 384 = 16$$
$$EG = 4$$

Remaining for BG of EC

$$BG = EC = \frac{20 - 4}{2} = 8$$

so

$$EC = 8 \text{ en } EB = 12$$

from which follows (Elem.I-47)

$$AE = 20 \text{ and } ED = 10$$

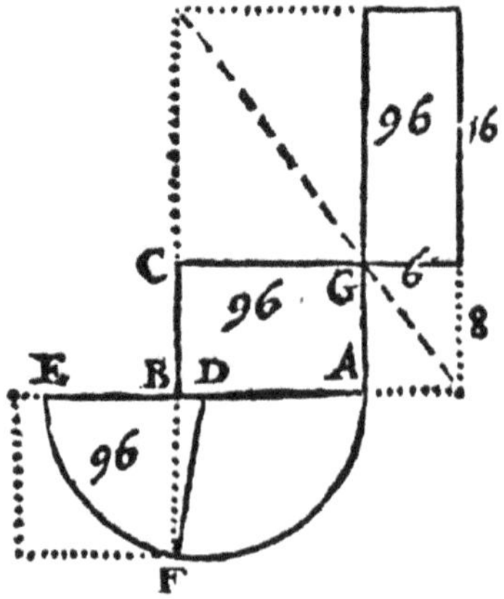

Figure 4.39: Question 27: alternative solution

Comment on Otherwise 1

In figure 4.39, D is the center point of AE. Since the triangles ABE and ECD are similar, so is

$$AB : BE = EC : DC \Leftrightarrow$$

$$BE \cdot EC = AB \cdot DC = 96 \text{ (Elem.VI-16)}$$

and given was

$$BE + EC = 20 \text{ (figure 2.38)}$$

Leading to the following system to be solved

$$\begin{cases} BE \cdot EC = 96 \\ BE + CE = 20 \end{cases}$$

In figure 4.39 the segments BE and EC from figure 4.38 are denoted, rather confusingly, as AB and BC. So, according to figure 4.39 the following system needs to be solved

$$AB + BC = 20 \text{ and } AB \cdot BC = 96$$

To do so, consider E on the extension of AB so that $BE = BC$, and so $AE = 20$, choose D on AE so that $AD = DE$, then $AD = DE = 10$. The circle with center D and radius AD intersects the perpendicular from B on AE in F, so $DF = 10$. According to Elem.IV-5 and III-31 the angle EFA

now is a right angle. From the similarity of triangle BEF and BFA and from Elem.VI-17 now follows

$$BF^2 = AB \cdot BE \text{ so } BF^2 = 96$$

because

$$AB \cdot BE = AB \cdot BC = 96$$

In triangle BDF we have

$$BD^2 = DF^2 - BF^2 \Leftrightarrow BD^2 = 100 - 96 \Leftrightarrow BD = 2$$

so

$$AB = 10 + 2 = 12 \text{ and } BE = 10 - 2 = 8 \text{ (figure 4.39)}$$

In figure 2.38 this implies that

$$BE = 12 \text{ and } EC = 8$$

This method is an example of §3.1.8, case 16.

$$\text{The } \textbf{system} \begin{cases} AB \cdot BC = 96 \\ AB + BC = 20 \end{cases}$$

is replaced by the equivalent **AH**-system and this being solved as explained in case 16 with the help of the geometric mean of AB and BC, i.e. $BF = \sqrt{96}$ in combination with **Q5**. While Cardinael does not discuss the construction but limits himself to the calculation, we here provide an outline of the construction

.

Firstly, the circle with diameter $AE = 20$ is constructed. Then B is constructed so that $BF = \sqrt{96}$. Next, the method of **Q5** is used for the right-angled triangleAFE.

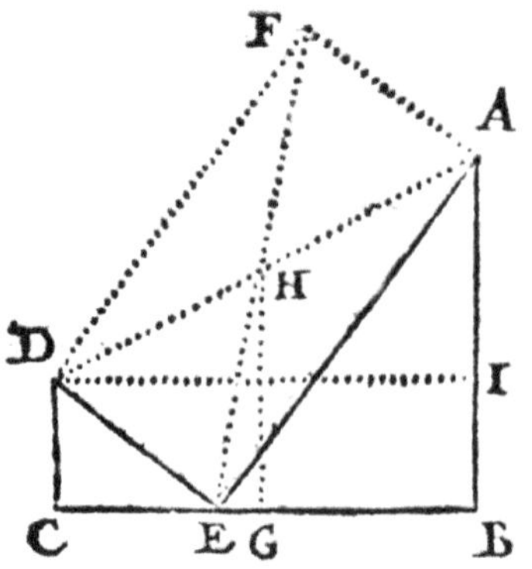

Figure 4.40: Question 27: second alternative solution

Comment on Otherwise 2

In figure 4.40, AED is a right angle, H is the center point of AD and F is on the extension of EH so that $EH = HF$. Hence, $AEDF$ is a rectangle with AD and EF as diagonals and H as its centre.

$$AB - CD = 16 - 6 = 10 \text{ and } ID = BC = 20$$

so according to Elem.I-47

$$AD = \sqrt{500} \text{ and } AH = HD = \sqrt{125}$$

and so

$$HE = \sqrt{125}$$

From $AB + DC = 16 + 6 = 22$ follows $HG = 11$ because H is the center point of AD and G the center point of BC. From $HG = 11$ and $HE = \sqrt{125}$ now follows $GE = 2$ and so

$$EC = CG - GE = 10 - 2 = 8 \text{ and } BE = BG + GE = 10 + 2 = 12$$

This is a fine application of Elem.IV-5 and III-31. Since the angle AED is a right angle, the foot of the ladders are on the circle with H as centre and HD as radius, so on the point of intersection of this circle with met BC.

Remarkable again, Cardinael fails to mention the existence and uniqueness of the solution, while the method of **Otherwise 2** however does give rise for this. From figure 4.40 it becomes directly obvious that the number of solutions depends on the existence of the number of intersecting points of of a circle with centre H and radius HD with BC. There is a simple relation between HD and the distance HG of H to BC

$$HD > HG \Rightarrow \text{ two solutions symmetrical on both sides of } G,$$
$$HD = HG \Rightarrow \text{ point } G \text{ the only solution, and}$$
$$HD < HG \Rightarrow \text{ there are no solutions}$$

Theory used

Answer (geometrical algebra, Pythagoras)

item Geometrical algebra, Elem.II-8

Otherwise 1 (proportions, Thales , Pythagoras)

Similarity, Elem.VI-16, VI-17; Elem.IV-5, Elem.III-31

§3.1.2 c: geometric mean, Elem.I-43

Q5

Otherwise 2 (Pythagoras , Thales)

Elem.IV-5 and III-31

Compare with

Q28

[Sitters 2008] for further details

Question 28

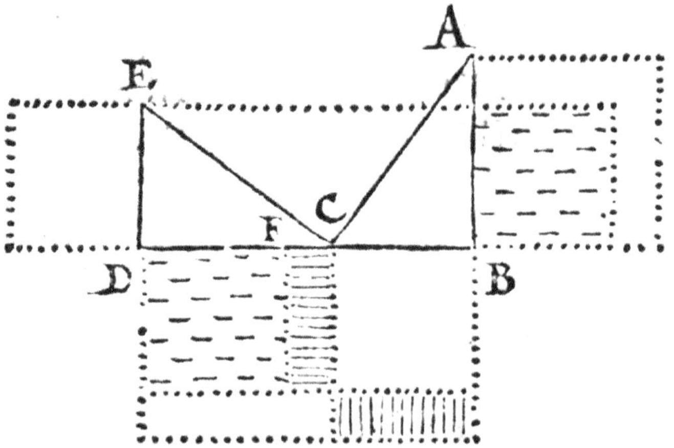

Figure 4.41: Subgroup 2, ladders of equal length from the same point against opposite walls

Comment on Answer

$$AC = EC$$

so

$$AB^2 + BC^2 = CD^2 + DE^2 \Leftrightarrow$$
$$CD^2 - BC^2 = AB^2 - DE^2 \Leftrightarrow$$
$$CD^2 - BC^2 = 175$$

The difference of the two squares on the left-hand side in the equation form a gnomon that can be transformed by turning the horizontally shaded part a quarter of a circle to a rectangle with length $BD = 35$ and breadth equal to

$$FC = DC - BC$$

So

$$FC \cdot 35 = 175 \Leftrightarrow FC = 5$$

and from this follows

$$BC + DF + 5 = 35 \Leftrightarrow 2BC = 30 \Leftrightarrow$$

$$BC = 15 \text{ and } DC = 20$$

and the length of the ladder will be

$$AC = \sqrt{225 + 400} = 25$$

General comment

The transition of the gnomon to the rectangle amounts to the identity from geometrical algebra

$$CD^2 - BC^2 = (CD + BC) \cdot (CD - BC)$$

as is mentioned in Elem.II ([Heath 1956], 383).

Theory used

Answer (geometrical algebra, Pythagoras)

Geometrical algebra, Elem.II ([Heath 1956], 383

Compare with

Q27;

[De Graaf 1706], 163; [De Graaf 1672], 63

Question 29

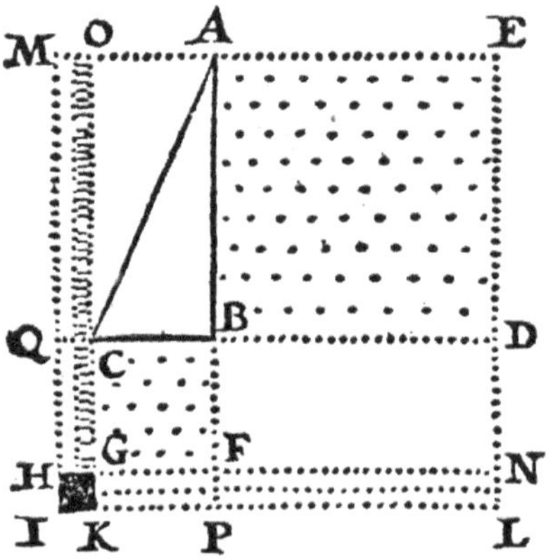

Figure 4.42: Subgroup 2, a right-angled triangle with sum of the right angle sides, the area and the hypotenuse known

Comment on Answer

In figure 4.42, O makes a rectangle of $ABCO$. Point M is on the extension of AO so that $OM = 2$. Point P is the center point of OM so that $OP = 1$. $ABDE$ and $BFGC$ are squares. In the rectangle $OGHM$, $OM = 2$ and $OG = AB + BC$. In order to calculate length and area, line segments AB and BC are represented by rectangles with one side with length 1. By $OP = 1$ Cardinael probably intends half of OM, so that the shaded strip along OG has the area

$$1 \cdot OC + 1 \cdot CG = 1 \cdot AB + 1 \cdot BC$$

So

$$1 \cdot AB + 1 \cdot BC + \text{opp.}ABC = 154 \Leftrightarrow$$

$$2 \cdot AB + 2 \cdot BC + \text{opp.}OABC = 308 \Leftrightarrow$$

$$\text{area } OMHGO + OABC = 308 \Leftrightarrow$$

$$\text{area } OMHGO + \text{ area } NLKGN + 2 \cdot OABC = 616 \Leftrightarrow$$

$$\text{area } OMHGO + \text{ area } NLKGN + OABC + BDNF = 616$$

Now by adding the three squares AB^2, BC^2 and HG^2 to the left-hand side of the equation, there arises the square $EMIL$ of which the area can be calculated since

$$AB^2 + BC^2 = AC^2 = 605$$

and

$$HG^2 = 4$$

because

$$HG = OM = 2$$

So

$$605 + 4 + 616 = EMIL \Leftrightarrow$$

$$EMIL = 1225 \Leftrightarrow$$

$$ME = \sqrt{1225} = 35$$

From this follows

$$OE = ME - OM = 35 - 2 = 33$$

$$AB + BC = 33$$

Having arrived at the two equations $AB + BC = 33$ and $AB^2 + BC^2 = 605$, we are back at the start of **Q14** to which Cardinael refers explicitly. In this question, AB and BC was determined in the following way:

$$(AB + BC)^2 - (AB^2 + BC^2) = 2 \cdot AB \cdot BC \Leftrightarrow$$

$$2 \cdot AB \cdot BC = 33^2 - 605 \Leftrightarrow$$

$$AB \cdot BC = 242$$

The system of equations consisting of

$$AB + BC = 33,\ AB^2 + BC^2 = 605 \text{ and } AB \cdot BC = 242$$

was solved in **Q14** in two different ways. In the underlying case of **Q29** he gives the results of applying the methods of **Q14** without explicit calculation.

In the underlying case of **Q29**, simpler and more general methods can be employed, as demonstrated in our General comment to **Q14** to solve **Q29** by using only the equations $AB + BC = 33$ and $AB \cdot BC = 242$. With these two equations one arrives at a **system I** of the form

$$\textbf{system I } (BA, BC, 242, 33) = \begin{cases} BA \cdot BC = 242 (= c) \\ BA + BC = 33 (= b) \end{cases}$$

with the condition $c \leq (\frac{1}{2}b)^2$ fulfilled since $242 \leq (\frac{33}{2})^2$.

In the General comment of **Q14**, this **system I** is solved in two ways.

The method of §3.1.8, case 16a (application of area) and of §3.1.8 case 16b (indirect solution via the equivalent **AH-system** $(DB, AC, \sqrt{242}, 33)$ with unknowns BA and BC) both lead to the following two solutions

$$BA = \frac{1}{2}b + \sqrt{(\frac{1}{2}b)^2 - c} \text{ and } BA = \frac{1}{2}b - \sqrt{(\frac{1}{2}b)^2 - c} \Leftrightarrow$$

$$BA = 22 \text{ and } BC = 11$$

or

$$BA = 11 \text{ and } BC = 22$$

In accordance with figure 4.42 Cardinael only gives the solution with $BA > BC$.

Theory used

Answer (geometrical algebra)

Q14;

§3.1.5 case 5 and §3.1.8 case 16

Compare with

Q5;

[Sitters 2008] for details.

Question 30

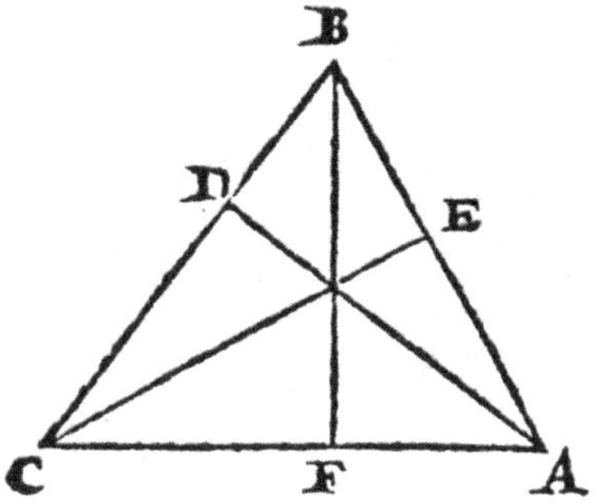

Figure 4.43: Subgroup 12, calculation of a triangle with the perpendiculars given

Comment on Answer

In the first of the two figures, figure 4.43, Cardinael draws the three perpendiculars through a single point without any further comment, while that is not a feature of the calculations below. From the similarity of triangles ADB and CEB follows

$$AD : EC = AB : BC$$

and from the similarity of triangles BFC and ADC follows

$$BF : AD = BC : AC$$

By substituting the given values in these equations follows

$$AB : BC = 60 : 64\frac{8}{13} \text{ (see figure 4.44)}$$

$$BC : AC = 56 : 60$$

The first equation can be expressed as

$$AB : BC = 52 : 56 \text{ (see figure 4.44)}$$

From the three equations follows

$$AB : BC : AC = 52 : 56 : 60 \text{ (figure 4.44, right)}$$

Assuming $AB = 52$, $BC = 56$ and $AC = 60$ and then applying the method of **Q1** one arrives at $AD = 48$. The real value of AD is given, being 60, so

$$AB = \frac{60}{48} \cdot 52 = 65, BC = \frac{60}{48} \cdot 56 = 70 \text{ and } AC = \frac{60}{48} \cdot 60 = 75$$

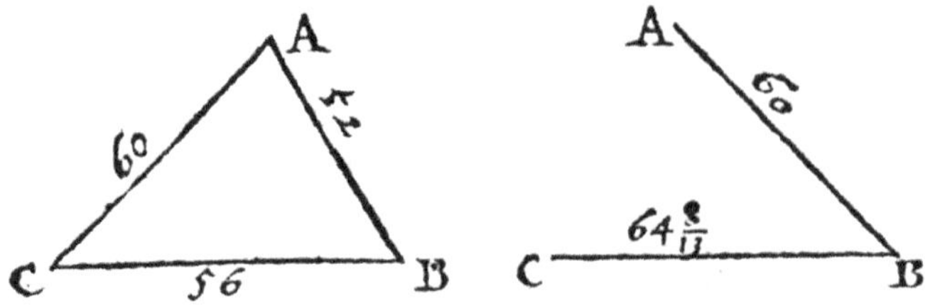

Figure 4.44: Question 30: solution by proportions

Comment on Otherwise

From

$$AB : BC = AD : EC$$

follows

$$AB : BC = \frac{1}{EC} : \frac{1}{AD} \tag{4.9}$$

and from

$$BC : AC = BF : AD$$

follows

$$BC : AC = \frac{1}{AD} : \frac{1}{BF} \tag{4.10}$$

From 4.9 and 4.10 follows

$$AB : BC : AC = \frac{1}{EC} : \frac{1}{AD} : \frac{1}{BF}$$

and substitution of the given values gives

$$AB : BC : AC = \frac{13}{840} : \frac{1}{60} : \frac{1}{56} = 13 : 14 : 15$$

If we assume that $AB = 13$, $BC = 14$ and $AC = 15$ then follows according to **Q1** that $AD = 12$. But because it is given that $AD = 60$ then

$$AB = 5 \cdot 13 = 65, \ BC = 5 \cdot 14 = 70 \text{ and } AC = 5 \cdot 15 = 75$$

General comment

It is remarkable that both in this Question as in the comparable **Q96** the solution is arrived at through the intermediate step of a triangle which is similar to the real solution, in the first case by calculation and in the second by construction .

Theory used

Answer (proportions, point multiplication)

Q1, Similarity of triangles, Regula Falsi

Otherwise (proportions, point multiplication)

Q1, Similarity of triangles, Regula Falsi

Compare with

Q96

Question 31

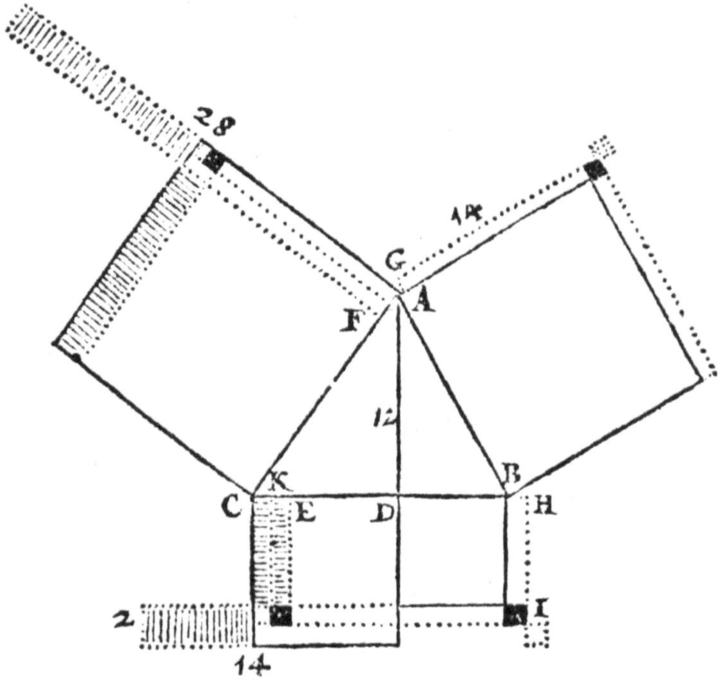

Figure 4.45: Subgroup 1, a triangle with base, perpendicular and sum of the other sides known

Comment on Answer

From **Q1**, $AF = AC - AB$ and $EC = DC - BD$ and the geometrical identity $a^2 - b^2 = (a+b) \cdot (a-b)$ follows that

$$AC^2 - AB^2 = DC^2 - BD^2 \Leftrightarrow$$

$$(AC + AB)(AC - AB) = (DC + BD)(DC - BD) \Leftrightarrow$$

$$28AF = 14EC \Leftrightarrow$$

$$AF = \frac{1}{2}EC$$

Let $AG = \frac{1}{2}AF$ and $BH = \frac{1}{2}EC$, then

$$AG = \frac{1}{2}AF = \frac{1}{4}EC = \frac{1}{2}BH$$

Firstly, the identity

$$BG^2 - (DH^2 + DA^2) = BH^2 - AG^2$$

can be proven as follows. From the figure it can be determined that $DH = 7, BG = 14$ and $AD = 12$.

$$\begin{aligned} BG^2 &= AB^2 + \text{area}_1 \\ &= AB^2 + 2BG \cdot AG - AG^2 \\ &= AB^2 + 28AG - AG^2 \end{aligned} \tag{4.11}$$

$$\begin{aligned} HD^2 + DA^2 &= DA^2 + DB^2 + \text{area}_2 \Leftrightarrow \\ &= AB^2 + \text{area}_2 \Leftrightarrow \\ &= AB^2 + 2DH \cdot HB - HB^2 \Leftrightarrow \\ &= AB^2 + 14HB - HB^2 \end{aligned} \tag{4.12}$$

and from 4.11 and 4.12 follows

$$\begin{aligned} BG^2 - (DH^2 + DA^2) &= 28AG - AG^2 - (14HB - BH^2) \\ &= BH^2 - AG^2 + 28AG - 14BH \\ &= BH^2 - AG^2 \end{aligned}$$

as

$$AG = \frac{1}{2}BH$$

Also

$$BG^2 - (DH^2 + DA^2) = 196 - (49 + 144) = 3$$

and therefore

$$BH^2 - AG^2 = 3$$

However, as previously stated

$$AG = \frac{1}{2}AF = \frac{1}{4}EC = \frac{1}{2}BH$$

so

$$BH = 1 \Rightarrow AG = \frac{1}{2} \Rightarrow BH^2 - AG^2 = \frac{3}{4}$$

From which follows (because of $BH^2 - AG^2 = 3$, so the real difference is 3) that

$$BH = \sqrt{\frac{3}{\frac{3}{4}}} = \sqrt{4} = 2$$

because

$$BH^2 - AG^2 = 3$$

Therefore the real difference is 3. Hence, BD, DC and AC can be simply calculated.

General comment

The use of a multiplication factor commences by posing $BH = 1$

$$BH = 1 \Leftrightarrow BH^2 - AG^2 = \frac{3}{4}$$

The real value of the difference is calculated as

$$BH^2 - AG^2 = 3$$

It is therefore concluded that the multiplication factor for squares is equal to $\frac{3}{\frac{3}{4}}$ from which follows, by posing $BH = 1$

$$BH^2 = 4 \cdot 1 \ \Leftrightarrow BH = 2$$

A solution without using a multiplication factor is equally possible. From

$$AG = \frac{1}{2}BH \text{ follows } BH : AG = 2 : 1$$

so

$$BH^2 : AG^2 = 4 : 1$$

which implies

$$BH^2 = 4 \cdot AG^2$$

So

$$BH^2 - AG^2 = 3 \cdot AG^2$$

and

$$3 \cdot AG^2 = 3 \ \Leftrightarrow AG = 1 \ \Leftrightarrow$$

$$BH = 2$$

In fact, this exercise treats the geometrical locus of the points A so that

$$AC + AB = 28$$

and that their distance to BC equal 12. So these are the two points of intersection of an ellipse with the principal axis along BC and a line parallel to BC on distance 12. Cardinael only considers the case where $BD < DC$ and neglects the solution of mirroring on the short axis $BD = 9$ and $DC = 5$.

Theory used

Answer (geometrical algebra, multiplication factor)

Geometric identity $a^2 - b^2 = (a + b) \cdot (a - b)$

Regula Falsi

Compare with

[Brasser 1663], 287; [Petri 1583], question 7: same question, but solved

by the Rule of Coss

Question 32

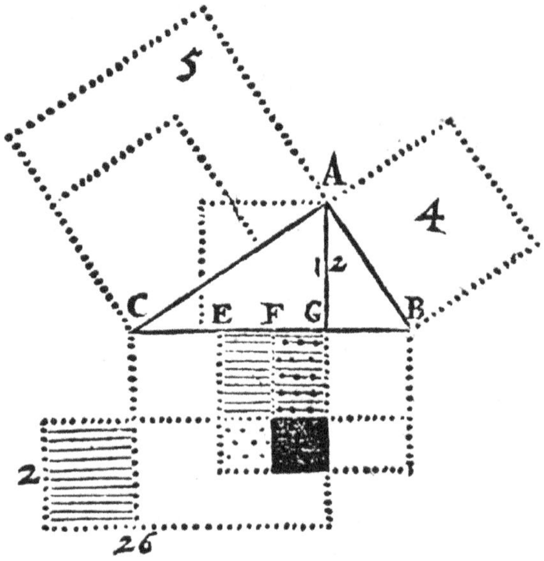

Figure 4.46: Subgroup 1, a triangle with its base, perpendicular and the ratio of the other sides known

Comment on Answer

From figure 4.46 it is obvious that the center point of BC is F. Thus, $CE = BG$ and so $GF = EF$.
Cardinael initially pays no attention to the given value $AG = 12$ and while the assumptions $GE = 2$ and $GF = 1$ are mentioned, they only serve as illustrative data.
He does use the identity from **Q1**, which is a variation of the Law of cosines.

$$AC^2 - AB^2 = GC^2 - GB^2$$

From $AC = \frac{3}{2}AB$ follows

$$AC^2 = \frac{9}{4}AB^2 \Leftrightarrow$$

$$AC^2 - AB^2 = \frac{5}{4} \cdot AB^2$$

and the same for

$$GC^2 - GB^2 = \frac{5}{4}AB^2$$

By cutting away the square $GB^2(= CE^2)$ from square GC^2, a gnomon results which can be transformed into a rectangle the usual way, with the long side 26 and the short side $2 \cdot GF$, hence with area $52 \cdot GF$.
This means that besides

$$GC^2 - GB^2 = \frac{5}{4} \cdot AB^2$$

we also have

$$GC^2 - GB^2 = 52 \cdot GF$$

and therefore

$$\frac{5}{4} \cdot AB^2 = 52 \cdot GF \Leftrightarrow AB^2 = 41\frac{3}{5} \cdot GF = 41\frac{3}{5} \cdot EF$$

From this moment on the given value $AG = 12$ comes into play, because this leads to

$$AG^2 + BF^2 = 313 \Leftrightarrow$$

$$AB^2 + BF^2 - BG^2 = 313 \tag{4.13}$$

From figure 4.46 it can be established for the gnomon $BF^2 - BG^2$ that

$$BF^2 - BG^2 = 2 \cdot GF \cdot BF - GF^2 = 26 \cdot GF - GF^2$$

from which follows by substitution in 4.13

$$AB^2 + 26 \cdot GF - GF^2 = 313$$

Substitution of $AB^2 = 41\frac{3}{5} \cdot GF$ leads to

$$67\frac{3}{5} \cdot GF - GF^2 = 313 \Leftrightarrow GF \cdot (67\frac{3}{5} - GF) = 313$$

So, this results in a (geometrical) equation as treated in §3.1.6 case 11 and in §3.1.11. Because $GF = EF$ in figure 4.46 and following the notation of Cardinael, this equation can be expressed as

$$EF \cdot (67\frac{3}{5} - EF) = 313$$

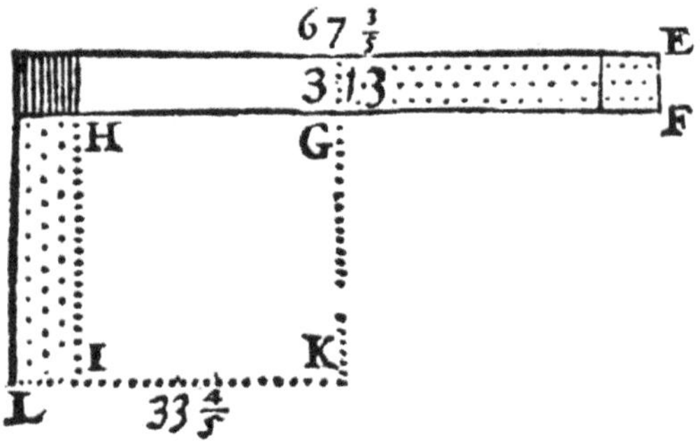

Figure 4.47: Question 32: solution

This equation is now solved geometrically by interpreting it as an elliptic application of rectangle $HFEQ$ with area 313 to the line segment PE of length $67\frac{3}{5}$, falling short by the square $RHQP$ with area PQ^2 (see our figure 4.48, which completes Cardinael's figure 4.47). In figure 4.47, M is the center point of $EP, QP = EF$. Here point G, as well as in Cardinael's figure 4.47 unfortunately has another denotation than in figure 4.46. The segment EF in figures 4.48 and 4.47 corresponds with the equal segment GF in figure 4.47 and thus also with the unknowns $GF = EF$ of the above geometrical equations.
The geometrical equation can be formulated as a **system I** (§3.1.6 case 11)

$$\textbf{system I}\ (QP, QE, 313, 67\frac{3}{5}) = \begin{cases} QP \cdot QE = 313 (= c) \\ QP + QE = 67\frac{3}{5} (= b) \end{cases}$$

with the condition fulfilled $c \leq (\frac{1}{2}b)^2$ because $313 \leq (33\frac{4}{5})^2$.

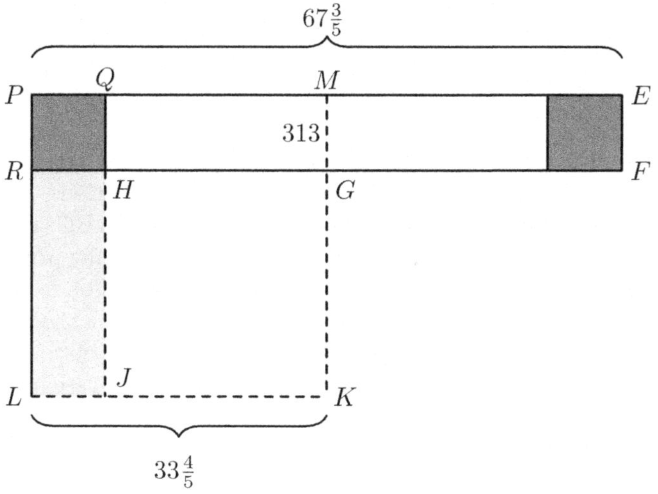

Figure 4.48: Cardinael's geometrical solution completed

For the two solutions we have calculated in §3.1.6 case 11

$$QP = \frac{1}{2}b - \sqrt{\left(\frac{1}{2}b\right)^2 - c} \text{ and } QE = \frac{1}{2}b + \sqrt{\left(\frac{1}{2}b\right)^2 - c} \Leftrightarrow$$

$$QP = 33\frac{4}{5} - 28\frac{4}{5} = 5 \text{ and } QE = 33\frac{4}{5} + 28\frac{4}{5} = 62\frac{3}{5}$$

or

$$QP = 62\frac{3}{5} \text{ and } QE = 5$$

In figure 4.46 these solutions correspond with

$$GF = EF = QP = 5 \text{ and } QE = 62\frac{3}{5}$$

or

$$GF = EF = QP = 62\frac{3}{5} \text{ and } QE = 5$$

In the underlying case, with the point G (in figure 4.46) on the hypothenuse BC between B and C, only the solution $GF = 5$ is valid.
This leads to

$$BG = 8 \text{ and } GC = 18$$

so

$$AB = \sqrt{208} \text{ and } AC = \sqrt{468}$$

General comment

From the geometrical equation for EF above follows as we have seen directly, that with $EF = 5$ also $EF = 67\frac{3}{5} - 5 = 62\frac{3}{5}$ is a solution. The solution $GF = 62\frac{3}{5}$ (figure 4.46) corresponds to the obtuse angle ABC, in which G is on the right side of B on the extension of CB. In this configuration, GF fits into the same equation as in the case of an acute angle ABC.

All the geometrical algebra used to arrive at the geometrical equation in the case of an acute angle remains valid also in the case of an obtuse angle ABC resulting in the same geometrical equation for GF. In that case, obviously one has to select the solution $GF = EF = 62\frac{3}{4}$.

It is easily understood that there are two different solutions depending on the geometrical loci. The points A with $AC : AB = 3 : 2$ and $BC = 26$ are situated on a circle with the centre on the extension of CB and the radius larger than 12. This circle is known in geometry as Apollonius's circle ([Molenbroek 1943] 199 – 200). The line through A parallel to BC and on distance 12 of BC shares with this circle two points of intersection and so there are two points A that satisfy.

Theory used

Answer (geometrical algebra, application of area, Pythagoras)

Elliptic application §3.1.6 case 11

Compare with

Q31, 33, 34;

[Molenbroek 1934] 199 – 200;

[Petri 1583], ex. 86, fol. 186^r;

[Schooten 1660], II, 72, voorstel 22;

[Sitters 2008], for details;

[Bos 2001], 88 Analysis 4.20

Question 33

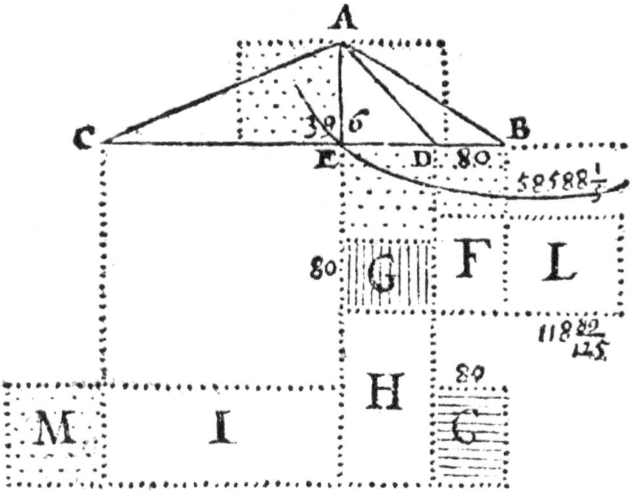

Figure 4.49: Subgroup 1, triangle with transversal , segments of the base and the proportion of the slanted sides known

Comment on Answer

Cardinael wrongly refers to BC in figure 4.49 where he means AC. E is the projection of A on BC

$$BD^2 = 6400 \text{ and } DA^2 = 27225 = DE^2 + EA^2$$

so

$$BD^2 + DE^2 + EA^2 = 6400 + 27225 = 33625$$

F is the rectangle $BD \cdot (BE - BD)$ and G is the rectangle $DE \cdot (BE - DE)$ so $F = G$.
If the rectangles F and G are added to both sides of this equation, then follows

$$AB^2 = 33625 + F + G \tag{4.14}$$

as

$$BD^2 + DE^2 + EA^2 + F + G = AB^2$$

From $AB : AC = 25 : 33$ follows that $AC^2 = 1\frac{464}{625}AB^2$. By multiplying both sides of the equation 4.14 with this factor we get

$$AC^2 = 58588\frac{1}{5} + 1\frac{464}{625}(F+G) \Leftrightarrow AC^2 = 58588\frac{1}{5} + F + G + \frac{464}{625}(F+G)$$

For $F+G$ we write $160 \cdot ED$ and with that the last equation becomes

$$AC^2 = 58588\frac{1}{5} + 160 \cdot ED + 118\frac{98}{125} \cdot ED \tag{4.15}$$

In the figure 4.49 we have $160 \cdot ED = F + G$ and $118\frac{98}{125} \cdot ED$ is represented as rectangle L with breadth ED and length $118\frac{98}{125}$ such that (4.15) can also be written as

$$AC^2 = 58588\frac{1}{5} + F + G + L \tag{4.16}$$

Further we know that

$$DC^2 = 396^2 = 156816$$

so

$$DC^2 + DE^2 + EA^2 = 156816 + 27225 = 184041$$

If we use M to denote the square DE^2, we may write

$$184041 = EA^2 + DC^2 + M \tag{4.17}$$

By elementary manipulation of the rectangles of figure 4.49, one can conclude that the right side of (4.17) can also be written in the form

$$AC^2 + M + G + H + I + M \tag{4.18}$$

By substituting (4.16) in (4.18), this last expression becomes (with $M = DE^2$)

$$58588\frac{1}{5} + F + G + L + M + G + H + I + M \Leftrightarrow$$

$$58588\frac{1}{5} + 160 \cdot ED + 118\frac{98}{125} \cdot ED + M + G + H + I + M \Leftrightarrow$$

$$58588\frac{1}{5} + 278\frac{98}{125} \cdot ED + 396 \cdot ED + I + M \Leftrightarrow$$

$$58588\frac{1}{5} + 278\frac{98}{125} \cdot ED + 396 \cdot ED + 396 \cdot ED$$

The last 3 terms taken together become $1070\frac{98}{125}\cdot ED$ such that the right-hand side of (4.18) will be

$$58588\frac{1}{5}+1070\frac{98}{125}\cdot ED$$

and for that (4.18) can be written as

$$184041=58588\frac{1}{5}+1070\frac{98}{125}\cdot ED$$

If we finally subtract $58588\frac{1}{5}$ from both sides, such that

$$125452\frac{4}{5}=1070\frac{98}{125}\cdot ED$$

from which follows

$$DE=117\frac{27}{169}$$

so

$$BE=197\frac{27}{169}\text{ and }EC=278\frac{142}{169}$$

F is the rectangle $BD\cdot(BE-BD)$ and G is the rectangle $DE\cdot(BE-DE)$, so $F=G$.

By applying Elem.I-47 to the triangles DEA, BEA and CEA we finally arrive at

$$AE=116\frac{31}{169}$$

and

$$AB=228\frac{11}{13}\text{ and }AC=303\frac{1}{13}$$

with their ratio 25:33 as required.

General comment

Cardinael's solution above is not constructive as it cannot be easily translated to a construction with compass and ruler, which would require the locus of the point E to be known. However, each step of the calculation can be interpreted as the calculation of area and can therefore be represented in figure 4.49. Also here, Cardinael does not mention the uniqueness of the solution. In this case the solution is unique if we limit ourselves to the upper half plane, limited by the line BC. The point A with distance 165 to D is on a circle with radius 165. The point A so that $AC : AB = 33 : 25$ is also on a circle with centre on the extension of CB. As already remarked in **Q32** this circle is known as Apollonius's circle ([Ogilvy 1969], 14-23). If point D is chosen between B and C, the circles have just one point of intersection, but if D is chosen on the extension of CB then there is no intersection point at all. So, after all the solution is unique.

If Cardinael's calculation is reformulated in exactly the same way but using parameters for the constants (see §3.2.2), this results in a formulation of Stewart's theorem. Given triangle ABC with $BD = a_1, DC = a_2, AD = d, BC = a$ and $AC : AB = R$ with $R^2 = 1 + r, AC = b$ and $AB = c$, it is required to know the length of DE expressed in a_1, a_2, a, d and R. The solution, following Cardinael's method in **Q33)** then becomes

$$DE = \frac{{a_2}^2 + d^2 - R^2({a_1}^2 + d^2)}{2 \cdot (a_1 R^2 + a_2)}$$

From this, together with

$$BE = a_1 + DE, CE = a_2 - DE, R = \frac{b}{c} \text{ and } AE^2 = d^2 - DE^2$$

the famous Stewart's theorem can be deduced (see §3.2.2 and [Molenbroek 1934])

$$ad^2 = a_1 b^2 + a_2 c^2 - a_1 a_2 a$$

Theory used

Answer (geometrical algebra)

Manipulation of rectangles and squares

Compare with

Q34;

Q68 for application of **Q33** and **Q34**;

[Molenbroek 1934], 183; 366;

[Sitters 2008] for more details

Question 34

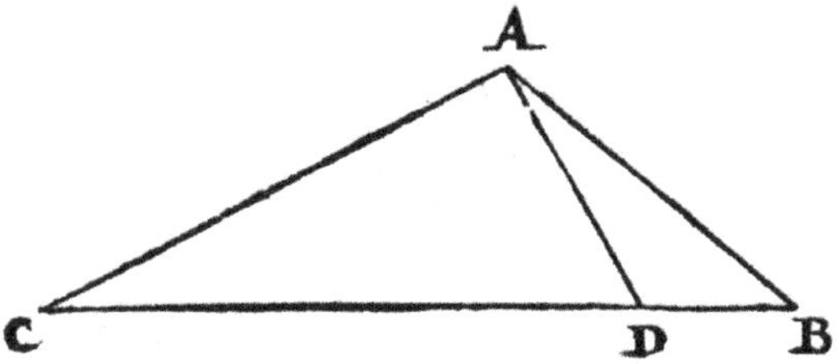

Figure 4.50: Subgroup 1, a triangle with its slanted sides and the ratio of the transversal to the segments of the base known

Comment on Answer

Cardinael exploits the property of a point multiplication of a triangle by a factor k where all mutual proportions remain equal. From **Q33** it is known for triangle ABC that

$$AB : AC = 25 : 33, BD : DC = 1 : 4\frac{19}{20}, BD : DA = 1 : 2\frac{1}{16}$$

and

$$AB = 218\frac{11}{13}, AC = 302\frac{1}{13} \text{ and } BD = 80$$

By multiplication with factor $k = \frac{1}{80}$, there results a triangle with with the same proportions but now with

$$AB = \frac{218\frac{11}{13}}{80} = 2\frac{179}{208} \text{ and } AC = 3\frac{807}{1040} \text{ and } BD = 1$$

By again multiplying this triangle by factor

$$\frac{25}{2\frac{179}{208}}(= \frac{33}{3\frac{807}{1040}}) = 8\frac{88}{119}$$

we get the required triangle with

$$AB = 25, AC = 33, BD = 8\frac{88}{119}$$

$$DC = 8\frac{88}{119} \cdot 4\frac{19}{20} = 43\frac{31}{119}, \; DA = 8\frac{88}{119} \cdot 2\frac{1}{16}$$

and in which also

$$BD : DC = 1 : 4\frac{19}{20} \text{ and } BD : DA = 1 : 2\frac{1}{16}$$

So

$$BC = BD + DC = 52$$

General comment

The calculation can be made more efficient by multiplying the triangle from **Q33** once with the factor

$$k = \frac{25}{218\frac{11}{13}} \left(= \frac{1}{80} \cdot \frac{25}{2\frac{179}{208}} \right)$$

as then arises a triangle ABC with $AB = 25$, while all proportions remaining equal, so we have

$$AC = 33, BD : DC = 1 : 4\frac{19}{20} \text{ and } BD : DA = 1 : 2\frac{1}{16}$$

as required.
That this solution also is unique can be deduced directly from the uniqueness of the solution of **Q33**, because two different solutions of **Q34** would by multiplication with the inverse of the factor lead to two different solutions of **Q33**.

An important conclusion from the two questions **Q33** and **Q34** is also that triangle ABC can be found with any given

$$AB, AC, BD : DC \text{ and } BD : DA$$

By supposing $BD = 1$, the lengths of DA and DC can be calculated and as the proportion $AB : AC$ is known by means of **Q33**, the sides AB and AC can be calculated for the case $BD = 1$. Because the real values of AB and AC are given, the multiplication factor is known to find the real lengths of BD, DA and DC.

Theory used

Answer (proportions, geometrical algebra)

Multiplication factor

Compare with

Q33;

Q68 for application of **Q33** and **Q34**

Question 35

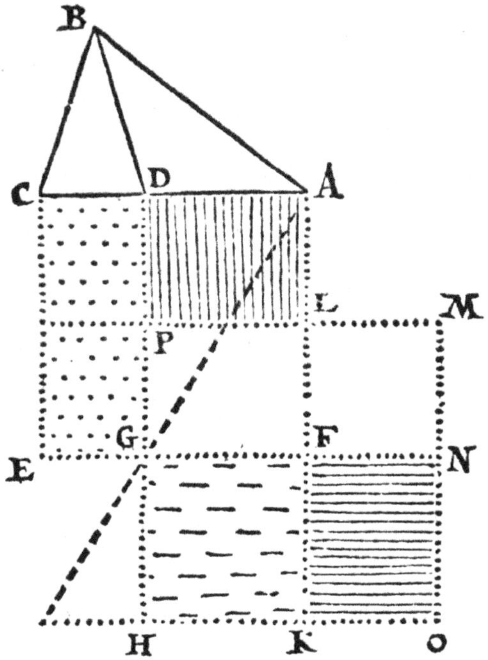

Figure 4.51: Subgroup 8, isosceles triangle with two angles of 72 degrees and the equal sides known

Comment on Answer

By constructing the bisector BD we arrive at two similar triangles BDC and ACB in which

$$\angle A = \angle ABD = \angle CBD = 36^\circ \text{ and } \angle C = 72^\circ$$

Therefore, BDC and ACB are similar. Because of this similarity, the proportion holds

$$CD : DB = BC : CA \Leftrightarrow CD : DA = DA : CA ...(1)$$

So AC is divided by D in extreme and mean ratio.

Let $ACEF$ be the square with area $EF^2 = 100$ and let G be on EF such that $GF = DA$ and $GH = GF$; P and L are the center points of DG and AF; M on the extension of PL such that $LF = LM$.

The configuration of squares and rectangles, which has arisen in this way and is shown in figure 4.51, can be interpreted as a special case of hyperbolic application of §3.1.5 case 10 (4), because AC is partitioned according to the Golden Section:

$$DC : AD = AD : AC$$

From the **Second consequence** of case 10, we have with $AC = 10$

$$AD = \frac{1}{2}AC \cdot (\sqrt{5} - 1) = 5(\sqrt{5} - 1) \text{ and } CD = \frac{1}{2}AC \cdot (3 - \sqrt{5}) = 5(3 - \sqrt{5})$$

from which follows

$$BC = AD = 5(\sqrt{5} - 1)$$

General comment

In the reverse way, it now becomes possible by means of the mean and extreme ratio (§3.1.5, case 10, **First consequence**) to construct angles of 36° and of 72° and so the construction of the pentagon and decagon become possible (**Q69**). Dividing a line segment AC in figure 4.51, by means of a point D according to the Golden Section (see §3.1.5, case 10, **First consequence**), a triangle ABC can be constructed, such that

$$DA = DB = BC$$

From the ratios $CD : DA = DA : CA$ follows that the triangles BDC and ACB are similar, because

$$CD : DB = BC : CA \text{ and } \angle BDC = \angle ACB$$

From their similarity further follows

$$\angle A = \angle DBC \text{ and } \angle A = \angle ABD \Leftrightarrow$$

$$BD \text{ is the bisector of } \angle ABC \Leftrightarrow$$

$$5 \cdot \angle A = 180 \Leftrightarrow$$

$$\angle A = 36^\circ \text{ and } \angle C = \angle B = 72^\circ$$

With the data of **Q35** not only the length of BC can be calculated but the line segment BC can also be constructed! The picture in **Q35** corresponds to the second picture in **Q25**, besides a mirroring in the line ML. In the same way is the picture of **Q35** in correspondence with figure 3.11, case 10, apart from a rotation over 90° to the left.

Theory used

Answer (proportions, hyperbolic identity)

§3.1.5 case 9: hyperbolic identity, §3.1.6 case 10

(special case: Golden Section)

Compare with

Q25: square $GPLI$ (from **Q25 Answer**) and square $FIHG$ (from **Q25 otherwise**) correspond to square $MPHO$ from **Q35**;

Q69

[Petri 1583], ex. 29, 30

[Brasser 1663], Questions 32, 33)

[Sitters 2008] §5.2.2.2: ex. 4, 11, 12, 13

Question 36

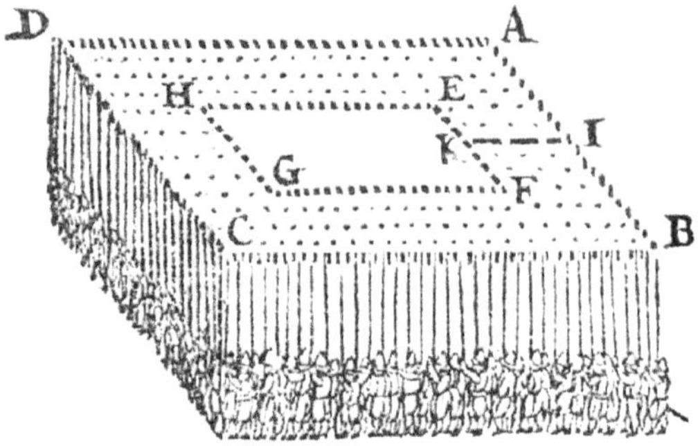

Figure 4.52: Subgroup 3, square battalion of musketeers encircled by lancers

Comment on Answer

Cardinael intends that the area of square $ABCD = 576$ and that the area of square $EFGH = 256$. Required is the width KI of the border area, which is the same width around the inner square. The outer square $ABCD$ contains $256 + 320 = 576$ soldiers. So the side of that square is $\sqrt{576} = 24$ soldiers long. The side of the inner square $EFGH$ is $\sqrt{256} = 16$ soldiers long. So there are $\frac{24-16}{2} = 4$ rows of lancers surrounding the musketeers.

Theory used

Answer (geometrical algebra)

Elementary manipulation of squares and rectangles

Compare with

Q37, 49, 50, 51

Question 37

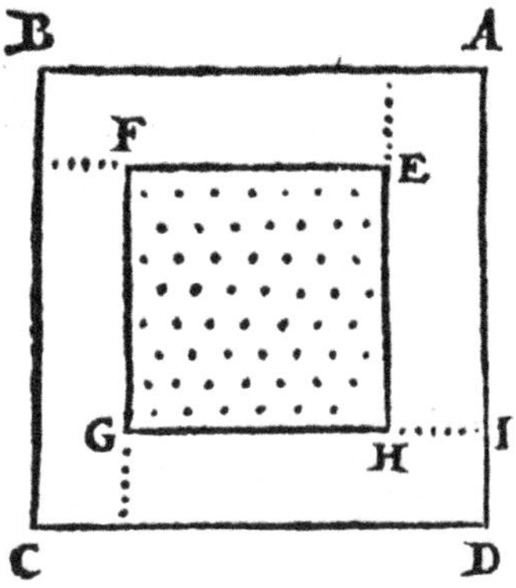

Figure 4.53: Subgroup 3, a squared battalion of 120 lancers covering and unknown number of musketeers

Comment on Answer

In geometrical terms, $HI = 3$ is the width of the border area and the area of square $ABCD-$ the area of square $EFGH = 120$. Required is the area of $EFGH$).
The area of lancers is divided into four equal rectangles each containing 30 lancers. As the breadth of these rectangles equals 3 (for example HI),the length must be 10 (for example AI). Then the length of the side of the square with musketeers is $10-3 = 7$(for example HE). The total number of marksmen therefore is 49.

Comment on Otherwise

The four squares on the four vertices together are 36. Subtracting this from 120 gives 84 for the sum of the four congruent rectangles all having breadth 3 and as length the side of the inner square (for example HE). Together they form a rectangle with one side 12, and the other side $\frac{84}{12} = 7$. So $HE = 7$ and the number of marksmen 49.

Theory used

Answer (geometrical algebra)

Elementary manipulation with squares and rectangles

Compare with

Q36, 49, 50, 51

Question 38

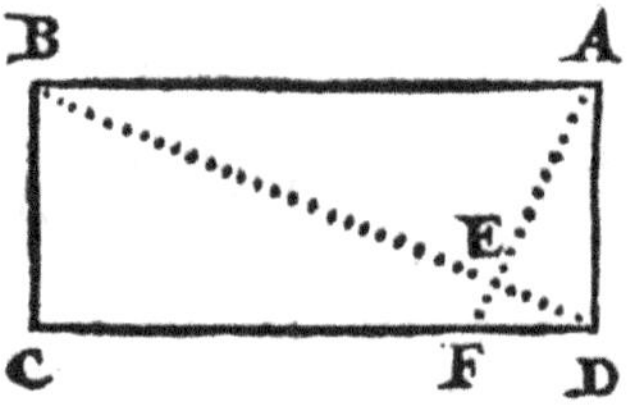

Figure 4.54: Subgroup 3, the area given of the rectangle on BE on the diagonal BD and of EF on the line AF perpendicular on BD

Comment on Answer

The solution depends on the iterative application of Elem.VI-8. The triangles FED, AED and AEB are similar, following the triplicate ratio ([Heath 1956], II, 114; 132)

$$EF : ED = ED : EA = EA : EB$$

Cardinael denotes EB as the third number of ED, meaning

$$EB = \left(\frac{ED}{EF}\right)^3 \cdot EF$$

Assuming that $EF = 1$ and $EB = 8$, instead of $EF = 2$ and $EB = 16$, the multiplication factor is $\frac{1}{2}$, one finds

$$8 = ED^3$$

so

$$ED = 2 \text{ and } EA = 4$$

The real values of the lengths thus are

$$ED = 4 \text{ and } EA = 8$$

So the length of diagonal BD is 20 and the area of $ABCD$ is

$$AE \cdot BD = 160$$

Comment on Otherwise

From

$$EF : ED = ED : EA = EA : EB$$

follows

$$EB \cdot EF = ED \cdot EA \text{ and } EA \cdot EF = ED^2$$

So

$$EB \cdot EF^2 = ED \cdot EA \cdot EF = ED \cdot ED^2 = ED^3$$

and

$$ED^3 = 16 \cdot 4 = 64 \Rightarrow ED = 4$$

From which follows

$$BD = 20 \text{ and } ABCD = 160$$

Theory used

Answer (proportions)

Similarity in right-triangles: Elem.VI-8

Triplicate ratio: [Heath 1956], II, 114; 132

[Dijksterhuis 1929-30], 82 – 83

Multiplication factor

Otherwise

Elem.VI-8;

Elem.VI-16 and 17

Compare with

Q39

Question 39

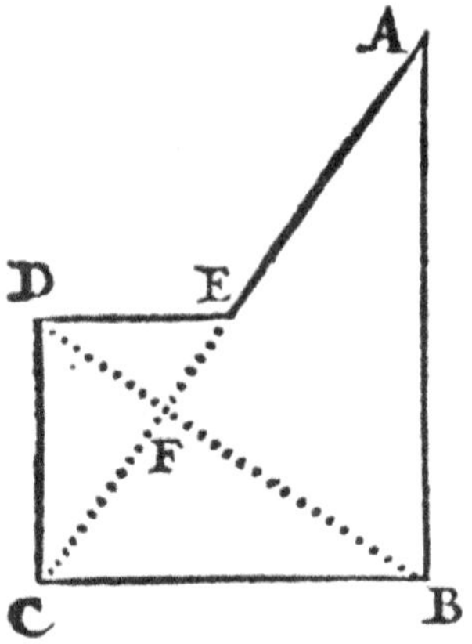

Figure 4.55: Subgroup 3, the area of a 'pentagon' calculated by the diagonal

Comment on Answer

$$FC^2 = FE \cdot FA$$

Without any explication, Cardinael states that FC is the geometrical mean of FE and FA. This indeed follows from the similarity of EFD, DFC, CFB and BFA from which follows the quadruplicate ratio

$$FE : FD = FD : FC = FC : FB = FB : FA$$

This means that FC is the geometrical mean between FE and FA ($FC^2 = FE \cdot FA$, as the quadruplicate ratio can be expressed in the form

$$(FD^2 = FE \cdot FC) \text{ and } (FC^2 = FD \cdot FB) \text{ and } (FB^2 = FC \cdot FA)$$

so

$$FC^2 \cdot FC^2 = (FD \cdot FB) \cdot (FD \cdot FB) = FD^2 \cdot FB^2$$
$$= (FE \cdot FC) \cdot (FC \cdot FA) = FC^2 \cdot (FE \cdot FA)$$

from which follows

$$FC^2 = FE \cdot FA$$

From $FE = 2$ and $AE = 30$ follows

$$FA = 32$$

so

$$FC^2 = 2 \cdot 32 = 64 \text{ and } FC = 8$$

Because FD is the geometrical mean of FE and FC (Elem.VI-17), we have:

$$FD^2 = FE \cdot FC = 2 \cdot 8 = 16$$

so

$$FD = 4$$

And for the same reason

$$FB = 16$$

The area of $AEDCB$ follows from

$$\frac{1}{2} \cdot FD \cdot EC + \frac{1}{2} FB \cdot AC = \frac{1}{2} \cdot 4 \cdot (2+8) + \frac{1}{2} \cdot 16 \cdot (32+8) = 20 + 320 = 340$$

Comment on Otherwise

Cardinael names, analogously to **Q38**, FA the fourth number in the progression of FD thus referring to the quadruplicate ratio (see **Answer**)

$$FE : FD = FD : FC = FC : FB = FB : FA$$

He then assumes, analogous to **Q38**,

$$FE = 1 \text{ and } FA = 16$$

so the multiplication factor is $\frac{1}{2}$ because the real values are 2 and 32. From this follows

$$FA = FD^4 \Leftrightarrow FD^4 = 16 \Leftrightarrow FD = 2$$

However, because of the used multiplication factor in reality

$$FD = 4$$

and so also FC, FB and FA are known and so is the answer.

Theory used

Answer (proportions)

Similarity in right-triangles: Elem.VI-8

Geometric Mean: Elem.VI-17

Otherwise (proportions)

Quadruplicate ratio (see **Q38**)

Multiplication factor

Compare with

Q38

Question 40

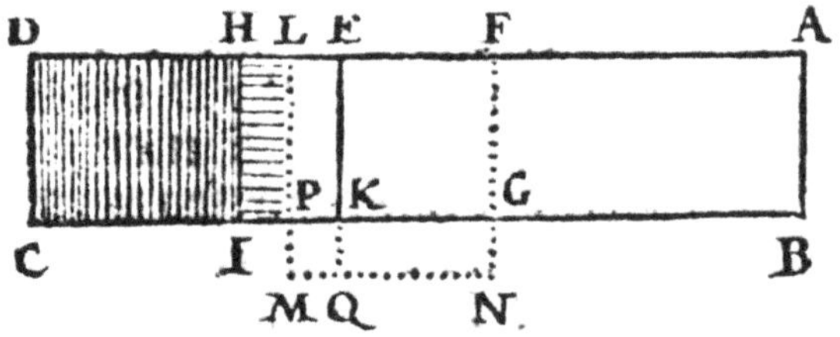

Figure 4.56: Subgroup 4, a rectangle with its area and the ratio between length and breadth known

Comment on Answer

Firstly, the geometrical configuration is given in which application of area as method of solution is possible. $FEKG$ is a square as $FE = \frac{1}{3} \cdot AE$ and $AE = 3 \cdot AB$. Then this configuration is interpreted as a hyperbolic application of area. Rectangle $FHIG$ is applied to $EH(= 4)$ exceeding by the square $FEKG$. This is exactly the situation of §3.1.5 case 9. Thus follows

$$FL^2 - EL^2 = FH \cdot FE \Leftrightarrow$$

$$FL^2 - 4 = FHIG \Leftrightarrow$$

$$FL^2 = 64 \Leftrightarrow$$

$$FL = 8 \text{ and } FE = 6$$

General comment

Algebraically one could pose $AB = x$ which results in the equation

$$x \cdot (3x + 12) = 180 \text{ corresponding with } AB \cdot AD = 180$$

The algebraic solution thus is analogous to the geometrical one

$$x \cdot (x + 4) = 60 :\leftrightarrow FHIG = \frac{1}{3} \cdot ABCD = 60$$

$$x \cdot (x + 4) + 4 = 64 \leftrightarrow FHIG = KPMQ = 60 + 4 = 64$$

$$(x + 2)^2 = 64 \leftrightarrow FLMN = 64$$

$$x + 2 = 8 \leftrightarrow FL = LM = \sqrt{64} = 8$$

$$x = 6 \leftrightarrow FL - EL = 8$$

thus

$$FE = 6$$

Theory used

Answer(geometrical algebra, hyperbolic application of area)

§3.1.5 case 9

Compare with

Q41, 42, 43, 44

Question 41

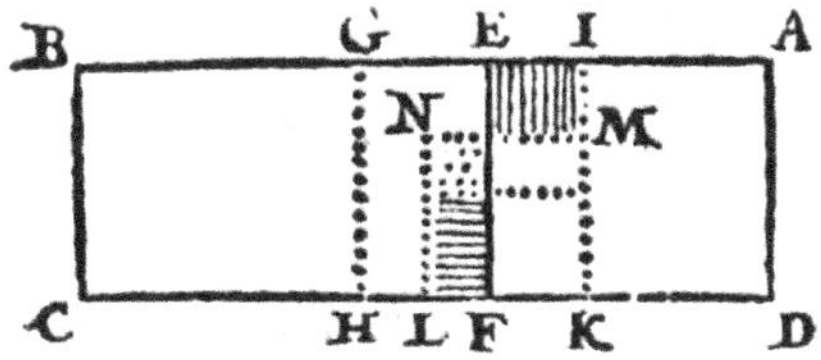

Figure 4.57: Subgroup 4, rectangle with the ratio of the sides and the ratio between area and length known

Comment on Answer

In figure 4.58 Cardinael assumes E from which follows

$$AEFD = 120$$

(2)...$EI = \frac{1}{3}EA$ and $GE = 6$, so $IGHK$ is a square with $IGHK = \frac{1}{3} \cdot 120 = 40$.
(3)...P is chosen such that $KP = FK$.
(4)...$IK = HK = FK + 6 = KP + 6 = KP + IP$, so $IP = 6$.
(5)...M is the center point of IP, so $IM = MP = 3$.
(6)...L is chosen such that $HL = LF = 3$.
(7)...N is chosen such that $NL = MK$.
(8)...S on FE such that $FS = KP$.

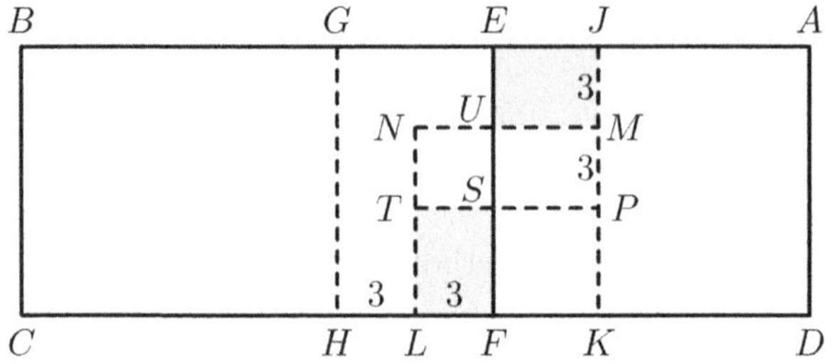

Figure 4.58: Solution to question 41

The solution can now be considered as an example of the hyperbolic identity, treated in §3.1.5 case 9. Rectangle $IEFK$ is applied to IP with quadratic excess $PSFK$. According to case 9 thus follows

$$KM^2 - PM^2 = KP \cdot KI \Leftrightarrow$$

$$KM^2 - 9 = IEFK \Leftrightarrow$$

$$KM^2 - 9 = 40 \Leftrightarrow$$

$$KM^2 = 49 \Leftrightarrow$$

$$KM = 7 \Leftrightarrow$$

$$KI = 10 \Leftrightarrow$$

$$AD = 10 \text{ and } AB = 30$$

General comment

Figure 4.58 demonstrates that it equally possible to use HP instead of rectangle $FI(= IEFK)$. In that case, the rectangle HP is applied to HF exceeding by the square $PSFK$ (the hyperbolic identity).

An algebraic solution would proceed as follows. Suppose $KM = x$, then

$$IEFK = 40 \Leftrightarrow$$

$$(x+3)(x-3) = 40 \Leftrightarrow$$

$$x^2 - 9 = 40 \Leftrightarrow$$

$$x^2 = 49 \Leftrightarrow$$

$$x = 7 \Leftrightarrow$$

$$KM = 7$$

$$KI = 10 \Leftrightarrow$$

$$AD = 10 \text{ and } AB = 30$$

Also here, there is a close correspondence between the geometrical and algebraic solution, requiring an equal amount of steps. The problem with the geometrical solution though is to find a suitable geometrical configuration, relating to the original geometrical formulation of the problem, so that the geometrical application – here application of area – can be used successfully. Cardinael often displays a great ingenuity in finding the optimal geometrical configuration. His solutions for the questions **Q1, 4, 5, 14, 40, 41, 42, 43, 48 50** and **51** are examples of such inventiveness.

Theory used

Answer (geometrical algebra, hyperbolic application)

§3.1.5 case 9

Compare with

Q40, 42, 43

Question 42

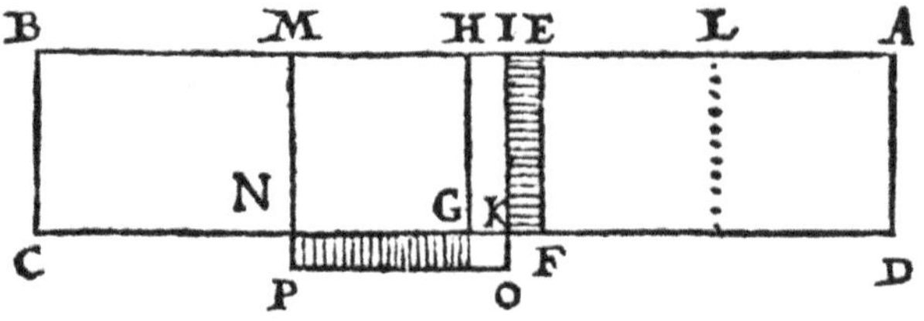

Figure 4.59: Subgroup 4, a rectangle with known length, breadth unknown, divided into a left part and a smaller right part

Comment on Answer

(1)...$BA = 40$ and $BE > EA$ so $BE > 20$.
(2)...EBCF= 192.
(3)...M is the center point of BE and L is the center point of EA, so $ML = 20$ and
$BE > EA \Rightarrow ME > EL \Rightarrow (ME > \frac{1}{2}ML$ and $EL < \frac{1}{2}ML)$.
(4)...$EMNF = 96$.
(5)...$EA = 2AD \Rightarrow EL = LA = AD$.
(6)...I is the center point of $ML \Rightarrow MI = IL = \frac{1}{2}ML = 10$.
(7)...$ME > MI$ and $EL < IL$ (follows from (3)), so
(8)...$(MI = IL$ and $EL = MN) \Rightarrow MI > MN$.
(9)...(H such that $MH = MN) \Rightarrow MH < MI$.
(10)...P and O such that $MP = MI = IO = 10$, so
(11)...$MN < MP$ and $IK < IO$ (K the intersection of IO and CD).
The perpendiculars from H and E on CD intersect CD in G and F.
$MH = MN \Rightarrow MH = EL \Rightarrow HI = IE = NP$.

It is possible to consider the solution as an example of the hyperbolic identity, treated in §3.1.5 case 9. Rectangle $MNFE$ is applied to HE with quadratic excess $MNGH$. Applying case 9 gives:

$$MI^2 - HI^2 = ME \cdot MH \Leftrightarrow$$

$$100 - HI^2 = EMNF \Leftrightarrow$$
$$100 - HI^2 = 96 \Leftrightarrow$$
$$HI = 2 \Leftrightarrow$$
$$AD = LE = LI - EI = 10 - 2 = 8$$

General comment

It is even possible to consider the solution as an example of the elliptic identity treated in §3.1.5 case 8. Rectangle $EMNF$ is applied to ML, falling short by the square LF. Applying case 8 gives:

$$LI^2 - EI^2 = EM \cdot EL \Leftrightarrow$$
$$100 - EI^2 = EMNF \Leftrightarrow$$
$$100 - EI^2 = 96 \Leftrightarrow$$
$$EI = 2 \Leftrightarrow$$
$$AD = LE = LI - EI = 10 - 2 = 8$$

Theory used

Answer (geometrical algebra, hyperbolic application of area)

§3.1.5 case 9, §3.1.5 case 8

Compare with

Q14: is a similar configuration. There is no evidence for Cardinael's choice for an elliptic or a hyperbolic identity. The calculation and the shading suggest the hyperbolic alternative

Q40, 41, 43;

Q51: here unequivocally the hyperbolic application is intended

Question 43

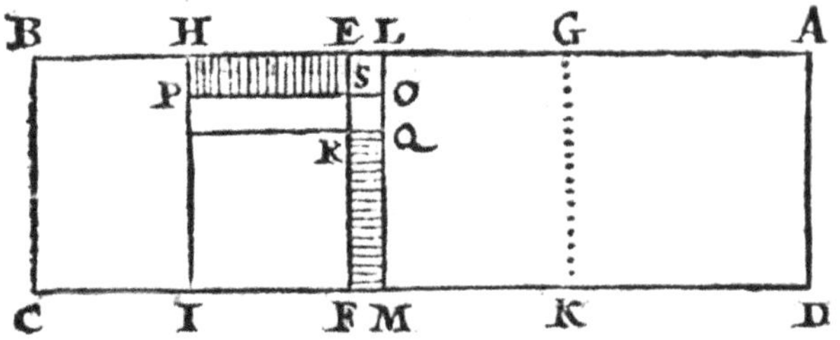

Figure 4.60: Subgroup 4, a rectangle with known length, breadth unknown, divided in left part and a larger right part

Comment on Answer

(1)...$AB = 40$ and $BE < EA$ so $BE < 20$.
(2)...$EBCF = 192$.
(3)...H is the center point of BE and G is the center point of EA, so $HG = 20$ and $BE < EA \Rightarrow HE < EG \Rightarrow (HE < \frac{1}{2}HG$ and $EG > \frac{1}{2}HG)$.
(4)...$EHIF = 96$.
(5)...$EA = 2AD \Rightarrow EG = GA = AD$.
(6)...L is the center point of $HG \Rightarrow HL = LG = \frac{1}{2}HG = 10$.
(7)...$HE < HL$ and $EG > LG$ (from 3), so
(8)...$HL = LG \Rightarrow HE < HL < EG$.
(9)...$EG = AD \Rightarrow HE < HL < AD$.
(10)...$HL - HE = EG - HL$.
(11)...T, P, R, S, Q and O are chosen on IH, FE and ML such that $IT = FR = MQ = HE$, $IP = FS = MO = HL = 10$ (so $IP^2 = 100$.)
From (10) and (11) follows that
(12)...$RTIF$, $MIPO$, $RQOS$ and $LESO$ are all squares with $RQOS = LESO$ because $HL - HE = AD - HL = ML - MO = MO - MQ$.

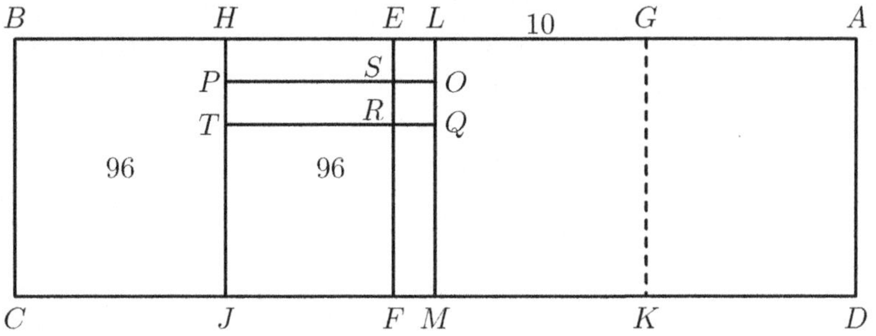

Figure 4.61: Completion of Cardinael's diagram of figure 4.60

Also now, the solution can be interpreted as a hyperbolic application of area as treated in §3.1.5 case 9. Rectangle $EHIF$ is applied to segment HT with quadratic excess the square $RTIF$. According to case 9 we have

$$IH \cdot IT = IP^2 - TP^2$$

$$96 = 100 - TP^2 \Leftrightarrow$$

$$TP = 2 \Leftrightarrow$$

$$AD = IH = IP + PH = IP + TP = 10 + 2 = 12$$

General comment

Also here, it is possible to consider the solution an example of the elliptic identity treated in §3.1.5 case 8. Rectangle $EHIF$ is then applied to segment HG falling short by the square $EGKF$. According to case 8 we have

$$HL^2 - EL^2 = EG \cdot EH \Leftrightarrow$$

$$100 - EL^2 = 96 \Leftrightarrow$$

$$EL = 2 \Leftrightarrow$$

$$AD = EG = HL + EL = 10 + 2 = 12$$

The use of the elliptic identity (§3.1.5 case 8) in the general comment of **Q42** and **Q43** yields in fact the two solutions of the same Question. These solutions are distinguished according to the condition $BE > EA$ in **Q42** and $BE < EA$

in **Q43**. Both Questions can be integrated as a single problem by means of **system I** (elliptic application of area) from §3.1.6 case 11, defined as

$$\textbf{system I}\ (EH, EG, 96, 20) = \begin{cases} EH \cdot EG = 96\ (= c) \\ EH + EG = 20\ (= b) \end{cases} \quad \text{with } c \leq \left(\frac{1}{2}b\right)^2$$

and with EH and EG as unknowns. Evidently, the condition $c \leq (\frac{1}{2}b)^2$ is satisfied as $96 < 10^2$.

Following the two solutions of case 11, derived by elliptic application, are found for **system I** $(EH, EG, 96, 20)$

Solution 1 (**Q42**, with the letters taken from **Q43**),
so $M \to H, I \to L, E = E, L \to G, N \to I, K \to M, F = F$:

$$EH > EG$$

$$EH = HL + EL \text{ and } EG = HL - EL \Leftrightarrow$$

$$EH = 10 + \sqrt{(\frac{1}{2}b)^2 - c} \text{ and } EG = 10 - \sqrt{(\frac{1}{2}b)^2 - c} \Leftrightarrow$$

$$EH = 10 = \sqrt{100 - 96} \text{ and } EG = 10 - \sqrt{100 - 96} \Leftrightarrow$$

$$EH = 12 \text{ and } EG = 8$$

Solution 2 (**Q43**: $EH < EG$)

$$EH = HL - EL \text{ and } EG = HL + EL \Leftrightarrow$$

$$EH = 10 - 2 = 8 \text{ and } EG = 10 + 2 = 12$$

Hence, this demonstrates how both Questions could have been treated as one single Question in which the point E on AB is a variable point.

Theory used

Answer (geometrical algebra, application of area)

§3.1.5 case 9, §3.1.5 case 8 (hyperbolic and elliptic identities)and §3.1.6 case 11 (elliptic application of area)

Compare with

Q40, 41, 42

Question 44

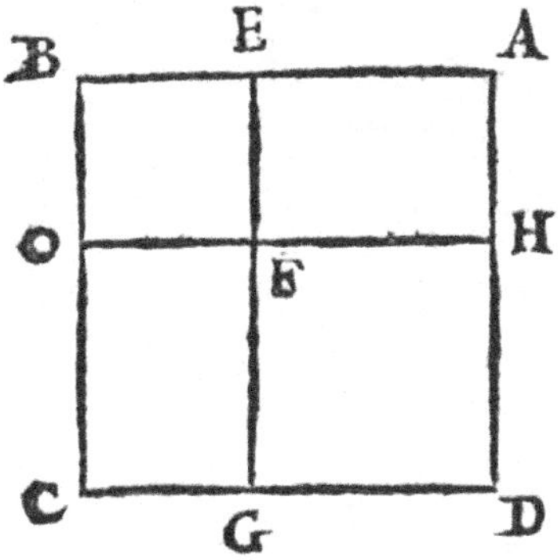

Figure 4.62: Subgroup 4, complement within a square of two connecting squares on the diagonal of that square

Comment on Answer

The equality $AEFH = FOCG$ in figure 4.62, follows from Elem.I-43, as

$$AEFH = 1200 \text{ and } AE + EF = 70 \Leftrightarrow$$

$$EB \cdot EA = 1200 \text{ and } EB + EA = 70$$

system I (see §3.1.6 case 11) can be applied as follows

$$\textbf{system I } (EB, EA, 1200, 70) = \begin{cases} EB \cdot EA = 1200 \ (= c) \\ EB + EA = 70 \ (= b) \end{cases} \quad \text{with } c \leq \left(\frac{1}{2}b\right)^2$$

and with EB and EA as unknowns.
The condition $c \leq (\frac{1}{2}b)^2$ is satisfied, because $1200 \leq 35^2$.

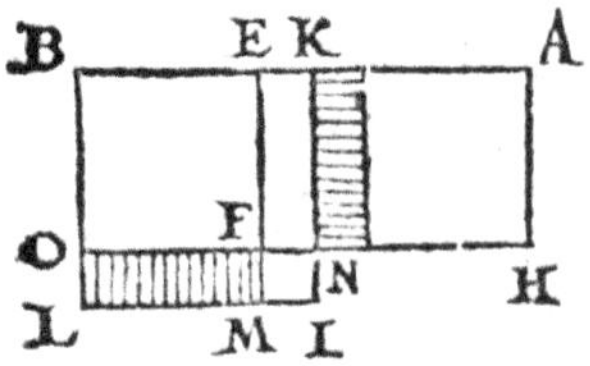

Figure 4.63: Question 44: solution

So according to case 11, using the elliptic application (see figure 4.61 and figure 4.63 in which the drawing of the square $KBLI$ illustrates the use of the elliptic identity) we have the two solutions

Solution 1 ($EB < BK$)

$$EB = BK - EK \text{ and } EA = BK + EK \Leftrightarrow$$

$$EB = 35 - \sqrt{35^2 - 1200} \text{ and } EA = 35 + \sqrt{35^2 - 1200} \Leftrightarrow$$

$$EB = 35 - 5 = 30 \text{ and } EA = 35 + 5 = 40$$

Solution 2 ($EB > BK$)

$$EB = BK + EK \quad \text{and } EA = BK - EK \Leftrightarrow$$

$$EB = 40 \text{ and } EA = 30$$

The second solution, $EB = 40$, as usual is not mentioned by Cardinael, though there remains a principal difference between the solutions

$$EBOF = 900 \text{ and } HFGD = 1600$$

and

$$EBOF = 1600 \text{ and } HFGD = 900$$

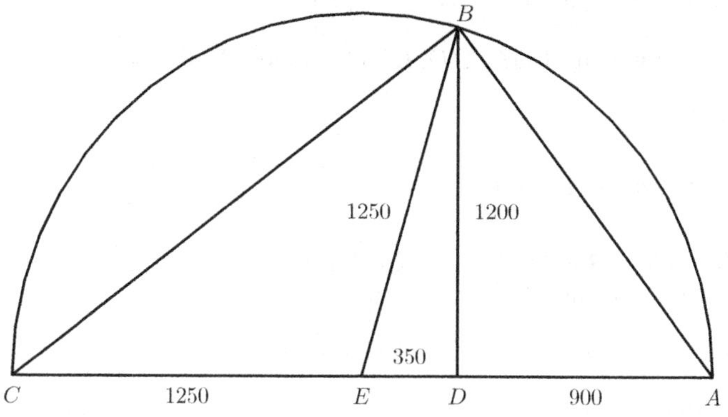

Figure 4.64: Solution to problem 44

Comment on Otherwise

As shown in figures 4.62 and 4.61, for the squares $EBOF$ and $HFGD$ we have

$$EBOF + HFGD = 2500$$

and for the rectangles $AEFH$ and $FOCG$ we have

$$AEFH = FOCG \Rightarrow 2AEFH = 4900 - 2500 = 2400 \Rightarrow AEFH = 1200$$

According to figure 4.62

$$EBOF : AEFH = AEFH : HFGD \Leftrightarrow$$

since $AEFH$ is the mean ratio between $EBOF$ and $HFGD$

$$EBOF \cdot HFGD = 1200^2$$

So the squares $HFGD$ and $EBOF$ meet the requirements of **system I**

$$\textbf{system I}\ (HFGD, EBOF, 1200^2, 2500) = \begin{cases} HFGD \cdot EBOF = 1200^2\ (= c) \\ HFGD + EBOF = 2500\ (= b) \end{cases}$$

with $HFGD$ and $EBOF$ as unknowns while the condition $c \leq (\frac{1}{2}b)^2$
is satisfied because $1200^2 \leq 1250^2$.

By renaming $HFGD = DC$ and $EBOF = DA$

$$\textbf{system I } (HFGD, EBDF, 1200^2, 2500)$$

can be formulated as

$$\textbf{system I } (DC, DA, 1200^2, 2500) = \begin{cases} DC \cdot DA = 1200^2 \; (= c) \\ DC + DA = 2500 \; (= b) \end{cases}$$

By solving this **system I** in the way of **Q5 Answer** as suggested by Cardinael, it becomes transformed into the equivalent

$$\textbf{AH-system} \quad (BD, CA, 1200, 2500)$$

as is demonstrated in §3.1.8 (figure 3.18. Here Cardinael's method is shown with a right-angled triangle CAB with right angle B and altitude $BD = \sqrt{c}$ and $CA = b$ as the hypothenuse in figure 4.64.

According to the equations §3.1.8 (3.17,3.18 we can write the two solutions as

Solution 1 $(DA > DA)$

$$DC = 1250 + \sqrt{1250^2 - 1200^2} \text{ and } DA = 1250 - \sqrt{1250^2 - 1200^2} \Leftrightarrow$$

$$DC = 1250 + \sqrt{122500} \text{ and } DA = 1250 - \sqrt{122500} \Leftrightarrow$$

$$DC = HFGD = 1600 \text{ and } DA = EBOF = 900$$

Solution 2 $(DC < DA)$

$$DC = 1250 - \sqrt{1250^2 - 1200^2} \text{ and } DA = 1250 + \sqrt{1250^2 - 1200^2} \Leftrightarrow$$

$$DC = HFGD = 900 \text{ and } DA = EBDF = 1600$$

Theory used

Answer (geometrical algebra, elliptic application of area)

Elem.I-43;

§3.1.6 case 11

Otherwise (geometrical algebra, Thales , Pythagoras)

3.1.8 case 11;

§3.1.8 case 16 (3.16)(3.16), (3.17 3.17, (3.18 3.18)

Q5

Compare with

Q14.

Question 45

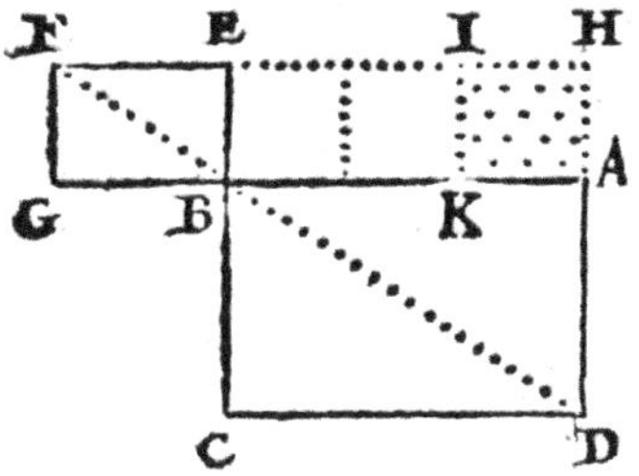

Figure 4.65: Subgroup 5, similar triangles contained in one rectangle

Comment on Answer

In rectangle FD we have

$$\text{rectangle } GC = ABEH$$

$$EBGF : ABEH = BG : AB$$

$$\text{rectangle } GC : ABCD = BG : AB$$

From this follows

$$EBGF : ABEH = \text{rectangle } GC : ABCD \Leftrightarrow$$

$$EBGF : ABEH = ABEH : ABCD$$

so $ABEH$ is the geometrical mean of $EBGF$ ($= 24$) and $ABCD$ ($= 96$). However it is known that

$$24 : 48 = 48 : 96$$

so the geometrical mean of 24 and 96 is 48. From this it can be concluded

$$ABEH = 48$$

I is chosen such that $HI = \frac{1}{3}AB$, so from $BE = \frac{1}{3}AB$ follows

$$HI = BE = AH$$

and therefore

$$\text{square } HIKA = \frac{1}{3}ABEH = 16 \Leftrightarrow$$

$$AH^2 = 16 \Leftrightarrow AH = 4 \Leftrightarrow \text{ (with } ABEH = 48)$$

$$AB = 12 \text{ and } BE = 4 \Leftrightarrow \text{ (with } ABCD = 96 \text{ and } EBGF = 24)$$

$$BC = 8 \text{ and } BG = 6$$

General comment

That $ABEH$, as the geometrical mean of $EBGF$ and $ABCD$, was also used in **Q14**.

Theory used

Answer (geometrical algebra, proportions)

§3.1.2 c: construction and calculation of the geometric mean

Compare with

Q14 Answer and **Q44 Otherwise**

Question 46

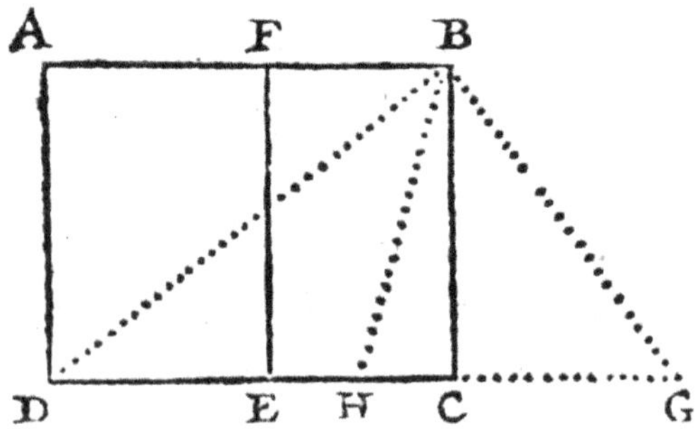

Figure 4.66: Subgroup 5, a rectangle with the same ratio as a truncated segment

Comment on Answer

From the data follows

$$DE : EF = CB : CD$$

By choosing CG equal to DE the ratio holds

$$CG : CB = DE : EF = CB : CD$$

So triangle GCB is similar to triangle BCD and angle DBG is a right angle (inversion of Elem.VI-8). If H is the center point of EC then H is also the center point of DG and because DBG is a right angle we have

$$DH = BH = GH \text{ (Elem.IV-5 and III-31)}$$

From $FBCE = 336$ and $AD = 24$ follows

$$EC = \frac{336}{24} = 14 \text{ and } EH = CH = 7$$

For triangle BCH holds

$$BH = \sqrt{BC^2 + CH^2} = \sqrt{625} = 25$$

so

$$HD = 25 \leftrightarrow$$

$$AB = DC = 25 + 7 = 32$$

Used

Answer (proportions, Pythagoras , Thales)

Inversion of Elem.VI-8 $BC^2 = GC \cdot DC \Leftrightarrow$ angle DBG is right

Elem.IV-5 and III-31

Question 47

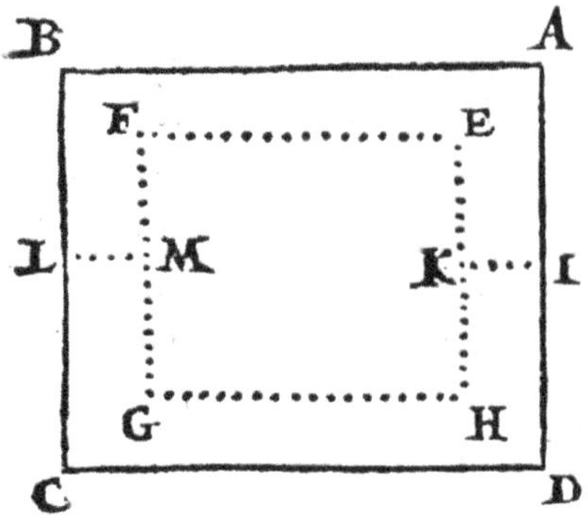

Figure 4.67: Subgroup 5, a right-angled garden in which is contained a similar flower-bed

Comment on Answer

$ABCD$ and $EFGH$ are similar, so for any given k holds that
$ABCD = k^2 \cdot EFGH$

so

$$k^2 = \frac{180}{80} \Rightarrow k^2 = 2\frac{1}{4} \Rightarrow k = 1\frac{1}{2}$$

By means of of Elem.V-17

$$1 : 1\frac{1}{2} = HG : (HG + 5) \Leftrightarrow \frac{1}{2} : 5 = 1 : HG \tag{4.19}$$

and in the same way with $HG + 5 = DC$

$$1 : 1\frac{1}{2} = HG : (HG + 5) \Leftrightarrow \frac{1}{2} : 5 = 1\frac{1}{2} : DC \tag{4.20}$$

From (4.19) follows $HG = 10$, so

$$EH = \frac{80}{10} = 8$$

and from (4.20) follows $AB = DC = 15$ and so $BC = 12$

Theory used

Answer(proportions)

Multiplication factor

Probably Elem.V-17

Question 48

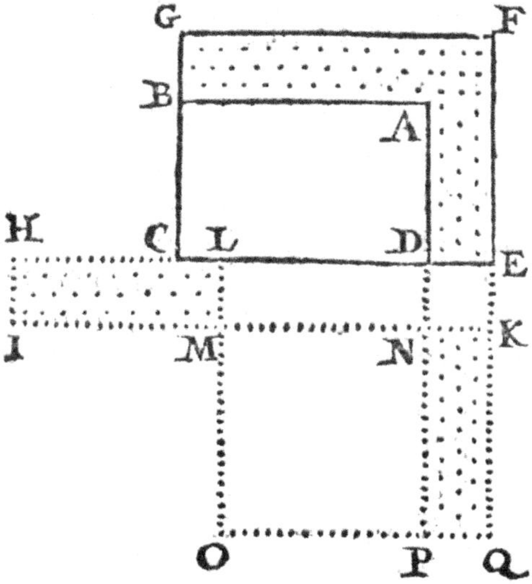

Figure 4.68: Subgroup 4, a farmland with a ditch of prescribed area

Comment on Answer

By re-arranging the rectangles through the linear concatenation of the border areas there arises a new rectangle $EHIK$, with unknown breadth EK, of which the area is known, as $EHIK$ equals the area of the required ditch, thus 120. $EHIK$ can be considered a hyperbolic application to HD exceeding by the square $EKND$. According to Elem.II-6 now we have:
$LE^2 - LD^2 = EHIK$ and $LE^2 - LD^2 = 120$ as $EHIK = 120$.
With $LD = 13$ follows $LE^2 = 289$ and so $LE = \sqrt{289} = 17$ and $ED = BG = LE - LD = 17 - 13 = 4$

General comment

This solution is the geometrical equivalent with the algebraic solution of the quadratic equation $x \cdot (26 + x) = 120$. By posing $BG = x$, the following correspondences hold

$$
\begin{array}{lll}
DE \cdot HE = 120 & \leftrightarrow & x \cdot (26 + x) = 120 \\
DE \cdot (26 + DE) = 120 & \leftrightarrow & x \cdot (26 + x) = 120 \\
LE^2 - LD^2 = 120 & \leftrightarrow & (x + 13)^2 - 13^2 = 120 \text{ (cutting off a square)} \\
LE^2 = 13^2 + 120 = 289 & \leftrightarrow & (x + 13)^2 = 289 \\
LE = 17 & \leftrightarrow & x + 13 = 17 \\
DE = 4 & \leftrightarrow & x = 4.
\end{array}
$$

The geometrical configuration in which the quadratic equation

$$DE \cdot (26 + DE) = 120$$

can be solved by a hyperbolic application of area is a natural interpretation of the given problem.

Despite the great similarity between the two solutions there are some essential differences. For many of the mathematicians at the beginning of the seventeenth century the geometrical interpretation had a more demonstrative validity than the algebraic one. In their epistemological considerations of what constituted a legitimate proof, Cardinael represented the more conservative group.

Theory used

Answer (geometrical algebra, hyperbolic application)

§3.1.6 case 12

Question 49

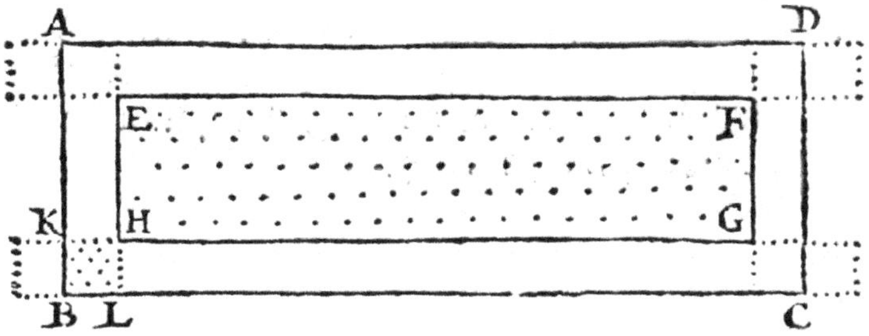

Figure 4.69: Subgroup 3, an unknown number of lancers cover a known number of musketeers

Comment on Answer

The Question can be formulated geometrically as follows. Given a rectangle $ABCD$ with $AD = 3AB$. If inside $ABCD$, a border area of constant breadth 4 is constructed, then the area of this border is 576. The Question is to calculate the area of the remaining rectangle $EFGH$.
If the border area, horizontally with length AD and vertically with length AB are subsequently laid down in a single row, there arises a rectangle with length

$$AD + DC + BC + AB = 2 \cdot BC + 2 \cdot AB$$

and breadth 4. The total area of this rectangle thus is

$$576 + 4 \cdot 16 = 640$$

The term $4 \cdot 16$ arises from counting the 4 vertices twice. Cardinael denotes the double-counted square in the picture as dotted areas. So

$$(2 \cdot AB + 2 \cdot BC) = 640 \Leftrightarrow AB + BC = 80$$

As $BC = 3 \cdot AB$

$$BC = 60 \text{ and } AB = 20$$

So

$$ABCD = 1200$$

and therefore there were

$$1200 - 576 = 624 \text{ lancers.}$$

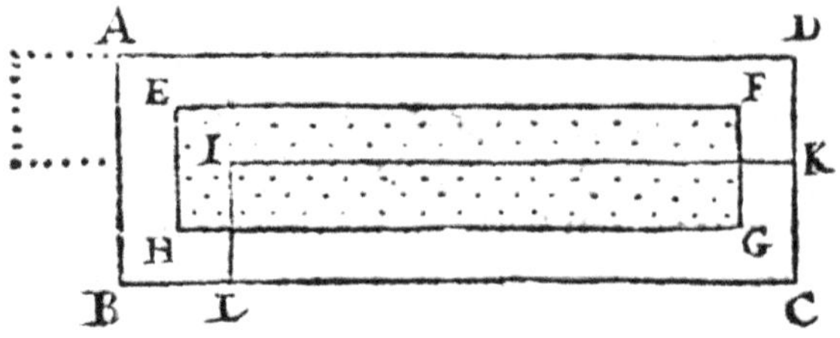

Figure 4.70: Question 49: alternative solution

Comment on Otherwise

In figure 4.70, $EFGH$ is shifted so that G meets with C and H with L. So now

$$BL = DK = 8$$

and also

$$ABCD - EFGH = 576$$

If the two border areas are placed behind each other without overlap, then there arises a rectangle with D as right vertex and the dotted strip at the left side, having area 576 and breadth 8. So the length is $\frac{576}{8} = 72$ and

$$AD + IL = 72$$

So

$$AD + AB = AD + IL + DK = 72 + 8 = 80$$

such that

$$4 \cdot AB = 80 \Leftrightarrow AB = 20 \text{ because } AD = 3 \cdot AB$$

From this follows

$$AD = 60 \text{ and } ABCD = 1200$$

General comment

If this problem would have been solved algebraically by posing $DC = x$ there arises the linear equation

$$64 + 2 \cdot (x - 8) \cdot 4 + 2 \cdot (3x - 8) \cdot 4 = 576$$

This implies a solution by means of geometrical algebra without application of area. The quadratic case occurs in the comparable **Q50**.

Theory used

Answer (geometrical algebra)

Ordinary elementary application of area

Compare with

Q50, 51

Question 50

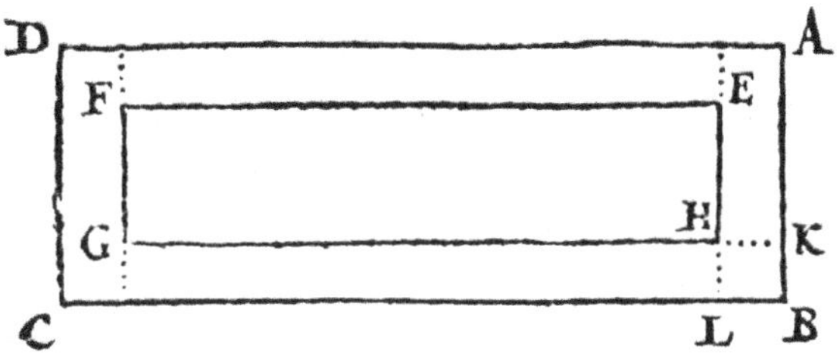

Figure 4.71: Subgroup 3, a known number of musketeers covered by an unknown number of lancers

Comment on Answer

The Question can be geometrically formulated as follows. Given a rectangle $ABCD$ with $AD = 3AB$. If a border area of constant breadth 4 is cut away, there remains a rectangle $EFGH$ with area 624. The Question is to calculate that border area.

Multiplying, $EFGH$ in horizontal direction by factor $\frac{1}{3}$, there arises $EHNM$ in figure 4.72 with area $\frac{1}{3} \cdot 624 = 208$. Though, Cardinael fails to mention that the multiplication takes place in horizontal direction, it is clearly shown in figures 4.71 and 4.72. Now by multiplying $ABCD$ in figure 4.71 also in horizontal direction with $\frac{1}{3}$, there arises $QPON$ as in figure 4.72.

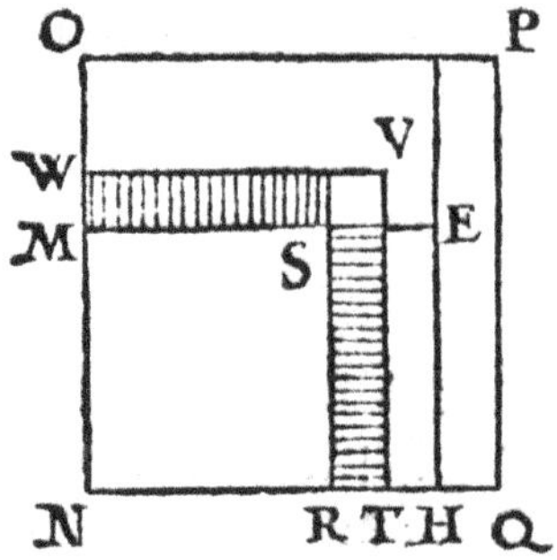

Figure 4.72: Question 50: solution

Let both figures come together in N so that NM is on NO.
Then $MO = 8$, as in vertical direction there was no multiplication and $QH = \frac{1}{3} \cdot 8 = 2\frac{2}{3}$, because MO is twice BK or HL.
Consider the square $RNMS$, then $HR = 5\frac{1}{3}$, because $OM = QR = 8$.
Choose T as the center point of RH, then

$$TH = TR = \frac{1}{2} \cdot 5\frac{1}{3} = 2\frac{2}{3}$$

and

$$TR^2 = 7\frac{1}{9}$$

so

$$EHNM + TR^2 = 208 + 7\frac{1}{9} = 215\frac{1}{9}$$

and

$$\text{square } TVWN = 215\frac{1}{9}$$

and

$$TN^2 = 215\frac{1}{9} \text{ and } TN = 14\frac{2}{3}$$

From

$$TN + TH + HQ = 14\frac{2}{3} + 2\frac{2}{3} + 2\frac{2}{3} = 20$$

follows

$$QN = 20, AB = 20 \text{ and } BC = 60$$

and therefore

$$ABCD = 1200 \text{ and the number of musketeers equals } 576$$

General comment

As opposed to **Q49** here the problems is a quadratic one. Assuming $AB = x$ the problem leads to the quadratic equation

$$(x - 8) \cdot (3x - 8) = 624$$

The horizontal multiplication with the factor $\frac{1}{3}$ is the intended configuration for the application of area as the solution method. The result of the multiplication is that $QPON$ becomes a square as in figure 4.72.
As $RNMS$ is also a square, it follows that

$$OM = QR$$

so

$$QR = 8$$

Here the method of §3.1.5 case 9 is used. Rectangle $EHNM$ (=208) is considered a hyperbolic application to the line segment of difference

$$HR(= 5\frac{1}{3})$$

exceeding by the square $RNMS$. According to §3.1.5 case 9 then

$$EHNM = TVWN - TR^2 \Leftrightarrow TVWN = 215\frac{1}{9} \Leftrightarrow TN = 14\frac{2}{3}$$

and so NQ and AB can be calculated.

Theory used

Answer (geometrical algebra, hyperbolic application of area)

§3.1.5 case 9

Compare with

Q49, 51

Question 51

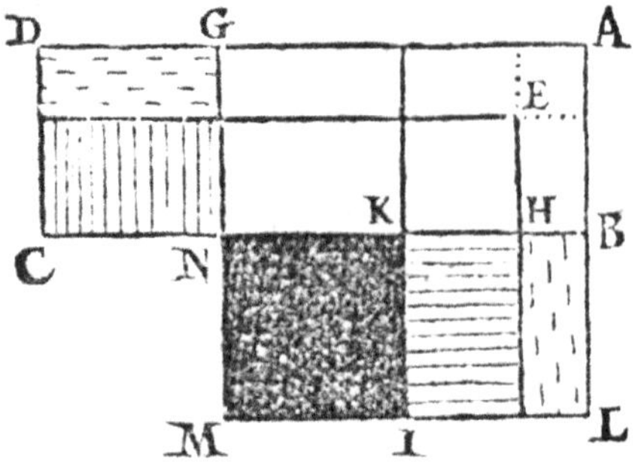

Figure 4.73: Subgroup 3: a given number of lancers covered by four rows of a known number of musketeers, the ratio of the sides of the rectangle unknown

Comment on Answer

The problem can be geometrically formulated as the translation of the inner rectangle to the vertex C. As in **Q49** otherwise, inside of rectangle $ABCD$ there is an extension along the succeeding sides AD and AB with constant breadth 8, so $BH = 8$. The area of the rectangle with diagonal CE is 624 and hence extension's breadth is 576. Required is the length of AB and AD.

The point K is selected so that $BK = AB$ and CK is the line segment of difference between AD and AB

$$CK = AD - AB$$

The center point of CK is N. Cardinael follows the method of §3.1.9 case 17 (ii) and **Q14 Otherwise**. As in **Q14** also here the sum and product of the two connected sides of rectangle $ABCD$ are known.
From the known values follows directly

$$AB \cdot AD = 576 + 624 = 1200$$

and

$$(AB + AD) \cdot 8 = 576 + 8^2 = 640 \Leftrightarrow$$

$$AB + AD = 80 \text{ (see } \mathbf{Q49}\text{)}$$

L is selected such that $BL = CN = \frac{1}{2}CK = DG$. Now we have

$$AG = NK + BK = \frac{1}{2}CK + AB$$

and

$$AL = AB + BL = AB + \frac{1}{2}CK$$

from which follows

$$AG = AL$$

But also

$$AG + AL = AD + AB = 80$$

so

$$2AG = 80 \Leftrightarrow$$
$$AG = 40 = BN$$

Now $ABCD$ can be interpreted as the hyperbolic application to CK exceeding by the square BK^2. According to §3.1.5 case 9 we have

$$BN^2 - KN^2 = BC \cdot BK \Leftrightarrow$$
$$1600 - KN^2 = 1200 \Leftrightarrow$$
$$KN^2 = 400 \Leftrightarrow$$
$$KN = 20$$

so

$$BK = BN - KN = 40 - 20 = 20$$

and therefore also

$$AB = BK = 20$$

and

$$AD = 80 - BK = 60$$

General comment

$$\text{The } \textbf{system I} = \begin{cases} AB \cdot AD = 1200 \\ AB + AD = 80 \end{cases}$$

is directly solved by hyperbolic application as described in §3.1.9 case 17 (ii) and in **Q14**.
The second solution, $AB = 60$ and $AD = 20$ is found by applying $ABCD$ to the line segment of the difference $AB - AD$ (see §3.1.9 case 17 (iiii)). As usual Cardinael does not mention the possibility of this solution.

Theory used

Answer (hyperbolic application)

§3.1.9 case 17 (*ii*)

Compare with

Q14 Otherwise

Q49, 50

Question 52

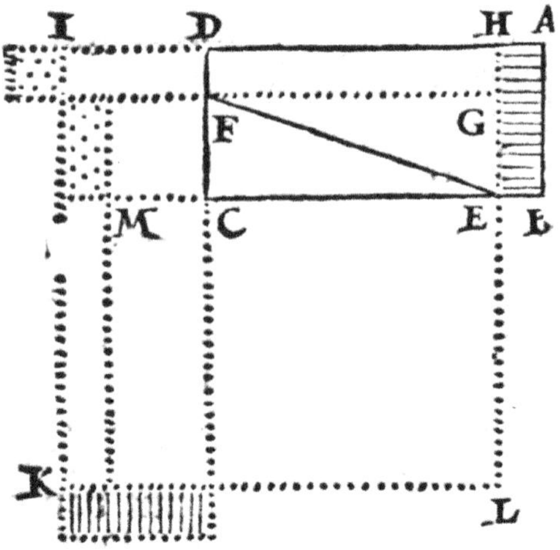

Figure 4.74: subgroup 3, a rectangle divided by a ditch

Comment on Answer

Rectangle $2 \cdot ABCD$ is divided symmetrically and appended to the two squares FC^2 and EC^2 such that a new square $HIKL$ is constructed. The shaded rectangle at vertex K is transferred to the rectangle along ID and the shaded rectangle $ABEH$ is divided over the two small dotted rectangles at I and M. The small square near I with area 4 remains. Therefore $HIKL = 365 - 4 = 361$.

Theory used

Answer (geometrical algebra)

Q14

Compare with

Q29 is just as **Q52** a generalisation of **Q14**

Question 53

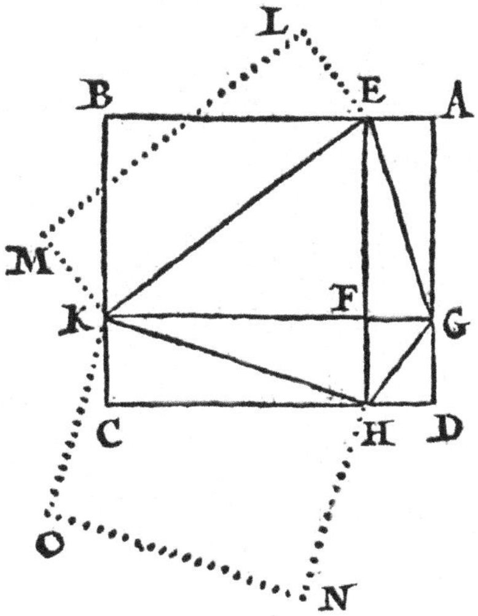

Figure 4.75: Subgroup 3, a circumscribed rectangle of a quadrangle, with their diagonals perpendicular to each other

Comment on Answer

The solution is based on the iterated application of Elem.VI-31. From the given ratio

$$FG : FH = FE : FK \text{ and so } FG : FE = FH : FK$$

follows

$$AEFG \approx FKCH \text{ and } GFHD \approx EBKF (\approx : \text{ similarity of figures})$$

By applying Elem.VI-31 to the rectangles $AEFG$ and $FKCH$ on the triangles EFK and GFH, follows for the sum $EKML$

$$EL = GH = 5$$

because $EKLM$ = rectangle on HG with $HG = 5$, so $EL = 5$ and area $EKLM = 5 \cdot 15 = 75$. Applying Elem.VI-31 to the rectangles $GFHD$ and $EBKF$ on the triangles HFK and GFE results for the sum $HKON$

$$HN = EG = \sqrt{90}$$

as $HKON$ = rectangle on EG with $EG = \sqrt{90}$, so is

$$HN = \sqrt{90} \text{ and area } HKON = \sqrt{90} \cdot \sqrt{160} = 120$$

From this follows the area of $ABCD$

$$ABCD = EKML + HKON = 75 + 120 = 195$$

The lengths of AB and BC can now be calculated. Firstly, the areas of $GFHD$ and $EBKF$, my means of their known proportion $5^2 : 15^2$ and hence in the same manner the areas of $AEFG$ and $FKCH$, by means of their proportion $(\sqrt{90})^2) : (\sqrt{160})^2$. Assuming that $FG = 1$ a straightforward calculation leads to the multiplication factor and so to the true lengths of AB and AD.

Comment on Otherwise

From the result $ABCD = 195$ obtained above, AB and AD can be calculated by an alternative method in directly posing $FG = 1$ and calculating $FE = 3$ by means of $FGH \approx FEK$ and calculating $FH = \frac{4}{3}$ and $FK = 4$ by $FGE \approx FHK$. This results in $EH = 4\frac{1}{3}$ and $GK = 5$ which gives

$$\text{area } ABCD = 21\frac{2}{3}$$

From this follows another multiplication factor $\frac{195}{21\frac{2}{3}}$ for the areas (squares) and so the actual values of EH and GK are found as

$$EH = 4\frac{1}{3} \cdot \sqrt{\frac{195}{21\frac{2}{3}}} = 13 \text{ and } GK = 5 \cdot \sqrt{\frac{195}{21\frac{2}{3}}} = 15$$

Theory used

Answer (geometrical algebra, similarity of polygons, and a generalization of the Pythagorean theorem)

Elem.VI-31

Multiplication factor and Regula Falsi

Otherwise

(Same as under Answer)

Question 54

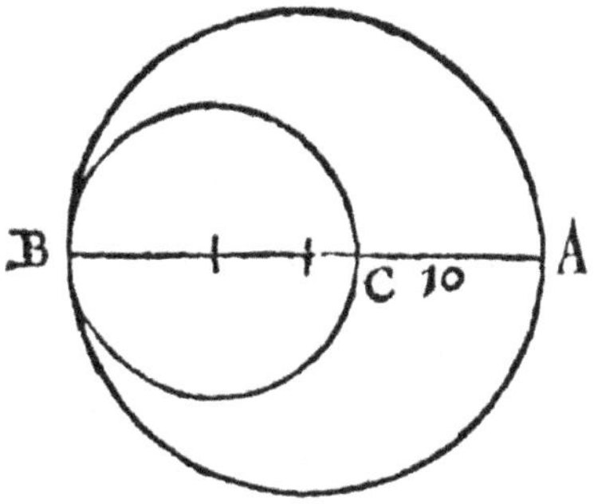

Figure 4.76: Subgroup 6, two circles with their difference in area and difference in diameters given

Comment on Answer

Cardinael consistently adopts the rational number $\frac{22}{7}$ for π. In figure 4.77, DE is the radius of the larger circle and EF the circumference, so

$$DE : EF = 1 : (2 \cdot \frac{22}{7})$$

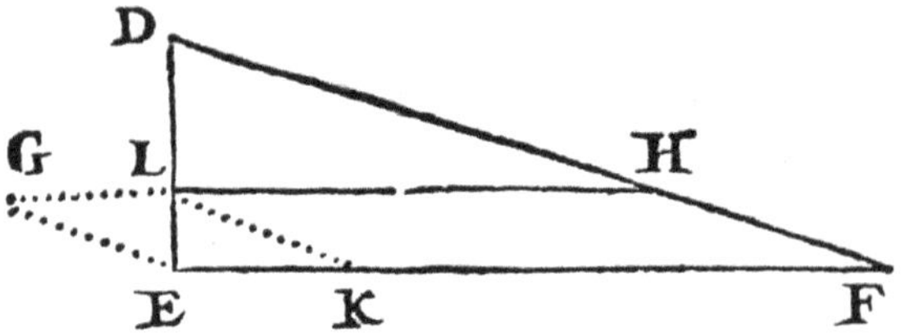

Figure 4.77: Question 54: solution

From $LE = 5$ follows that DL is the radius of the smaller circle, so that

$$DEF = \frac{1}{2} \cdot DE \cdot EF = \frac{1}{2} \cdot DE \cdot 2 \cdot \frac{22}{7} \cdot DE = \frac{22}{7} \cdot DE^2 = \text{ area big circle}$$

and

$$DLH = \text{ area small circle}$$

and therefore

$$ELHF = 361\frac{3}{7}$$

From $\frac{EF}{DE} = 2 \cdot \frac{22}{7}$ and therefore also $EF = \frac{44}{7} \cdot DE$, follows

$$EK = \frac{44}{7} \cdot LE = \frac{44}{7} \cdot 5 = \frac{220}{7} \Leftrightarrow$$

$$LEK = \frac{220}{7} \cdot \frac{5}{2} = \frac{550}{7}$$

and because ELG is constructed as a triangle congruent to LEK, we have

$$ELG = \frac{550}{7}$$

From this follows

$$EGHF = 361\frac{3}{7} + 78\frac{4}{7} = 440$$

But we also have

$$EGHF = 5 \cdot EF$$

so

$$EF = \frac{440}{5} = 88 \Leftrightarrow$$

$$\frac{44}{7} DE = 88 \Leftrightarrow$$

$$DE = 14 \text{ and } AB = 28 \text{ and } BC = 18$$

General comment

With some imagination this problem can be considered as the problem of boundary extensions applied to circles instead of quadrangles (§3.2.2). The calculation is a typical application of elementary geometrical algebra. Imagining the areas of both circles as similar right triangles with the sides of the right angle R and $2\pi R$, respectively r and $2\pi r$, there arises as the difference a trapezium with known height and area. Then the two radii can easily be calculated by pasting the triangle ELG ($LE = 5$ and $LG = EK = 2 \cdot \frac{22}{7} \cdot 5$) to LE, congruent with the known triangle LEK. So the trapezium $ELHF$ is transformed into the parallelogram $EGHF$.

Cardinael's use of triangle here is probably inspired by Archimedes *Measurement of the Circle*, where he considers the area of a circle as right triangles with the right-angle sides as the radius and the circumference of the circles. The difference between two circles then is the difference of two similar triangles, being a trapezium. This trapezium then is transformed into a parallelogram.

A solution in Cardinael's usual way, as in **Q1**, also would have been possible. The gnomon

$$(\sqrt{\frac{22}{7}} \cdot DE)^2 - (\sqrt{\frac{22}{7}} \cdot DL)^2 = 361\frac{3}{7}$$

can be is transformed to the rectangle

$$(\sqrt{\frac{22}{7}} \cdot DE + \sqrt{\frac{22}{7}} \cdot DL) \cdot (\sqrt{\frac{22}{7}} \cdot DE - \sqrt{\frac{22}{7}} \cdot DL) = 361\frac{3}{7} \Leftrightarrow$$

$$5 \cdot \sqrt{\frac{22}{7}} \cdot (\sqrt{\frac{22}{7}} \cdot DE + \sqrt{\frac{22}{7}} \cdot DL) = 361\frac{3}{7} \Leftrightarrow$$

$$5 \cdot \frac{22}{7} \cdot (DE + DL) = 361\frac{3}{7} \Leftrightarrow$$

$$DE + DL = 361\frac{3}{7} \cdot \frac{7}{110}$$

And with the help of $DE - DL = 5$ follows

$$DE = 14 \text{ and } DL = 9$$

and so

$$AB = 28 \text{ and } BC = 18$$

Theory used

Answer (proportions, geometrical algebra)

Archimedes: circumference and area of a circle with $\pi = \frac{22}{7}$
Similarity, congruence

Compare with

Q1

[Petri 1583] Ex.41

Question 55

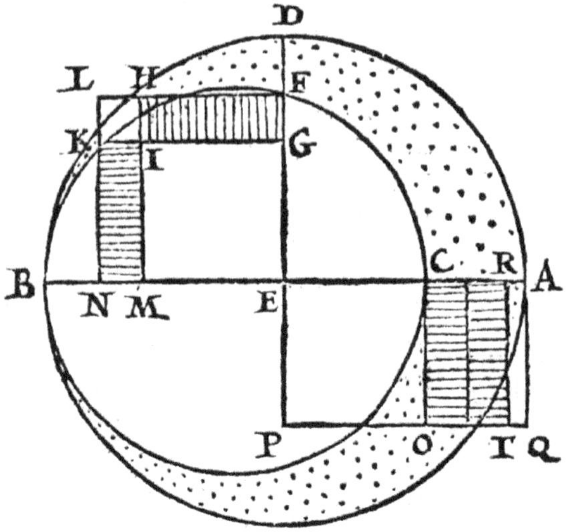

Figure 4.78: Subgroup 6, two circles with the difference of the diameters known and the width of the 'lunar' difference known

Comment on Answer

Given $AC = 9$ and $FD + 5$, choose G on ED such that $EG = EC$. From this follows

$$FG = DG - FD = 9 - 5 = 4$$

Choose P on the extension of DE such that

$$EP = EG = EC$$

Because

$$FG = IK = 4 \text{ and } GI = IM = EP$$

the two rectangles $FHIG$ and $IKNM$ can be drawn within $ACOQ$ such that the rectangle $ARTQ$ with $AR = 1$ remains. From

$$\text{square } ECOP = \text{square } GIME$$

and

$$EAQP = \text{square } EFLN$$

as EF is the mean proportional between EC and EB so square $EFLN = EC \cdot EB = EC \cdot A =$ rectangle $EAQP$, so follows

square $IHLK =$ rectangle $ARTQ \Leftrightarrow$ (with $IH = HL = FG = 4$ and $AR = 1$)

$$4^2 = 1 \cdot AQ \Leftrightarrow$$

$$AQ = EP = EC = 16$$

So

$$AE = EC + AC = 16 + 9 = 25$$

which means that

$$\begin{aligned} \text{radius small circle } &= \tfrac{1}{2}(EC + EB) = \tfrac{1}{2}(16 + EB) \\ &= \tfrac{1}{2}(16 + AE) \\ &= \tfrac{1}{2} \cdot (16 + 25) \\ &= 20\tfrac{1}{2} \end{aligned}$$

General comment

Though the solution can be considered as a successful application of geometrical algebra, an algebraic solution is clearly superior here. The algebraic translation of $EF^2 = EC \cdot EB$ is

$$(R - 5)^2 = (R - 9) \cdot R \text{ (with } AE = R)$$

from which follows, by cancelling out the squares at both sides of the equation,

$$R = 25$$

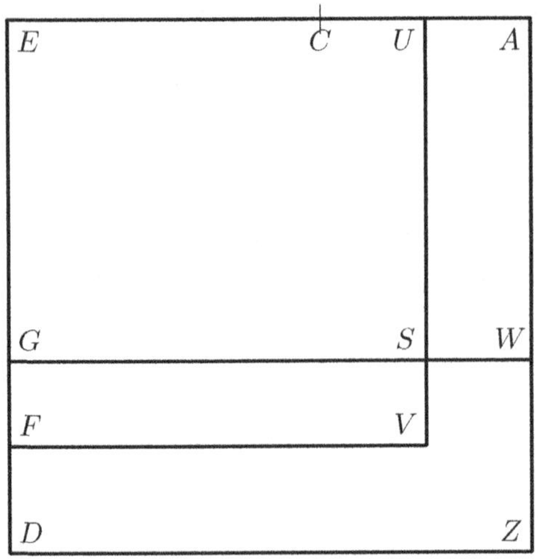

Figure 4.79: Q55 figure 2

A pure geometrical solution of the last equation with the help of simple geometrical algebra now becomes available (figure 4.79) by considering in the vertex E of the square $EAZD$ with sides of length R, the square $EUVF$ with sides of length $R-5$ and the rectangle $EAWG$ with sides $EA = R$ and $EG = R-9$, from which follows $UA = 5$ and $GF = 4$. Hence, we have point C on EA such that $UC = 4$, so that $AC = 9$ and $EC = EG$ and point S is the intersection of UV and WG.

Segment EA with points C and U corresponds with EA in figure 4.79 where U is the center point of CR. Segment ED with points G and F corresponds with ED in figure 4.79. With EF as the mean proportional in figure 4.79 of EC and EB follows

$$EF^2 = EC \cdot EB \Leftrightarrow \text{ (in Cardinael's figure ??)}$$

$$EF^2 = EG \cdot EA \Leftrightarrow \text{ (in figure 4.79)}$$

$$EUVF = EAWG \Leftrightarrow$$

$$GSVF = UAWS \Leftrightarrow$$

$$4(R-5) = 5(R-9) \Leftrightarrow$$

$$R = 25$$

Theory used

Answer

§3.1.2 d: $EF^2 = EC \cdot EB$, the perpendicular from the right-angled vertex is the geometric mean of the segments of the hypotenuse

Archimedes: area circle $= \frac{22}{7} \cdot (\text{radius})^2$.

Question 56

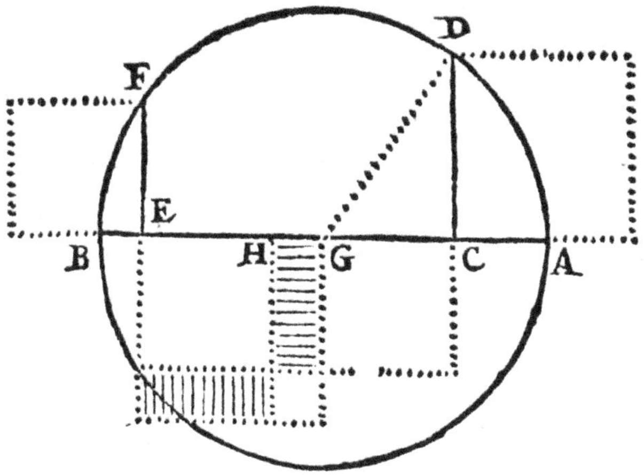

Figure 4.80: Subgroup 6, two chords known, perpendicular on the diameter of a circle and at a given distance of each other

Comment on Answer

G is the centre point of the circle with H chosen so that $EH = GC$.
So

$$EG^2 - CG^2 = (GF^2 - EF^2) - (GD^2 - CD^2) =$$

$$(GD^2 - EF^2) - (GD^2 - CD^2) = CD^2 - EF^2 = 24^2 - 18^2 = 252$$

Also

$$CE^2 = (CG + EG)^2$$

$$= CG^2 + 2 \cdot CG \cdot EG + EG^2 = 42^2 = 1764 \text{ (Elem.II-4)}$$

So

$$
\begin{gathered}
CE^2 - (EG^2 - CG^2) = 1512 \Leftrightarrow \\
CG^2 + 2 \cdot CG \cdot EG + EG^2 - EG^2 + CG^2 = 1512 \Leftrightarrow \\
2 \cdot CG^2 + 2 \cdot CG \cdot EG = 1512 \Leftrightarrow \ (\S 3.1.5 \text{ case } 5) \\
2 \cdot CG \cdot (CG + EG) = 1512 \Leftrightarrow \\
2 \cdot CG \cdot 42 = 1512 \Leftrightarrow \\
CG = \frac{1512}{84} = 18 \text{ so } DG = \sqrt{24^2 + 18^2} = 30
\end{gathered}
$$

Comment on Otherwise

From

$$CD^2 - EF^2 = 252$$

and

$$
\begin{aligned}
CD^2 - EF^2 &= (CD^2 - CG^2) - (GF^2 - EG^2) \\
&= EG^2 - CG^2 \text{ as } GD = GF \\
&= CE \cdot GH \text{ because of } \S 3.1.5 \text{ case } 7
\end{aligned}
$$

follows

$$
\begin{gathered}
CE \cdot GH = 252 \Leftrightarrow \\
GH = \frac{252}{CE} = \frac{252}{42} = 6
\end{gathered}
$$

so

$$CG = \frac{CE - GH}{2} = \frac{42 - 6}{2} = 18$$

from which follows again

$$DG = 30$$

General comment

The alternative method leads to a more direct solution then **Answer**. It depends on exactly the same application of geometrical algebra as used in **Q1 Answer**. From the solution in **Answer** it is known that

$$EG^2 - CG^2 = CD^2 - EF^2 = 24^2 - 18^2 = 252$$

The gnomon $EG^2 - CG^2$ with area 252 is replaced by the equal rectangle with sides CE and GH (§3.1.5 case 4 and §3.1.5 case 7) with given $CE = 42$. So

$$GH = \frac{252}{42} = 6$$

Theory used

Answer (geometrical algebra, Pythagoras)

Elem.II-3 and §3.1.5 case 5

Otherwise

(Same as **Answer**)

§3.1.5 case 4 and §3.1.5 case 7

Compare with

Q1 Answer.

Question 57

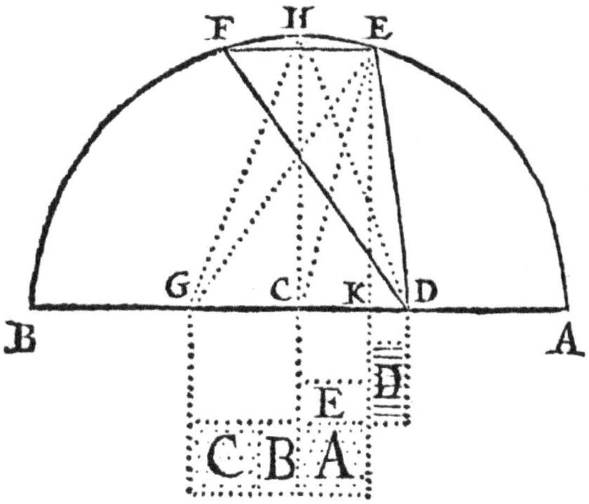

Figure 4.81: Subgroup 6, a semi-circle with an inscribed triangle with as top a point on the diameter and as base a chord parallel with that diameter

Comment on Answer

Cardinael commences the construction of the required triangle DEF with sides $DE = 10, DF = 12$ and EF the chord parallel with AB. Strictly speaking the existence of this triangle DEF should have been proven first. One way to proceed would be as follows.

From **Q20** follows the existence of the triangle GED with $GD = 8$ (C is the center point of GD), $GE = 12, DE = 10$ and $CE = \sqrt{106}$, so E is a point on the semi-circle AB. Mirroring GE on CH, the perpendicular on AB, gives the line segment DF with $DF = GE = 12$. As such, the existence of triangle DEF is demonstrated. From $DC = CG = 4$ and **Q1** follows that $KC = 2\frac{3}{4}$ and so $EF = 5\frac{1}{2}$.

Taking for granted the existence of a triangle DEF with $DE = 10$ and $EG = 12$, he proceeds by calculating DC and GC based on the equality

$$DH^2 + GH^2 = DE^2 + GE^2 \tag{4.21}$$

He provides a proof of (4.21) by means of elementary geometrical algebra on figure 4.81 $C = A = KC^2$, $B = E = D$, GE is the result of mirroring DF in CH and C, A, B, E and D are rectangles indicated in the drawing.
The right-hand member of (4.21) is reduced to

$$DE^2 + GE^2 = DK^2 + KG^2 + 2 \cdot EK^2$$

Subtracting $2 \cdot KC^2 + 2 \cdot EK^2$ leads to

$$DK^2 + KG^2 + 2 \cdot EK^2 - (2 \cdot KC^2 + 2 \cdot EK^2) =$$

because $KG^2 - 2 \cdot KC^2 = KG^2 - C - A = KC^2 + CG^2 + E + B$

$$DK^2 + KC^2 + CG^2 + E + B =$$

and since $DK^2 + KC^2 + E + B = DK^2 + KC^2 + E + D = DC^2$

$$DC^2 + CG^2 = 2DC^2$$

If the same expression $2 \cdot KC^2 + 2 \cdot EK^2$ is also subtracted from the left-hand side member of (4.21) this leads to

$$DH^2 + GH^2 - (2KC^2 + 2EK^2) =$$

because

$$HC^2 = EC^2 = KC^2 + EK^2$$

$$DH^2 + GH^2 - 2HC^2 =$$
$$DC^2 + GC^2 + 2HC^2 - 2HC^2 = DC^2 + GC^2 = 2DC^2$$

With this the equality of both members of (4.21) are proven.
From (4.21) follows

$$DH^2 + GH^2 = DE^2 + GE^2 = 100 + 144 = 244 \Leftrightarrow$$
$$DH^2 = GH^2 = 122 \Leftrightarrow$$
$$DH^2 - HC^2 = 122 - 106 = 16 \Leftrightarrow$$
$$DC^2 = 16 \Leftrightarrow$$
$$DC = GC = 4$$

By means of **Q1**, (see the same calculation preceding (4.21) where **Q20** was used) follows $KC = 2\frac{3}{4}$ and so $EF = 5\frac{1}{2}$.

General comment

In the proof of (4.21), the lengths of the line segments do not matter. This implies a proof for the following proposition.

Proposition: Given a circle with centre C, diameter AB, H on the circle with $HC \perp AB$ and points D and G on AB so that $DC = GC$ then E on the semi-circle with projection K between G and $D \Rightarrow$

$$GE^2 + DE^2 = \text{ const. } = GH^2 + DH^2$$

The equality (4.21)

$$DH^2 + GH^2 = GE^2 + DE^2$$

can be proved more directly by using Stewart's theorem (see §3.2.2). Stewart's theorem leads to following identities for triangle DEG (with $GC = DC = a$)

$$2a \cdot EC^2 = a \cdot GE^2 + a \cdot DE^2 - 2a^3 \Leftrightarrow$$
$$GE^2 + DE^2 = 2 \cdot EC^2 + 2a^2 \Leftrightarrow$$
$$GE^2 + DE^2 = 2 \cdot (HC^2 + a^2) \Leftrightarrow$$
$$GE^2 + DE^2 = 2 \cdot DH^2 = DH^2 + GH^2$$

by which (4.21) is proven.

At the end of **Answer**, Cardinael refers to **Q20** and Elem.II-10. He probably intended Elem.II-9 and II-10 in the numbering of [Heath 1956], implying that the Question can be solved by using only **Q20** or by (4.21). Cardinael provides a proof in his particular way, but this also easily follows from Elem.II-9 and Elem.II-10.

Now, the existence of triangle DEF can be proven by the following general theorem If G and D are on AB such that $GC = CD$ and E is a point on the semicircle AB, then we have (with H on the semicircle and $CH \perp AB$)

$$GE^2 + DE^2 = GH^2 + DH^2$$

Proof: We assume that the projection K of E on AB is located between G and D (in the opposite case the proof is similar but depends on Elem.II-10 rather then II-9). According to Elem II-9 we have:

$$GK^2 + KD^2 = 2GC^2 + 2CK^2$$

From figure 4.81 follows

$$\begin{aligned} GE^2 + DE^2 &= GK^2 + EK^2 + KD^2 + EK^2 \\ &= GK^2 + KD^2 + 2EK^2 \\ &= GK^2 + KD^2 + (2CE^2 - 2CK^2) \\ &= 2GC^2 + 2CK^2 + 2CE^2 - 2CK^2 \text{ (Elem.II-9)} \\ &= 2GC^2 + 2CH^2 \\ &= GH^2 + DH^2 = 2GH^2 \end{aligned}$$

This leads to the following equivalence :

$$GE^2 + DE^2 = 2GH^2 \Leftrightarrow E \text{ is on the semicircle } AB \tag{4.22}$$

In the particular case of **Q57** we have:

$$CH = \sqrt{106}$$

when G and D are chosen so that

$$GC = CD = 4$$

then follows from figure 4.81

$$2GH^2 = 2(GC^2 + CH^2) = 2(16 + 106) = 244$$

and therefore

$$GE^2 + DE^2 = 244 \Leftrightarrow E \text{ is on the semicircle } AB \tag{4.23}$$

From (4.23) follows the existence of triangle GED with E on the semicircle AB, with $GC = CD = 4$, $GE = 12$ and $DE = 10$.

Mirroring GE on CH gives the line DF with $DF = 12$, implying that we have got a triangle DEF as required with $DE = 10, DF = 12$ and a chord EF parallel to AB. Hence, the existence of triangle DEF is proven.

Calculation of EF

Given the existence proof of triangle DEF, a simple calculation suffices to obtain the length of EF

From $DG = DC + GC = 8, GE = 12$ and $DE = 10$
follows with the help of **Q1** that:

$$DK = \frac{5}{4} \Rightarrow KC = 2\frac{3}{4} \Rightarrow EF = 5\frac{1}{2}$$

We finally can figure out what Cardinael must have had in mind, given that he mentions three distinct approaches for solving **Q57**):

1. With the help of **Q20** triangle GDE immediately becomes available and using **Q1** all necessary calculation could be done, including the existence proof of DEF as we provided in **Comment on Answer**).

2. With the help of Elem.II-9 or Elem.II-10 a more general method of calculation was possible and leads to the general equivalence :

   $$GE^2 + DE^2 = a = GH^2 + DH^2 \Leftrightarrow E \text{ on the semicircle } AB$$

   In the underlying case:

   $$GE^2 + DE^2 = 244 \Leftrightarrow E \text{ on the semicircle } AB$$

   Also here, the existence proof and calculation of EF are straightforward, as we showed above in **General Comment**).

3. His own solution, where he takes more or less for granted the existence of triangle DEF, and gives a handcrafted proof of (4.21) for the special case of **Q57** without the use of Elem.II-9,10.

From a strict mathematical point of view method 2 is clearly superior to methods 1 and 3.

Theory used

Answer

Q20 for the proof of the existence;

Q1

Compare with

Q20 and Elem.II-9,10

Q70

Question 58

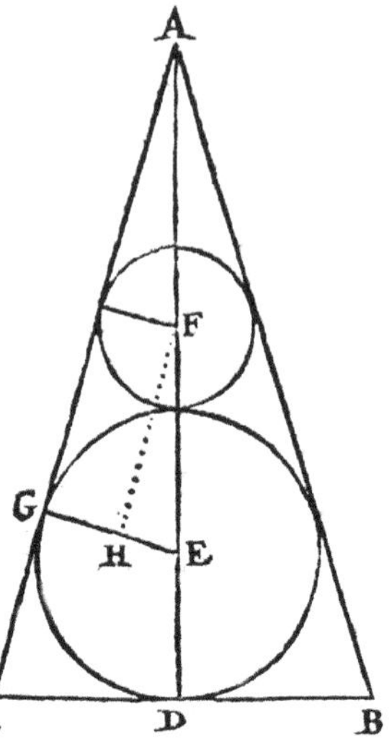

Figure 4.82: Subgroup 6, an isosceles triangle with two inscribed circles

Comment on Answer

From figure 4.82 it is clear that

$$EH = 6 - 4 = 2 \text{ and } EHF = 6 + 4 = 10$$

From the similarity of the triangles EHF and EGA follows

$$EH : EF = EG : EA \Leftrightarrow$$

$$2 : 10 = 6 : EA \Leftrightarrow$$

$$EA = 30$$

So

$$AD = EA + ED = 30 + 6 = 36$$

Also

$$AG^2 = EA^2 - EG^2 \Leftrightarrow$$

$$AG^2 = 864 \Leftrightarrow AG = \sqrt{864}$$

Further we have

$$AG : AE = AD : AC \Leftrightarrow$$

$$\sqrt{864} : 30 = 36 : AC \Leftrightarrow$$

$$AC = AB = \sqrt{1350}$$

from which follows

$$BD = \sqrt{AB^2 - AD^2} = \frac{1}{2} \cdot \sqrt{216}$$

so

$$BC = \sqrt{216}$$

Theory used

Answer

Similarity

Question 59

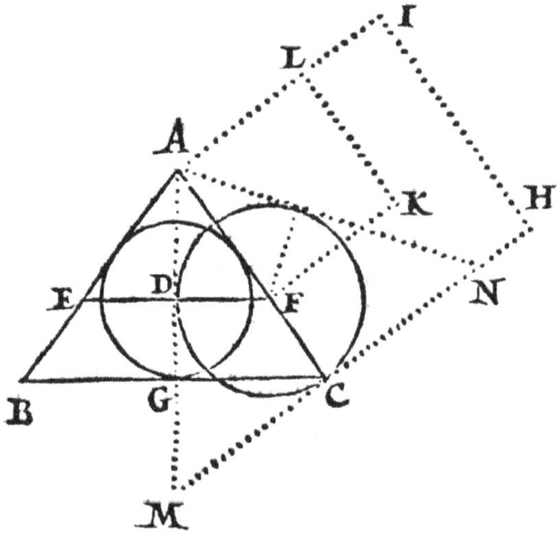

Figure 4.83: Subgroup 6, an isosceles triangle with an inscribed circle

Comment on Answer

The areas of similar figures as in figure 4.83 are in the same ratio as the squares of the corresponding sides, so $AF^2 = \frac{25}{64} \cdot AC^2 = 625$ from which follows

$$AF = 25 \text{ and } FC = 15$$

MN is perpendicular to AC and M is on the extension of AG, so that $MC = NC$ so AC is bisector of angle MAN. So the triangles ABC and AMN are similar. From the ratio

$$GD : DA = CF : FA$$

with GD the radius of the inscribed circle of ABC follows that FC is the radius of the inscribed circle of AMN and thus $FD = FC = 15$. CG (and therefore also BC) can be determined by means of the similarity of the triangles AFD and ACG.

General comment

Cardinael is not using geometrically algebra here. The use of the squares $AFKL$ and $ACHI$ has the intention to geometrically justify every step. Surprisingly however, FD is calculated in an unusual way by means of triangle AMN, similar to ABC. One would have expected the use of the more evident equality of the angles CDF and DCF, while CD is the bisector of angle GCF and EF is parallel with BC, so that angle GCD is equal to angle CDF, and so arrive at the equality of FD and FC.

Theory used

Answer

Similarity

Inscribed circle

Elem.VI-20

Question 60

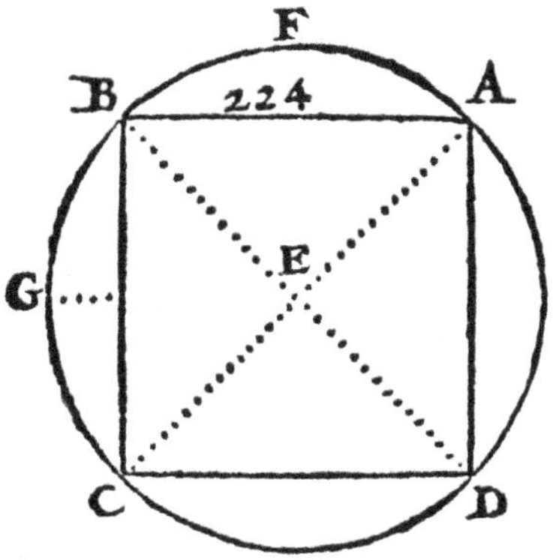

Figure 4.84: Subgroup 6, a circle with an inscribed square and the size of the remaining segments of the circle known

Comment on Answer

Cardinael consistently uses $\frac{22}{7}$ for π. We here use π as a short notation for the fraction $\frac{circumference\ circle}{diameter\ circle}$.If assumed is that $AC = 1$ then follows

$$AEBF = \frac{1}{4} \cdot \pi \cdot (\frac{1}{2})^2 = \frac{1}{4} \cdot \frac{22}{7} \cdot \frac{1}{4} = \frac{11}{56}$$

and that

$$AEB = (\frac{1}{2})^3 = \frac{1}{8}$$

This means that from $AC = 1$ for segment ABF follows

$$ABF = \frac{11}{56} - \frac{7}{56} = \frac{4}{56} = \frac{1}{14}$$

In reality however $ABF = 224$. From the proportion

$$\frac{1}{14} : 1^2 = 224 : AC^2$$

follows

$$AC = 56$$

Theory Used

Answer

Archimedes: area circle $= \frac{22}{7} \cdot r^2$;

Similarity and multiplication factor.

Question 61

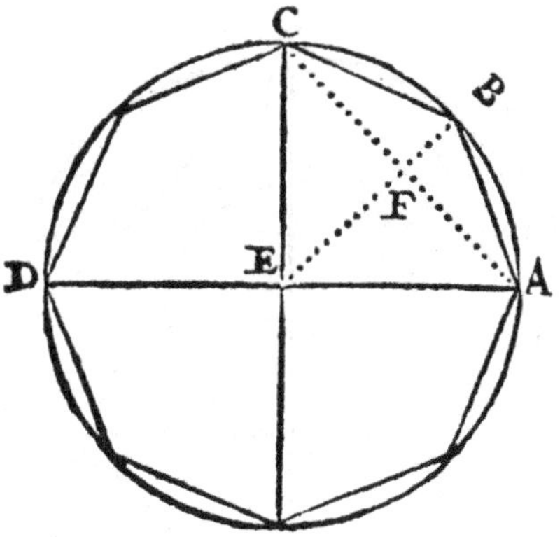

Figure 4.85: Subgroup 6, a circle with an inscribed octagon with known sides

Comment on Answer

Posing that $AE = 1$ then

$$AB = \sqrt{2 - \sqrt{2}}$$

So there is a multiplication factor that is equal to

$$\frac{\sqrt{72 - \sqrt{2592}}}{\sqrt{2 - \sqrt{2}}} = 6$$

Therefore, the real size of AE is 6 and the diameter is

$$AD = 12$$

Theory used

Answer

multiplication factor

Question 62

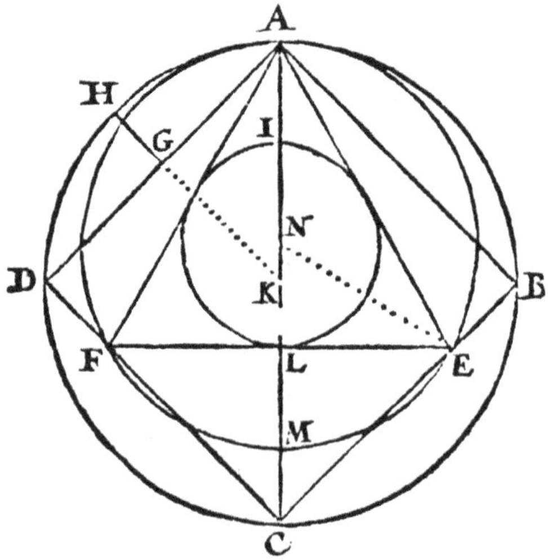

Figure 4.86: Subgroup 6, a circle with an inscribed square itself containing an inscribed isosceles triangle

Comment on Answer The equality relations $LF = LC$ and $AG = GK$ are direct consequences of the fact that AGK and FLC are right-angled triangles. Since $HK \perp AD$ and $\angle GAK = 45°$ and because $AC \perp FE$ and owing to $FE \parallel DB$ we have $\angle CFL = 45°$. The equality relation $NL = \frac{1}{3}AL$ follows from the property of the three medians in a triangle: they meet in a single point and divide each other in the ratio 2 : 1. While this property of triangles is nowhere explicitly proved by Euclid, it was known to Archimedes.

Also Cardinael uses this property without giving any reference, shows that such knowledge, which can simply be proven, was part of the mathematical toolbox of a reckoning masters, such as Cardinael.

The calculation is interesting for its handling of fractions of irrational binomials. Cardinael obviously wants to eliminate irrational terms in the denominator, probably based on rules for dealing with irrational binomials ([Heath 1956], vol. 3, X, 83 ff.) as for instance

$$FE = \frac{16\sqrt{2}}{1+\sqrt{3}} = \sqrt{512 - \sqrt{196608}}$$

Cardinael's method can be simplified as follows. First, suppose of all $AD = 2$

$$AD = 2 \Rightarrow AG = 1 \Rightarrow KA = \sqrt{2} \Rightarrow$$

$$HG = \sqrt{2} - 1$$

As we know that $HG = \sqrt{32} - 4$, the multiplication factor is

$$\frac{\sqrt{2} - 1}{\sqrt{32} - 4}$$

from which follows the true value of AD

$$AD = 2 \cdot \frac{\sqrt{32} - 4}{\sqrt{2} - 1} = 8 \tag{4.24}$$

Now, suppose that $FE = 2$

$$\begin{aligned} &FE = 2 \Rightarrow FL = 1 \Rightarrow (LC = 1 \text{ and } AL = \sqrt{3}) \Rightarrow \\ &AC = AL + LC = 1 + \sqrt{3} \Leftrightarrow \\ &AD = \tfrac{1+\sqrt{3}}{\sqrt{2}} \end{aligned} \tag{4.25}$$

From (4.24) and (4.25) follows the multiplication factor

$$\frac{1 + \sqrt{3}}{8\sqrt{2}}$$

so the true value of FE will be

$$FE(= AE = AF) = 2 \cdot \frac{8\sqrt{2}}{1 + \sqrt{3}} = 8\sqrt{2}(\sqrt{3} - 1)$$

which is written by Cardinael in the form:

$$8\sqrt{2}(\sqrt{3} - 1) = 8\sqrt{6} - 8\sqrt{2} = \sqrt{(8\sqrt{6} - 8\sqrt{2})^2}$$

$$= \sqrt{512 - 128\sqrt{12}} = \sqrt{512 - \sqrt{196608}}$$

Hence follows the diameter of the inscribed and circumscribed circle

$$\begin{aligned} \text{diameter inscribed circle} &= NL \\ &= \tfrac{1}{3}AL = \tfrac{1}{3} \cdot \sqrt{3} \cdot FL \\ &= \tfrac{1}{3}\sqrt{3} \cdot \tfrac{1}{2}FE = \tfrac{1}{3}\sqrt{3} \cdot \tfrac{1}{2} \cdot 8\sqrt{2}(\sqrt{3} - 1) \\ &= \tfrac{4}{3}\sqrt{6}(\sqrt{3} - 1) \end{aligned}$$

which is written by Cardinael in the form

$$NL = \sqrt{\frac{16}{9}(18 + 6 - 12\sqrt{3})}$$
$$= \sqrt{42\frac{2}{3} - \sqrt{1365\frac{1}{3}}}$$

$$\begin{aligned} \text{diameter circumscribed circle} &= NE = NA = 2 \cdot NL \\ &= \sqrt{170\tfrac{2}{3} - 4\sqrt{1365\tfrac{1}{3}}} \\ &= \sqrt{170\tfrac{2}{3} - \sqrt{21845\tfrac{1}{3}}} \end{aligned}$$

General comment Remarkably in **Q62** is the reason why the solution is unique. There is an infinity of possible circles going through the three the points H, A and D so that $HG = \sqrt{32} - 4$. By adding the condition that AD also is a side of inscribed square makes that there is only one circle that satisfies this condition! An algebraic solution makes this clear immediately Pose

$$AK = x \; K \text{ center of the circle through} A, H \text{and } D)$$

As the circle with K as centre and AK as radius is the circumscribed circle of square $ADCB$, it follows that so

$$KG = \frac{1}{2}x\sqrt{2}$$

Because

$$KH = x$$

follows from this:

$$x - \frac{1}{2}x\sqrt{2} = \sqrt{32} - 4 \Leftrightarrow x = 4\sqrt{2}$$

Theory used

Answer (Regula Falsi, multiplication factor)

Repeated application of the Regula Falsi

The obvious insight that in an equilateral triangle the centroid and the centre of the circumscribed and inscribed circle are the same

Compare with

[Dijksterhuis 1929-30], 2, 184-197

[Heath 1956], vol.3 X, 83 ff.: binomial forms

[Sitters 2008] for a more extensive treatment

Question 63

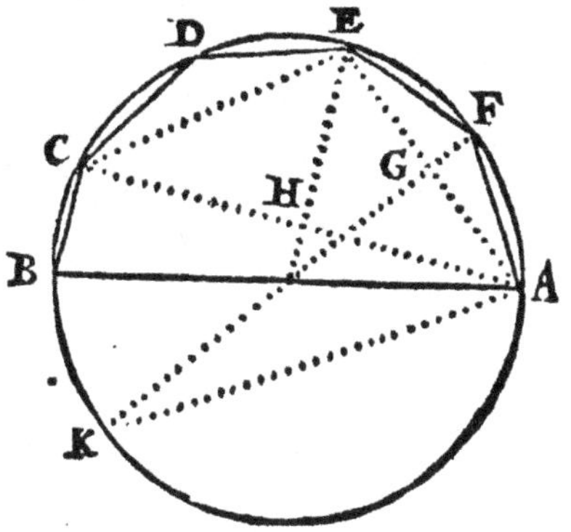

Figure 4.87: Subgroup 6, a semicircle with known radius with four chords given and an unknown remainder

Comment on Answer

K is chosen such that KF is a diameter. From the similarity of the triangles KAF and AGF follows

$$KF : AF = AF : FG \Leftrightarrow (\text{with } KF = AB = 50)$$

$$50 \cdot FG = AF^2 \Leftrightarrow (\text{with } AF = \sqrt{250})$$

$$FG = \frac{250}{50} = 5$$

For triangle AGF holds

$$AG = \sqrt{AF^2 - FG^2} = \sqrt{250 - 25} = 15$$

and

$$AE = 30 \text{ and also } EC = 30$$

A similar calculation can be done for EH and AH. In triangle LAE, with L (not drawn) as the intersection point of the extension of EH with the circumference of the circle, follows from the similarity of triangles LAE and AHE:

$$LE : AE = AE : EH \Leftrightarrow \text{ (with } LE = AB = 50)$$

$$50 \cdot EH = EA^2 \Leftrightarrow \text{ (with } EA = 30)$$

$$EH = \frac{900}{50} = 18$$

In triangle AHE

$$AH = \sqrt{AE^2 - EH^2} = \sqrt{900 - 324} = 24$$

so that

$$AC = 48$$

For that reason we have in triangle ABC

$$BC = \sqrt{50^2 - 48^2} = 14$$

Theory used

Answer

Elem.III-21, 29: AE and FG are perpendicular and bisect each other

Elem.III-31 and IV-5: angles KAF and LAE are right angles

Elem.VI-8: KAF and AGF are similar just as LAE and AHE

Question 64

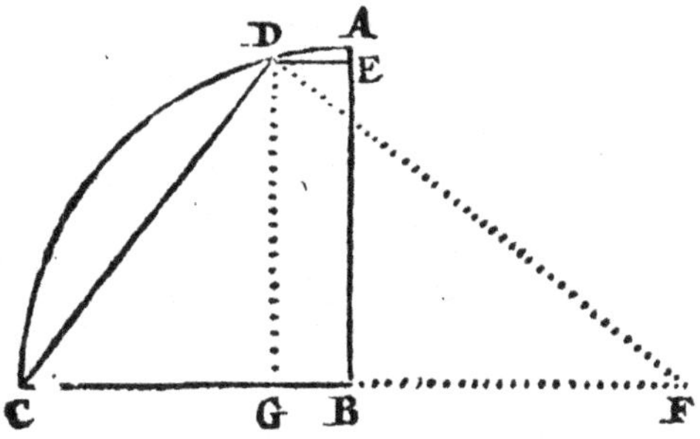

Figure 4.88: Subgroup 6, a quadrant of a circle

Comment on Answer

If F is chosen on the extension of CB so that $FB = BC$ then B is the centre of the circle through C, D, A and F and angle FDC is a right angle and

$$FG = FB + BG = BC + ED = CD$$

As angle FDC is a right angle

$$CG : CD = CD : FC \Leftrightarrow (\text{with } CD = FG)$$

$$CG : FG = FG : FC$$

So FC is divided by G in mean and extreme ratio and

$$FG = 10 - \sqrt{20}$$

This corresponds exactly with the configuration of **Q25**. Using that solution we can conclude

$$CG = \sqrt{180} - 10 \text{ and } FC = \sqrt{80}$$

from which follows

$$BC = \sqrt{20}$$

Theory used

Answer (similarity)

Elem.III-31: angle FDC is a right angle

Elem.VI-8: similarity in right-angled triangle

Q25

Question 65

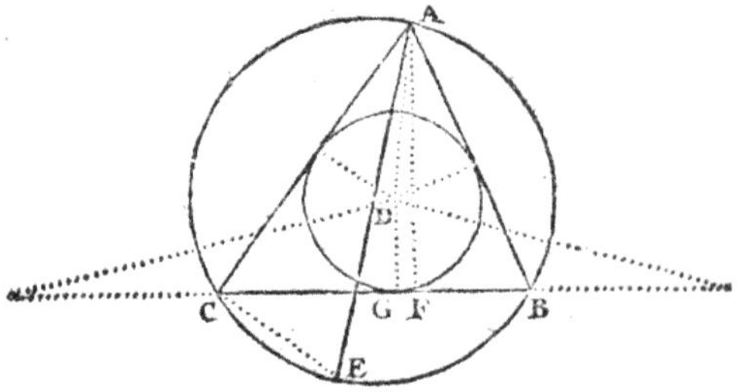

Figure 4.89: Subgroup 6, a triangle with an inscribed and circumscribed circle

Comment on Answer

To find the diameter AE of the circumscribed circle according to **Q1**, the length 12 of the perpendicular AF is calculated. As triangle EAC and triangle BAC are at the same base AC with E and B on the same side of AC, from Elem.III-21 follows that the angles AEC and ABC are equal according. The angles ACE and AFB are both right angles and so BAF and EAC are also equal. So the triangles ACE and AFB are similar. From

$$AF : AB = AC : AE\text{and } AF = 12, AB = 13, AC = 15$$

then follows

$$AE = 16\frac{1}{4}$$

for the diameter of the circumscribed circle.

The diameter of the inscribed circle follows from the area of triangle ABC, which is equal to 84. This doubled gives 168. By adding all three sides together, that is 42, and dividing 168 by 42, this results in 4 for GD, half the diameter, so the diameter of the inscribed circle is 8.

General comment

By the pragmatic choice of the perpendicular AF of triangle ABC and the diameter AE of the circumscribed circle as useful lines the calculation of AE

by means of the similarity of the triangles AFB and ACB is only a question of technique. From the figure 4.89 follows directly the formula for the radius $R(=\frac{1}{2}\cdot AE)$ of the circumscribed circle using the similarity of ABF and AEC. So

$$AF : AB = AC : AE \Leftrightarrow$$

$$\frac{h_a}{c} = \frac{b}{2\cdot R} \Leftrightarrow R = \frac{abc}{4O}$$

Calculating DG, the radius r of the inscribed circle, Cardinael proceeds in the usual way: considering the area of triangle ABC as the sum of the three kites with diagonals DB, DA and DC follows

$$\text{area } \triangle\, ABC = DG \cdot \left(\frac{AB + BC + CA}{2}\right) \Leftrightarrow$$

$$DG = \frac{2\cdot area \triangle\, ABC}{AB + BC + CA} \Leftrightarrow r = \frac{2O}{a+b+c}$$

Concerning the drawing 4.89, although the sizes do not fit, the intend was probably to suggest that the base CB is extended on both sides with the adjacent slanted sides CA and BA, so that the triangle with this extension as basis, D as top and height DG, has the same area as triangle ABC, being the right-hand member of the first equation.

Theory used

Answer (similarity, Thales)

Q1

Elem.III-21; Elem.III-31; Elem.IV-5

Compare with

[Schooten 1660], II, 86 proposal 36

Question 66

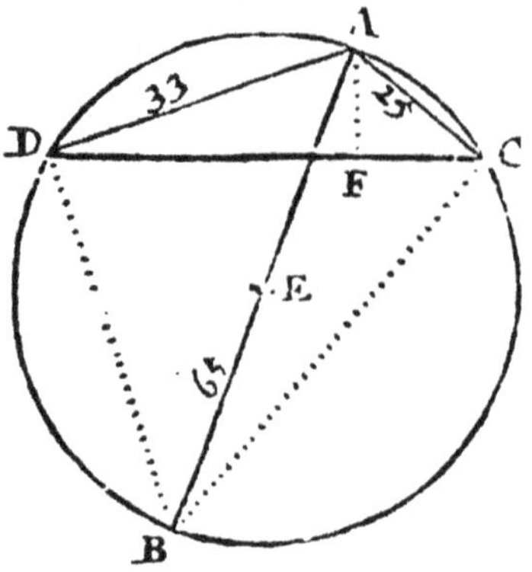

Figure 4.90: Subgroup 7, a triangle with two known sides and a given circumscribed circle

Comment on Answer

F is the base of the perpendicular from A on DC. Triangles ADB and AFC are similar as angles ADB and AFC are right angels and angle ACF is equal to angle ABD (same arch, see **Q65**). So

$$AC : AB = AF : AD \Leftrightarrow$$

$$25 : 65 = AF : 33 \Leftrightarrow$$

$$AF = 12\frac{9}{13}$$

For triangle AFC holds

$$CF = \sqrt{AC^2 - AF^2} = \sqrt{25^2 - (12\frac{9}{13})^2} = 21\frac{7}{13}$$

In the same way for AFD holds

$$FD = \sqrt{AD^2 - AF^2} = \sqrt{33^2 - (12\frac{9}{13})^2} = 30\frac{6}{13}$$

So

$$CD = CF + FD = 52$$

The calculation method for AF by means of the similarity of AFC with the right-angled triangle ADB is also applied in **Q65**, where AFB is similar to the right-angled triangle ACE.

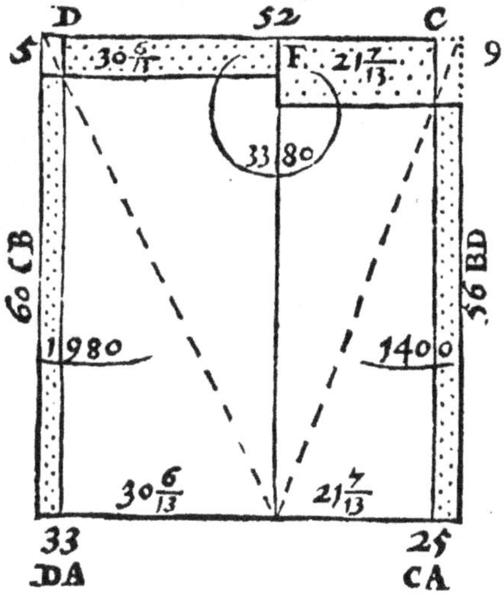

Figure 4.91: Problem 66: alterative solution

Comment on Otherwise 1

For the right-angled triangles ADB and ACB holds

$$BD = \sqrt{AB^2 - AD^2} = \sqrt{65^2 - 33^2} = 56$$

and

$$BC = \sqrt{AB^2 - AC^2} = \sqrt{65^2 - 25^2} = 60$$

From the similarity of triangles ACB and AFD follows

$$AB : BC = AD : FD \Leftrightarrow$$

$$65 : 60 = 33 : FD \Leftrightarrow$$

$$FD = 30\frac{6}{13}$$

The similarity of triangles ABD and ACF leads to

$$AB : BD = AC : CF \Leftrightarrow$$

$$65 : 56 = 25 : CF \Leftrightarrow$$
$$CF = 21\frac{7}{13}$$

So

$$DC = FD + CF = 52$$

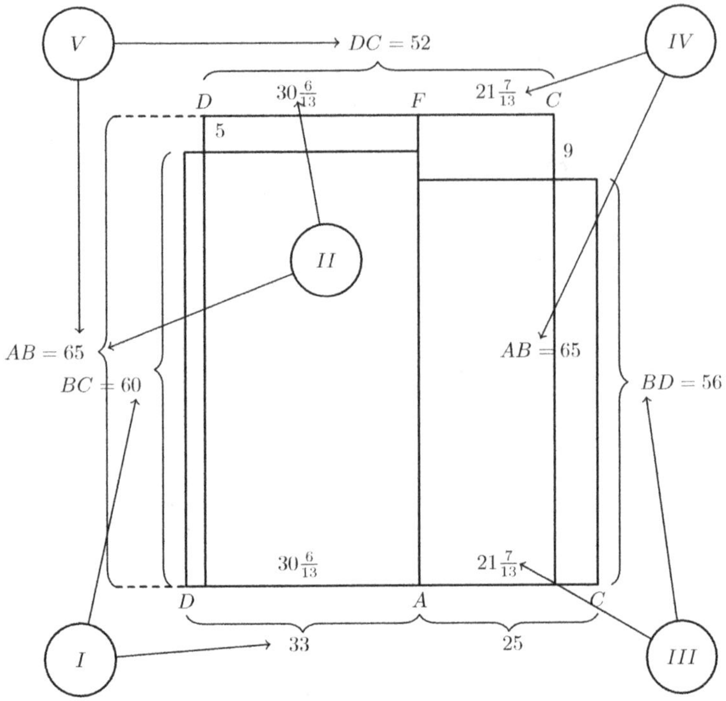

Figure 4.92: Question 66: elaborating on Cardinael's picture

Comment on Otherwise 2

First BC and BD are calculated with the help of Thales and the Pythagorean theorem, as was done in **Otherwise 1**

$$BC = 60 \text{ and } BD = 56$$

So

$$AC \cdot BD = 25 \cdot 56 = 1400$$

$$AD \cdot BC = 33 \cdot 60 = 1980$$

and therefore

$$AD \cdot BC + AC \cdot BD = 3380 \tag{4.26}$$

and

$$DC = \frac{3380}{AB} = \frac{3380}{65} = 52$$

Here Ptolemy's theorem is used:

$$AD \cdot BC + AC \cdot BD = AB \cdot DC$$

The proof of Ptolemy's theorem for this special case in which the diagonal of the quadrilateral inscribed in the circle is also the diameter of the circumscribed circle is proven by Cardinael. He uses the similarities of ACB and AFD and of ABD and ACF, proven and used in **Otherwise 1**.
The figure accompanying this special case of Ptolomy's theorem is figure 4.91. However, for elucidating Cardinael's rather cryptic figure we elaborated the drawing as figure 4.92. The different rectangles that Cardinael uses are indicated in the following way

$$AD \cdot BC = I$$
$$AB \cdot FD = II$$
$$AC \cdot BD = III$$
$$AB \cdot CF = IV$$
$$AB \cdot DC = V$$

Figure 4.92 shows directly in a pure geometrical way that

$$II + IV = V \tag{4.27}$$

This amounts to the distributive law for addition as it can be formulated as

$$AB \cdot FD + AB \cdot CF = AB \cdot DC$$

in which

$$FD + CF = DC$$

The intentions of figure 4.91 seems to be the geometrical illustration of this law and also of Elem.I-43 by means of the dotted diagonals, as explication below.

The outline of Cardinael's proof of Ptolemy's theorem is as follows. From the similarity of ACB and AFD follows

$$I = II \tag{4.28}$$

From the similarity of ABD and ACF follows

$$III = IV \tag{4.29}$$

Adding (4.28) and (4.29) gives

$$I + III = II + IV \tag{4.30}$$

Substitution of (4.27) in the left-hand member of equation (4.30) gives

$$I + III = V$$

Written out this can be formulated as:

$$AD \cdot BC + AC \cdot BD = AB \cdot DC \text{ (Ptolemy's theorem)}$$

from which follows

$$DC = \frac{AD \cdot BC + AC \cdot BD}{AB} = \frac{3380}{65} = 52$$

Similarity of ACB and AFD gives

$$AB : BC = AD : FD \Leftrightarrow$$

$$AD \cdot BC = AB \cdot FD \text{ see (4.28)}$$

Similarity of ABD and ACF gives

$$AB : BD = AC : CF \Leftrightarrow$$

$$AC \cdot BD = AB \cdot CF \text{ see (4.29)}$$

The equalities

$$I = II (4.28) \text{ and } III = IV (4.29)$$

are accentuated by Cardinael as two dotted lines in his figure 4.91. According to Elem.I-43 the two dotted rectangles in the left enclosing rectangle with sides 65 and 33 are equal, from which follows

$$I = II$$

and for the same reason we have in the right enclosing rectangle with sides 25 and 56 (see figure 4.91 and figure 4.92)

$$III = IV$$

Adding (4.28) and (4.29) leads to

$$AD \cdot BC + AC \cdot BD = AB \cdot FD + AB \cdot CF$$

so

$$AD \cdot BC + AC \cdot BD = AB \cdot (FD + FC) \Leftrightarrow$$
$$AD \cdot BC + AC \cdot BD = AB \cdot DC$$

and by this Ptolemy's theorem is proven, at least for this special case.

In **Otherwise 2** Cardinael has derived by means of **Otherwise 1** a special application of Ptolemy's theorem

$$AC \cdot BD + AD \cdot BC = AB \cdot DC$$

It appears that Cardinael has introduced this third demonstration anticipating the general case of Ptolemy's theorem which will be proven in a similar way in **Q68**, though by means of a construction line AG, corresponding to the perpendicular AF in **Q66**) which is not a perpendicular.

Theory used

Answer (Thales , similarity, Pythagoras)

Elem.III-21, III-31, I-47

Otherwise 1 (Thales , Pythagoras , similarity)

Elem.III-31, I-47, III-2

Otherwise 2 (Thales , Pythagoras , similarity, Ptolemy)

Elem.III-31, I-47, III-21

Compare with

Q65: calculation of AF;

Q68: general proof for the Ptolemy' theorem

Question 67

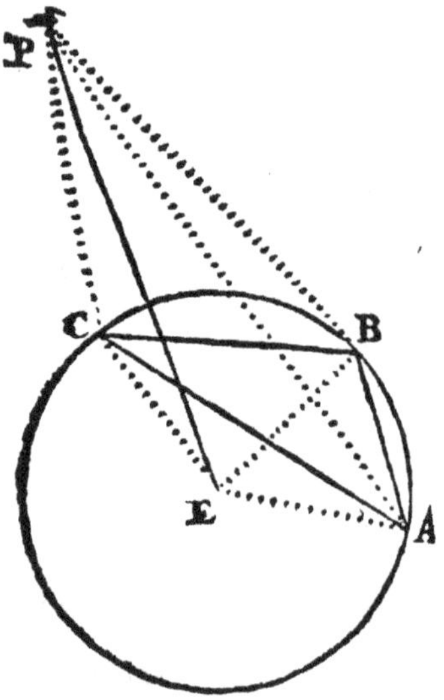

Figure 4.93: Subgroup 13, three marksmen with the same distance from a parrot

Comment on Answer

Because the marksmen all have the same distance from the parrot

$$PA = PB = PC$$

and so the triangles PEA, PEB and PEC are congruent, with PE being the distance of the parrot to earth. Therefore

$$EA = EB = EC$$

so E is the centre of the circumscribed circle of ABC. By means of **Q65** the radius of the circumscribed circle can be calculated

$$EA = EB = EC = 65$$

Applying Elem.I-47 in for example triangle EBP gives

$$BP = \sqrt{EB^2 + PE^2} = \sqrt{65^2 + 156^2} = 169$$

Theory used

Answer (congruency, Pythagoras)

Q65, Elem.I-47

Compare with

Q76

Question 68

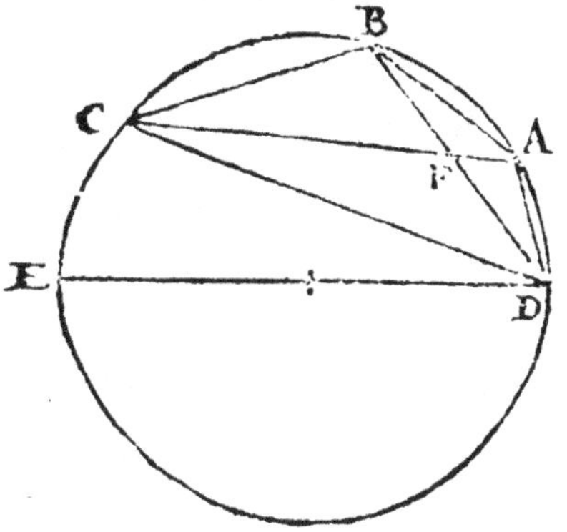

Figure 4.94: Subgroup 7, a circle with sides of an inscribed quadrilateral known

Comment on Answer

By assuming $FA = 1$, Cardinael calculates the lengths of the diagonals AC and BD using the similarity of triangles AFD and BFC and triangles BFA and CFD. By assuming $FA = 1$ the proportions remain unchanged, as for example

$$\frac{FC}{FD} = \frac{BC}{AD} = \frac{33}{16} \text{ and so on}$$

The length of FB follows from

$$\frac{FB}{FA} = \frac{FC}{FD}$$

and by means of

$$FA = 1 \text{ and } FB = \frac{FC}{FD}$$

so

$$FB = \frac{33}{16} = 2\frac{1}{16}$$

In the same way are calculated

$$FD = 2\frac{2}{5} \text{ and } FC = 4\frac{19}{20}$$

In the same way are calculated

$$FD = 2\frac{2}{5} \text{ and } FC = 4\frac{19}{20}$$

so

$$AC = 5\frac{19}{20} \text{ and } BD = 4\frac{37}{80}$$

Using Ptolemy's theorem, the multiplication factor can now be calculated. If $FA = 1$, then

$$AC \cdot BD = 5\frac{19}{20} \cdot 4\frac{33}{80} = 26\frac{883}{1600}$$

However, the real value of this product follows from Ptolemy's theorem

$$AC \cdot BD = AD \cdot BC + AB \cdot DC = 528 + 1500 = 2028$$

From this follows

$$\text{the square of the multiplication factor} \;=\; \frac{2028}{26\frac{883}{1600}} \Leftrightarrow$$

$$\text{(square because } AC \cdot BD = \text{ product of lengths)}$$

$$\text{the multiplication factor} \;=\; \sqrt{\frac{2028}{26\frac{883}{1600}}} = \sqrt{\frac{1040}{19}} = 8\frac{88}{119}$$

and therefore also

$$AF = 8\frac{88}{119},\ AC = 5\frac{19}{2} \cdot 8\frac{88}{119} = 52 \text{ and } BD = 4\frac{37}{80} \cdot 8\frac{88}{119} = 39$$

With $AB = 25, BC = 33$ and $AC = 52$ the radius of the circumscribed circle follows from **Q65**

$$R = \frac{25 \cdot 33 \cdot 52}{4 \cdot O} = \frac{25 \cdot 33 \cdot 52}{4 \cdot 330} = 32\frac{1}{2}$$

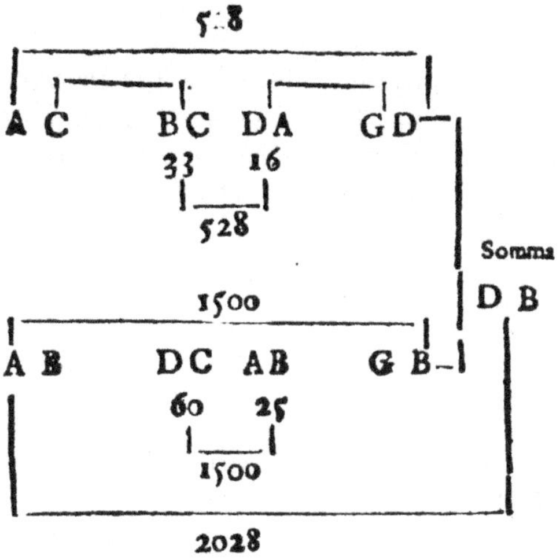

Figure 4.95: Problem 68: alternative solution

Comment on Otherwise

A proof without using Ptolemy's theorem is also feasible. Based on the properties of an inscribed quadrilateral in a circle BAC the ratios hold $FA : FC = 1 : 4\frac{19}{20}$ and $FA : FB = 1 : 2\frac{1}{16}$ from the similarity of triangles ADF and BCF. Also given is $AB = 25$ and that $BC = 33$, being the exact same data of **Q34**, which was solved by means of **Q33** using a multiplication factor. AC and FB so can be calculated and hence FD and the diameter ED. This proof of Ptolemy's theorem, or of 'Ptolemaeus and Copernicus' as referred to by Cardinael, is as yet given completely. Here, as in **Q66**, Cardinael dwells on the application of the distributive law with respect to addition. He does not rely on a geometrical explication as in figure 4.91 in **Q66** despite the reference to it, but instead uses a schematic depiction of the ratios as in figure 4.95, demonstrating that

$$AC \cdot GD = BC \cdot AD$$

added to

$$AC \cdot GB = DC \cdot AB$$

gives the Ptolemy's theorem

$$AC \cdot DB = BC \cdot AD + DC \cdot AB$$

General Comment

This last part where Cardinael provides a proof of Ptolemy's theorem, shows the amount of effort Cardinael has to go through, as most of his contemporaries, for justifying steps which basically are simple formal manipulations, as the use of the distributive law with respect to addition (see **Q66 Otherwise 2**).

Theory used

Answer (similarity)

Properties of the quadrilateral inscribed in the circle with the help of Elem.III-21

Multiplication factor with Regula Falsi and Ptolemy's theorem

Q65

Otherwise

Using **Q34** and **Q33**

Q65

For the proof of Ptolemy's theorem:

Elem III-21;

Elem.VI-16 and Elem.II-1 (geometrical algebra)

Q65

Compare with

Q66

Question 69

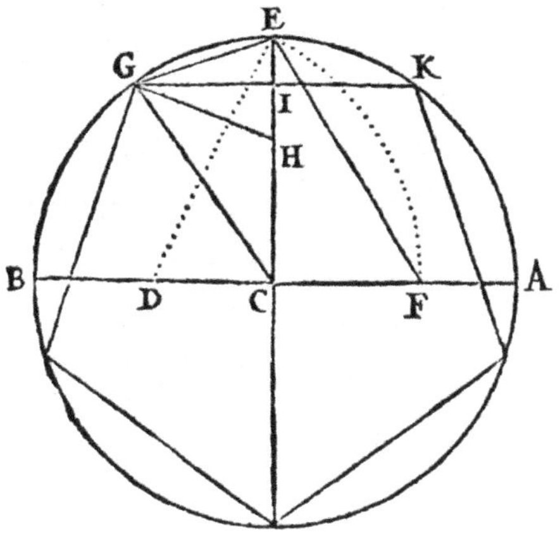

Figure 4.96: Subgroup 8, a pentagon and a decagon in a circle with a given diameter

Comment on Answer

In figure 4.96, Cardinael starts with the construction and calculation of a point F on AC such that

$$DE = DF \text{ (with } D \text{ the center point of } BC)$$

From this construction follows

$$FC = \sqrt{125} - 5$$

assuming that AC is divided by F in mean and extreme ratio, which indeed follows from Elem.VI-30 or from Elem.II-11 or from §3.1.5 case 10 (figure 3.11). The conclusion, referring to **Q35**, that $FC = EG =$ side decagon is not fully correct. In **Q35** the reverse has been proven. In an isosceles triangle with two angles of 72°, a bisector from a vertex of the shortest side divides an opposite side in a mean and extreme ratio. However, as shown in the **General comment** of **Q35**, the converse also can easily be proven. By $CH = CF$ in figure 4.96, CE is divided by H in mean and extreme ratio and the triangle CEG with

$$CH = HG = GE$$

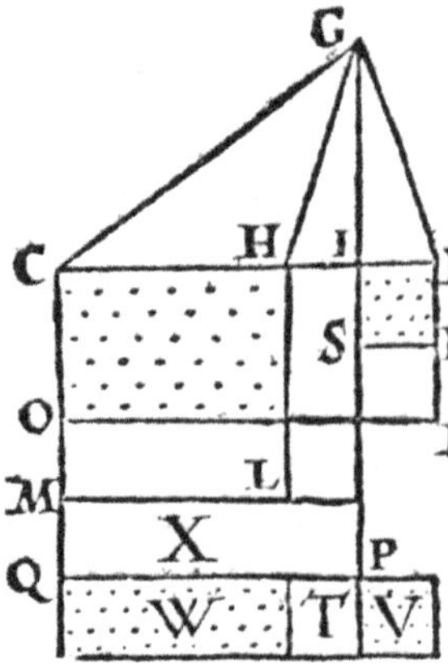

Figure 4.97: Problem 69: demonstration

is then an isosceles triangle with $\angle CGE = \angle CEG = 72°$ and $\angle ECG = 36°$, so $GE = FC = \sqrt{125} - 5$ is the side of a decagon.

To prove $GK = 2GI = FE$ (with GK the side of the pentagon) Cardinael proceeds as follows

$$GK = FE \Leftrightarrow$$
$$GK^2 = EG^2 + CG^2 \Leftrightarrow$$
$$4GI^2 = EG^2 + CG^2 \Leftrightarrow$$
$$EI^2 + IC^2 = 2GI^2$$

so in order to show that $GK = FE$, it suffices to prove this last equality which Cardinael does in two steps.

*First step

For triangle ECG of figure 4.97 holds

$$\angle CEG = \angle CGE = 2 \cdot \angle ECG$$

do follows from **Q35**

$$EH : HC = HC : EC$$

so

$$HC^2 = EH \cdot HC$$

or

$$HCML = ECON$$

as

$$HC^2 = HCML \text{ and } EC \cdot EH = ECON$$

Because

$$HC = EG = HG$$

this leads to

$$\begin{aligned} ECON - GI^2 &= HCML - GI^2 \\ &= HC^2 - GI^2 \\ &= HG^2 - GI^2 \\ &= EI^2 \\ &= EISR \end{aligned}$$

so

$$ECON - EISR = GI^2 \Rightarrow$$

$$NRSICON = GI^2$$

***Second step**

As remarked above, what needs to be proven is

$$EI^2 + IC^2 = 2GI^2$$

Subtracting from both members HC^2 gives

$$EI^2 + \text{ gnomon } IPQMLH = 2GI^2 - HC^2 \Leftrightarrow$$

(because $GI^2 - HCML = EI^2$)

$$EI^2 + \text{ gnomon } IPQMLH = GI^2 - EI^2 \Leftrightarrow$$

$$2EI^2 + \text{ gnomon } IPQMLH = GI^2 \Leftrightarrow \text{ (see figure 4.97)}$$

$$GI^2 = V + T + W + X \text{ (with } W + X = \text{ gnomon } IPQMLH)$$

From figure 4.97 also follows

$$V + T + W + X = NRSICON$$

so what needs to to be proven:

$$GI^2 = NRSICON$$

Which is precisely the result of I, so proven is

$$EI^2 + IC^2 = 2GI^2$$

from which follows

$$EI^2 + GI^2 + IC^2 + GI^2 = 4GI^2 \Leftrightarrow \text{ (Elem.I-47)}$$

$$EG^2 + CG^2 = 4GI^2 \Leftrightarrow$$

$$EG^2 + CG^2 = GK^2 \text{ (because } GK = 2 \cdot GI) \Leftrightarrow$$

$$\text{(side decagon)}^2 + \text{ (radius circumscribed circle)}^2 = \text{ (side pentagon)}^2$$

For the side of the pentagon we find in this way:

$$GK = \sqrt{EG^2 + CG^2} = \sqrt{(\sqrt{125} - 5)^2 + 10^2} = \sqrt{125 - 10\sqrt{125} + 125} =$$

$$= \sqrt{250 - 50\sqrt{5}} = 5\sqrt{10 - 2\sqrt{5}}$$

General comment

Remarkable here is that Cardinael derives a general formula for the relation between the side of the pentagon and the decagon. This formula was also known to Euclid.

The calculation of the side of the decagon was prepared by means of **Q35**. Using **Q1** and figure 4.97, the calculation of GK, the side of the pentagon is easily accomplished. However, Cardinael delivers more than than what was required by deriving by means of geometrical algebra, the relation between the side of the pentagon, the side of the decagon and the radius of their circumscribed circle.

Theory used

Answer (geometrical algebra, Golden Section)

Elem.II-11 of VI-30

§3.1.5 case 10: hyperbolic application (special case: Golden Section)

Elem.I-32

Compare with

Q35

[Petri 1583], ex.29 f. 164^v

[Brasser 1663], question 32

[Ceulen 1615], 253:3

[De Graaf 1672], 158 prop. 136

Elem.IV-11 (same exercise but solved without numbers and similarities, this appears in Elem.VI)

[Sitters 2008], §5.2.2.2 ex. 4, 11, 12, 13

Question 70

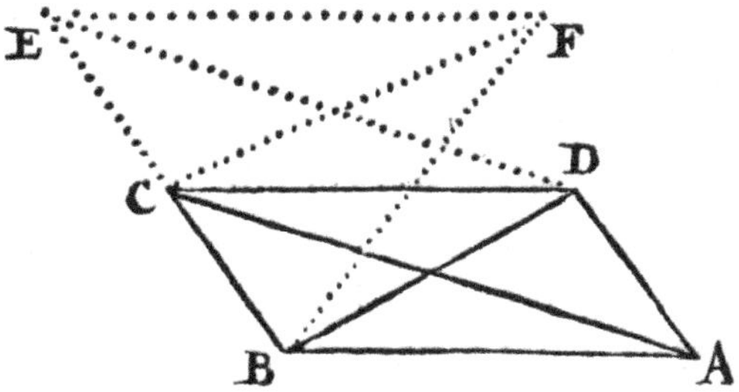

Figure 4.98: Subgroup 9, a quadrangle with a pair of opposite sides of equal length given, $\sqrt{106}$, the diagonals given with lengths 12 and 10 and the other pair of opposite sides of equal length

Comment on Answer

From the given data follows that $ABCD$ is a parallelogram. BC is extended to E so that $EC = BC$. So $ACED$ is a parallelogram with $AC = DE = 12$. From **Q57** follows

$$BD^2 + DE^2 = \text{ constant}$$

if D is on a circle with centre C and radius CD. Point F is chosen such that

$$CF = CD \text{ and } CF \perp BE$$

Then it follows that

$$BF^2 + EF^2 = BD^2 + DE^2 = 100 + 144 = 244 \Leftrightarrow$$

$$2 \cdot BF^2 = 244 \text{ (want } BF = EF)$$

In triangle BCF we have

$$BC^2 = 122 - 106 = 16 \Leftrightarrow$$

$$BC = 4$$

General comment

It is not that obvious that there exists of parallelogram $ABCD$ which satisfies the requirements of **Q70**. The calculation demonstrates that for the existence of such parallelogram, it is necessary and sufficient that the the triangle BDE exists with $BD = 10, DE = 12, DC = \sqrt{106}$ and $BC = EC$. That this is the case, follows from **Q20**.

The solution could have followed **Q20** where a triangle BCA appeared with slanted sides 10 and 12 and D as the center point of BC so that $DA = \sqrt{106}$. Then followed $BD = EC = 4$ corresponding with $BC = CE = 4$ in the case of this **Q70**. In **Q57** the same triangle appears.

Theory used

Answer (congruency of triangles)

Q57, Q20

Compare with

Q20

Question 71

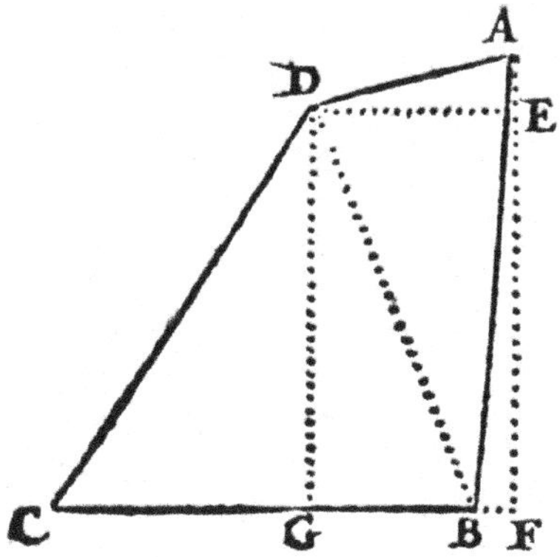

Figure 4.99: Subgroup 9, a quadrangle with three sides and the two diagonals known

Comment on Answer

Following **Q2**, AF can be determined in triangle ABC as $AF = 13\frac{11}{13}$ DG is calculated in triangle BCD following **Q1**: $DG = 12\frac{12}{13}$, resulting in

$$EA = AF - EF = 13\frac{11}{13} - 12\frac{12}{13} = \frac{12}{13}$$

And also

$$BG = \sqrt{14^2 - DG^2} = \sqrt{14^2 - \left(12\frac{12}{13}\right)^2} = 5\frac{5}{13}$$

and

$$FB = \sqrt{AB^2 - AF^2} = \sqrt{192\frac{1}{4} - \left(13\frac{12}{13}\right)^2} = \frac{19}{26}$$

so

$$FG = ED = \frac{19}{26} + 5\frac{5}{13} = 6\frac{3}{26}$$

so that finally

$$AD = \sqrt{(6\frac{3}{26})^2 + (\frac{12}{13})^2} = \sqrt{38\frac{1}{4}}$$

Theory used

Answer

Q1 and **Q2**

Question 72

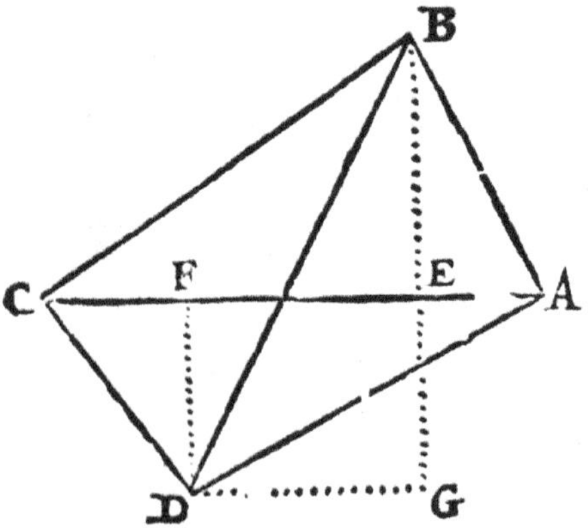

Figure 4.100: Subgroup 9, a quadrangle with its four sides and one diagonal known

Comment on Answer

Following **Q1**, AE and FC can be determined

$$AE = 10 \text{ and } FC = 12$$

From this follow the perpendiculars EB and FD

$$EB = 24 \text{ and } FD = 16 \text{ and so } BG = 40$$

Also

$$EF = AC - (AE + FC) = 42 - 22 = 20$$

and

$$GD = 20$$

Resulting in

$$BD = \sqrt{GD^2 + GB^2} = \sqrt{2000}$$

Theory used

Answer

Q1

Question 73

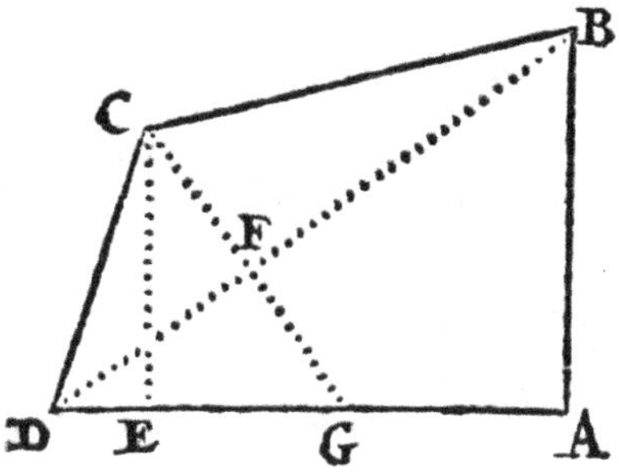

Figure 4.101: Subgroup 9, a quadrangle with its four sides known and a right angle

Comment on Answer

The perpendicular from C to BD intersects BD in F and AD in G. BD can be calculated in the right-angled triangle BAD

$$BD = 56$$

Following **Q1**, DF and FC can be calculated as

$$DF = 20 \text{ and } FC = 15$$

From the similarity of triangles DFG and DAB follows

$$DA : AB = DF : FG \Leftrightarrow 44\frac{4}{5} : 33\frac{3}{5} = 20 : FG$$

and so

$$FG = 15 \text{ and } GC = 15 + 15 = 30$$

From the similarity of BDA and GCE follows

$$BD : AD = GC : EC \Leftrightarrow$$

$$56 : 44\frac{4}{5} = 30 : EC$$

so

$$EC = 24$$

Theory used

Answer (similarity)

Q1

Compare with

Q74: analogous exercise

Q84 with application of **Q73**

Question 74

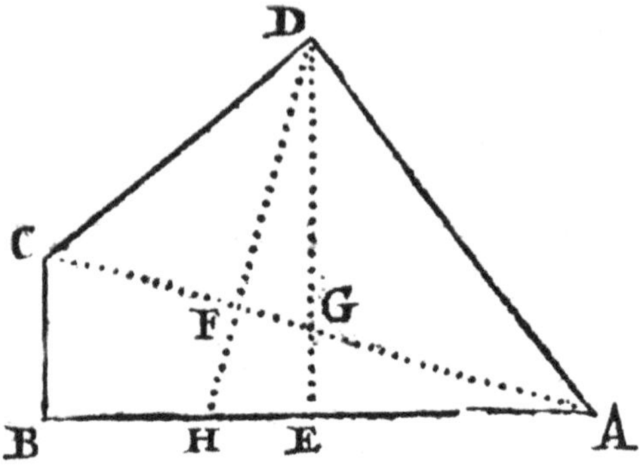

Figure 4.102: Subgroup 9, a quadrangle with its four sides known and a right angle

Comment on Answer

The perpendicular from D on CA intersects CA in F and AB in H. First AC is calculated

$$AC = 50$$

then AF and DE according to **Q1**

$$AF = 32 \text{ and } DF = 24$$

From the similarity of triangles CAB and DGF then follows

$$AB : BC = DF : GF \Leftrightarrow$$

$$48 : 14 = 24 : GF \Leftrightarrow$$

$$GF = 7$$

From the similarity of CAB and DGF follows

$$AB : AC = DF : GD \Leftrightarrow$$

$$48 : 50 = 24 : GD \Leftrightarrow$$

$$GD = 25$$

And further

$$AG = AF - GF = 32 - 7 = 25$$

From the similarity then of ABC and AEG follows

$$AC : CB = AG : GE \Leftrightarrow$$

$$50 : 14 = 25 : GE \Leftrightarrow$$

$$GE = 7$$

so that

$$DE = 25 + 7 = 32$$

Otherwise

Also here AC, AF and FD are first determined. Hence follows from the similarity of ABC and AFH that $FH = 9\frac{1}{3}$ from which follows

$$DH = 24 + 9\frac{1}{3} = 33\frac{1}{3}$$

From the similarity of ABC and DEH follows finally

$$DE = 32$$

Theory used

Answer (similarity)

Q1

Otherwise (similarity)

Compare with

Q73 (same type of exercise), **Q72** (**Q74** is derivable from **Q72**)

Q85: there **Q74** applied

Question 75

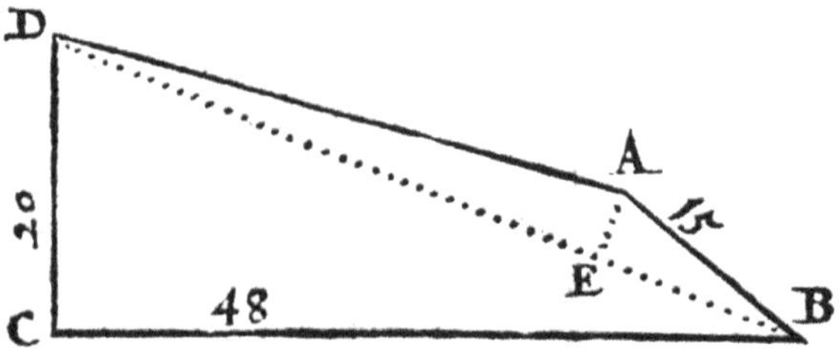

Figure 4.103: A quadrangle with a right angle, with area and three sides given

Comment on Answer

The area of BCD is 480. Therefore

$$\text{area } ABC = 576 - 480 = 96$$

In ABD

$$BD = 52$$

so

$$AE = \frac{96}{\frac{1}{2} \cdot BD} = \frac{96}{26} = 3\frac{9}{13}$$

In AEB follows from this

$$BE = 14\frac{7}{13}$$

and so follows

$$DE = 52 - 14\frac{7}{13} = 37\frac{6}{13}$$

Result: for triangle AED holds

$$AD = \sqrt{(37\frac{6}{13})^2 + (3\frac{9}{13})^2} = \sqrt{1417}$$

Theory used

Answer

Formula for area of triangle

Question 76

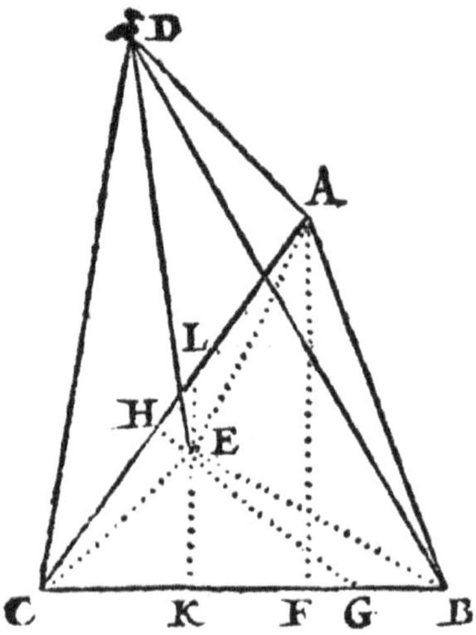

Figure 4.104: Subgroup 13, a tetrahedron with four congruent planes and three pairs of edges with three marksmen on known distances from a parrot and at the same distance from each other

Comment on Answer

The perpendicular from D on BC intersects BC in K.
The perpendicular from A on BC intersects BC in F.
The perpendicular from D on AC intersects AC in H.
Point L is on AC such that LK is the perpendicular of L on BC.
Point G is on BC such that GH is the perpendicular of G on AC.
Point E is the intersection of LK and GH.

From these notational conventions follows that DE is the altitude from D on ABC, because CA is perpendicular DH and GH, so DE is perpendicular CA. But also CB is perpendicular to DK and LK, so DE is perpendicular to CB. From this it follows that DE is perpendicular to the two intersecting lines CA and CB in the plane ABC and therefore DE is the height of the parrot above the plane ABC. According to **Q1**

$$DK = 120, KC = 50, KB = 90, AF = 120, FC = 90 \text{ and } FB = 50$$

so

$$FC : AF = KC : LK \Leftrightarrow$$
$$90 : 120 = 50 : LK \Leftrightarrow$$
$$LK = 66\frac{2}{3}$$

Again according to **Q1**

$$HC = 66$$

from which results

$$FC : AF = HC : GH \Leftrightarrow$$
$$90 : 120 = 66 : GH \Leftrightarrow$$
$$GH = 88$$

and also

$$FC : CA = CH : GC \Leftrightarrow$$
$$90 : 150 = 66 : GC \Leftrightarrow$$
$$GC = 110$$

and for that

$$KG = 110 - KC = 110 - 50 = 60$$

and also, from the similarity of the triangles AFC and GKE

$$AF : FC = KG : EK \Leftrightarrow$$
$$120 : 90 = 60 : EK \Leftrightarrow$$
$$EK = 45$$

From this follows

$$EG = 75 \text{ and } EH = 88 - 75 = 13$$

Applying Elem.I-47 to EKB gives

$$BE = \sqrt{EK^2 + KB^2} = \sqrt{10125}$$

and

$$CE = \sqrt{4525} \text{ and } AE = 85$$

And so the distances of the marksmen to E are known.
To find ED, AE^2 is subtracted from AD^2

$$ED = \sqrt{AD^2 - AE^2} = \sqrt{12375} \approx 111\frac{1}{4}$$

or

BE^2 is subtracted from BD^2 and then the root is taken

or

CE^2 is subtracted from CD^2 and the root is taken,
the result is always $\sqrt{12375}$.

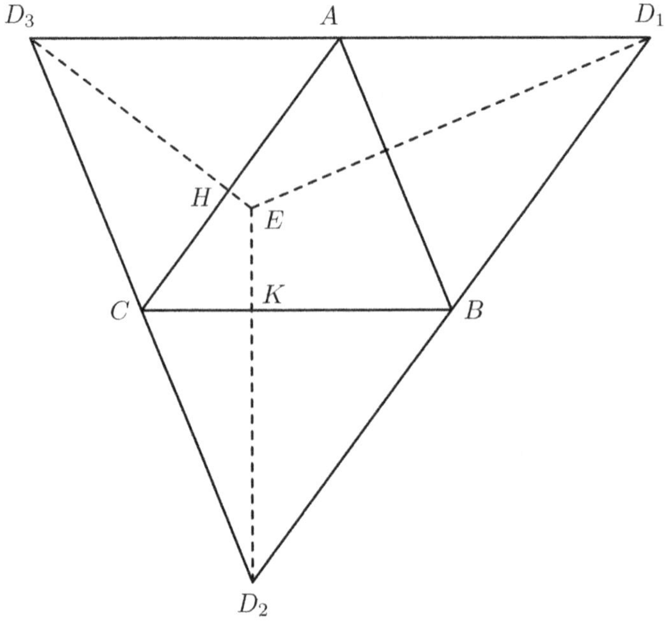

Figure 4.105: Question 76: solution

General comment

In general, it is not self-evident that with given lengths of AD, BD and CD there exists a point D which satisfies the requirements. In this case of the parrot we can take for granted that such a point D exists because the distances can be actually measured. For completeness two proofs of the existence of point D are provided.

1. If the virtual tetrahedron $DABC$ is unfolded, by turning the given faces DAB, DBC and DAC around the edges AB, BC and AC until they all fall into the same plane determined by ABC, figure 4.105 results. The triangles obtained by this process are designated as D_1AB, D_2BC and D_3AC.

   From the data follows that the four triangles ABC, D_1AB, D_2BC and D_3AC are all congruent to each other. A simple calculation of angles learns that the newly arisen figure 4.105 is a triangle with sides $D_1D_2 = 300$, $D_2D_3 = 260$ and $D_1D_3 = 280$.

   From the parallelism of BC an D_1D_3, AC with D_1D_2 and of AB with D_2D_3 follows that the lines D_2K and D_3H are perpendiculars in triangle $D_1D_2D_3$ which intersect in E, so also line D_1E is a perpendicular in triangle $D_1D_2D_3$.

   Without filling in the details, simple reasoning shows that by turning back the triangles D_1AB, D_2BC and D_3AC the three points D_1, D_2 and D_3 will meet each other in one single point D, such that DE is the altitude of the real existing tetrahedron $DABC$.

2. A somewhat easier proof makes only use of one triangle (face), for example D_2BC, lying in the plane ABC. From $D_2K = DK = 120$, $AF = 120$ and $KF = FC - KC = 90 - 50 = 40$ follows that $D_2A = \sqrt{240^2 + 40^2} \approx 243$.

   If D_2BC is turned around BC until D_2 again falls into the plane ABC on the line through A parallel to BC, then $D_2A = KF = 40$. So while turning point D_2, because of continuity, there must have been a point where $D_2A = 140$ holds. This was the required position of point D, as then $DB = 150$, $DC = 130$ and $DA = 140$.

The three three ways in which Cardinael finally calculates ED seem to be inspired by similar considerations as those above to demonstrate the existence of an unique point D or may be motivated by Cardinael's doubts on the existence of point D.

Theory used

Answer

Q1;

Similarity of right-angled triangles

Congruency of triangles

Compare with

Q67

Question 77

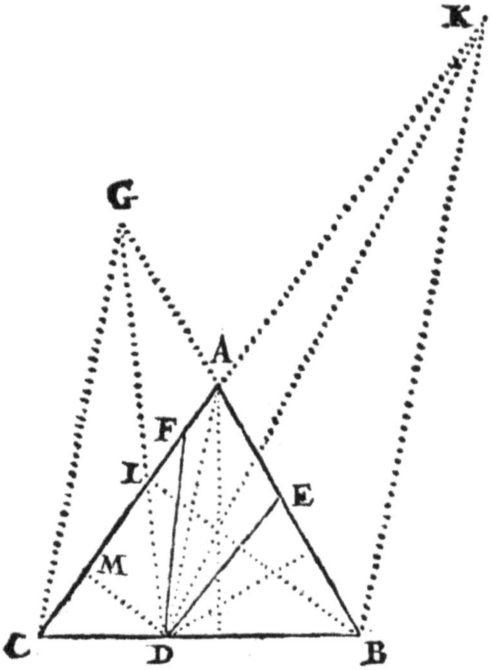

Figure 4.106: Subgroup 11, partitioning a triangle in three parts

Comment on Answer

By drawing the line segments CG and BK parallel to DA there arise two triangles ADG and ADK

$$ADG = ADC \text{ and } ADK = ADB$$

From this follows

$$BDG = BDA + ADG = BDA + ADC = ABC$$

and

$$CDK = CDA + ADK = CDA + ADB = ABC$$

The required line segments are DF and DE with F and E such that CF and BE are equal to a third part of respectively CK and BG (Elem.VI-9).
From the similarity of triangles ADC and KBC follows that

$$KC = \frac{15}{6} \cdot AC = 35$$

and so

$$CF = \frac{1}{3} \cdot 35 = 11\frac{2}{3}$$

In the same way BE can be calculated

$$BE = 7\frac{2}{9}$$

The solution can be generalised to partitioning the triangle ABC into an arbitrary number of parts. A considerable advantage of this approach to the geometer is that it is constructive. The required partitioning can be constructed without any calculation, as it does not matter for where point D is chosen on BC. For example, if D is on BC so that the area $ADB > \frac{2}{3}\cdot$ area ABC (so $CD < \frac{1}{3} \cdot BC$), then by using Cardinael's construction , the points E and F can be found on BA so that $DBE = \frac{1}{3} \cdot ABC$, $DEF = \frac{1}{3} \cdot ABC$ and $DFAC = \frac{1}{3} \cdot ABC$.

Comment on Otherwise
The perpendicular BL on AC can be calculated according to **Q1**: $BL = 12$. From the similarity of the triangles BLC and DMC, with M being the projection of D on AC, follows that $DM = 4\frac{4}{5}$.

As the area of triangle $ABC = 84$, the area of triangle FDC equals 28, so that from $DM = 4\frac{4}{5}$ follows $CF = 11\frac{2}{3}$ and in the same way $BE = 7\frac{2}{9}$.
The solution is clearly not constructive, but again D can be an arbitrary point on BC.

Theory used

Answer

Elem.I-37, VI-1

Otherwise

Q1

Question 78

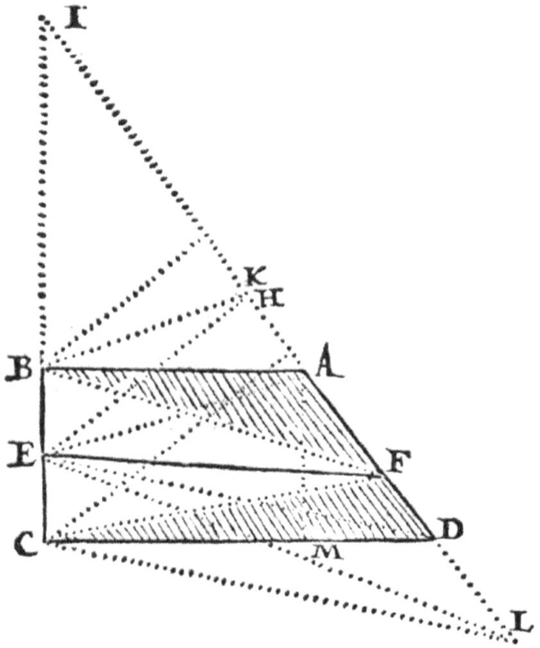

Figure 4.107: Subgroup 11, partitioning of a right-angled trapezium into two equal parts from the center of the right angled side

Comment on Answer

The lines CB and DA intersect each other in I. The perpendicular from E on DI intersects DI in H. The perpendicular of A on CD intersects CD in M. From the similarity of DMA and ABI follows that $BI = 32$ and $EI = 40$. From the similarity of DMA and EHI follows dat $EH = 24$.

K is choosen on AI so that AEB and AEK have the same double area (so according to Elem.I-37, BK is parallel to EA)

$$AB \cdot BE = 24 \cdot 8 = 192$$

Because the height of AEK is equal to $EH = 24$ it follows that the base

$$AK = \frac{192}{24} = 8$$

In the same way DEC is replaced by the equal triangle DEL, with CL parallel to ED. The area of these triangles is equal to

$$EC \cdot DC = 8 \cdot 36 = 288$$

As the height of DEL is equal to $EH = 24$ it follows for the base that

$$DL = \frac{288}{24} = 12$$

Therefore, triangle KEL with base $KL = 40$ and height EH has the same area as $ABCD$. If F is the center point of segment KL (so $FL = 20$) then $DF = 20 - 12 = 8$. F is the required point as LFE equals half of $ABCD$ and also

$$LFE = DFE + DEL = DFE + DEC = DFEC$$

from which follows

$$DFEC \text{ is equal to half } ABCD$$

The method aplied shows some analogy with the method of **Q77**. In **Q77** are constructed from the given starting point D of the unknown dividing lines DE and DF the triangles BDG and CDK, having the same area as the original triangle ABC.

In **Q78** is constructed from the given starting point E of the unknown dividing line EF the triangle KEL having the same area as the original quadrilateral $ABCD$.

Points K and L can be found or constructed as the intersecting points of the lines through B and C parallel to AE and DE with the line ID. Hence, also F can be constructed as the center point of KL. Cardinael also uses this construction in **Q90**, being a generalisation (construction) of this **Q78**. Uncharacteristically, Cardinael does not refer to the propositions of the Elements that he has used (see below).

Though it is not clear from the drawing, points H and K are two different points! As they are not shown as separated points, the lines EH and EK also are drawn as one single line.

Comment on Otherwise

The triangles BEF and CEF have the same area because they have in common the base and the height. So also the (**shaded**) triangles ABF and DCF have the same area. As angles FAB and FDC are each others supplement, we have (Elem.VI-15)

$$\begin{aligned} AB \cdot AF = DC \cdot FD \Leftrightarrow \\ AB : DC = FD : AF \Leftrightarrow \\ (AB + DC) : (FD + AF) = AB : FD = DC : AF \Leftrightarrow \\ 60 : 20 = 24 : FD = 36 : AF \end{aligned} \tag{4.31}$$

so

$$FD = 8 \text{ and } AF = 12$$

The alternative method is much simpler. Also here, there is no reference to the propositions used (see below).

The question can again be solved in a purely constructive way. Firstly, equation (4.31) is derived in a constructive way, without any calculation, with by means of Elem.VI-15, given the condition that E is the center point of BC. Further we have in (4.31) the segments $AB + BC$, $FB + AF$, AB and BC geometrically given from the data. The construction of AF and FD amounts to an application of Elem.VI-12, or the construction of the fourth proportional. As already been remarked, here the position of E as the center point of BC is a necessary condition because of the use of Elem.VI-15.

Theory used

Answer

Elem.I-37, VI-1

Otherwise

Elem.VI-1, VI-15 last part (variant for the case of two angles being each others supplement), V-18, VI-12

Compare with

Q77, 79

Q90 construction of the general partitioning case of any given convex quadrangle and E being an arbitrarily chosen point on a sid)

Question 79

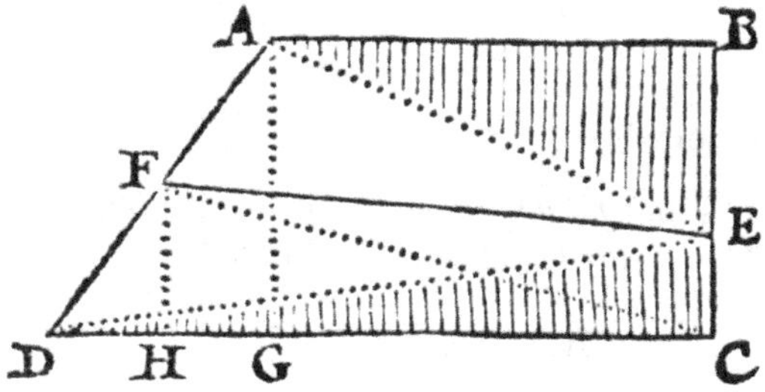

Figure 4.108: Subgroup 11, partitioning a right-angled trapezium from the centre of the slanted side into two parts

Comment on Answer

From the similarity of the triangles FDH and AGD follow FH and DH

$$FH = 8 \text{ and } HD = 6$$

From

$$\text{area } CDF = \frac{1}{2} \cdot FH \cdot DC = 4 \cdot 36 = 144$$

and so

$$\text{area } ECF = \frac{1}{2} \cdot ABCD - CFD = 240 - 144 = 96$$

Resulting in

$$EC = \frac{96}{HC} = \frac{96}{30} = 6\frac{2}{5} \text{ and } EB = 9\frac{3}{5}$$

This solution is clearly not constructive (compare with **Q77 Otherwise**). Point F can be an arbitrary point on AD.

Comment on Otherwise 1

It is possible, as in **Q78 Answer**, to replace $ABCD$ by a triangle with F at the top and the base along BC. Sych base is fixed by the intersection points K and L on the lines through A and D and parallel to FB and FC with the line BC. Point E then is the center point of KL.

This method is analogous to **Q78 Answer**. So the solution is constructive with the position of F on AD being fully arbitrary.

Comment on Otherwise 2

As the areas of EFA and EFD are equal, the same holds for ABE and DCE. As angles B and C are equal, following Elem.VI-15

$$AB \cdot BE = DC \cdot EC \Leftrightarrow$$

$$AB : DC = EC : BE \Leftrightarrow$$

$$(AB + DC) : (EC + BE) = AB : EC = DC : BE \Leftrightarrow$$

$$60 : 16 = 24 : EC = 36 : BE$$

so

$$EC = 6\frac{2}{5} \text{ and } BE = 9\frac{3}{5}$$

The same method as in **Q78 Otherwise**, so the solution is constructive and F must be the center point of AD.

Theory used

Answer

Similarity

Otherwise 1

Elem.I-37, I-38, VI-1

Otherwise 2

Elem.VI-1, VI-15, V-18, VI-12

Compare with

Q78, 90

Question 80

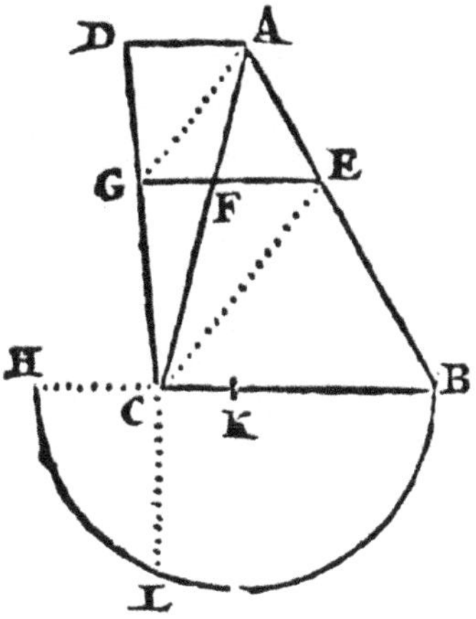

Figure 4.109: Subgroup 10, partitioning a trapezium such that the greatest trapezium has the same area as the greatest triangle and for that same reason the smallest trapezium has the same area as the smallest triangle

Comment on Answer

Firstly, the geometrical mean of BC and AD is constructed and calculated by choosing CH equal to AD and drawing the circle with centre K, the center point of BH. The point of intersection of the perpendicular from C on BH with the circle is L. The geometrical mean is equal to CL. So, for CL holds

$$HC : CL = CL : BC \Leftrightarrow$$

$$CL^2 = AD \cdot BC \Leftrightarrow$$

$$CL^2 = 8 \cdot 18 = 144 \Leftrightarrow CL = 12$$

That EG indeed is the geometrical mean between AD and BC, follows from the equality of triangle ABC and quadrilateral $BEGC$. Hence follows that AFE is equal to FGC and so CAE is equal to CGE. According to Elem.I-37

(Cardinael probably means Elem.I-39) AG and EC are parallel. So AGD and EGC are similar triangles just as AEG and EBC. So

$$AD : EG = AG : EC \text{ and } AG : EC = EG : BC$$

From this follows

$$AD : EG = EG : BC$$

so EG is the geometrical mean of AD and BC and we have $EG = CL = 12$.

Also

$$BC : BE = EG : EA \Leftrightarrow$$
$$BC : EG = BE : EA \Leftrightarrow$$
$$18 : 12 = BE : EA \Leftrightarrow \text{ (with } BE + EA = 25)$$
$$30 : 25 = 12 : EA \Leftrightarrow$$
$$EA = 10 \text{ so } BE = 15$$

In the same way GD and CG can be calculated

$$CG = 13 \text{ and } GD = 8\frac{1}{3}$$

General comment

The solution also leads to a construction method of the required trapezium $BEGC$, as $EG = CL$ with EG parallel to BC can be simply constructed by choosing a point L' on BC such that $CL = CL'$ and drawing a line through L' parallel to DC.

When this exercise is slightly reformulated as follows

> replace a given triangle ABC (calculation or construction) by an equal trapezium $BEGC$ with the same base, and with one base angle in common with a base angle of the triangle and with the other base angle given,

then it becomes clear that the inverse

> replace a given trapezium (calculation or construction) by an equal triangle with the same base and with one base angle in common with a base angle of the trapezium

can be performed in the same way. The inverse then also is a completion of the comparable **Q88**. In the case of the inverse, the equal replacing triangle has the base and one basis angle in common with the original trapezium. In the case of **Q88** the replacing triangle has both base angles in common with the trapezium.

Theory used

Answer

Elem.VI-13 and VI-17: geometrical mean;

Elem.I-39: triangles with equal area and same base

Compare with

Q86 otherwise: here **Q80** is applied without explicitly been noticed

Q87: use of **Q80** without explicitly been noticed

Q88

Q91: here too an important application of **Q80** without being mentioned by Cardinael

Q94: here Cardinael refers to **Q80**

[Sitters 2008], §5.2.2.2 ex. 14

Question 81

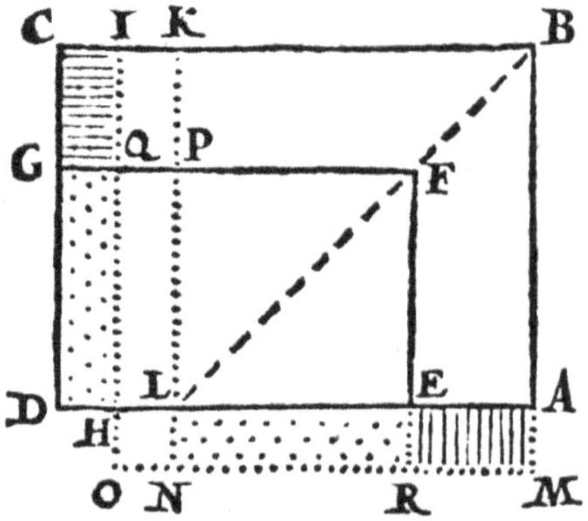

Figure 4.110: Subgroup 11, partitioning a rectangle into two equal parts by means of a constant width area along the two succeeding sides

Comment on Answer

The bisector of angle B intersects AD in point L. A simple calculation leads to

$$\text{area } BCDA = 12 \cdot 16 = 192$$

All sides of the inscribed square $BALK$ are 12, so

$$\text{area } BALK = 144$$

For the circumscribed square $BMOI$, with O on the extension of BL, M on the extension of BA and I on BC between K and C such that $AM = IC = 2$, it holds

$$ICGQ = AMRE$$

which means that $BMOI$ has a constant boundary area equal to the area of $BCDA$

$$\text{boundary area of } BMOI = \text{boundary area of } BCDA = \frac{1}{2} \cdot 192 = 96$$

The area of the square $BMOI$ can be calculated as

$$BMOI = 14^2 = 196$$

By truncating the constant boundary area of 96 from $BMOI$ in figure 4.110 there results for the square $FROQ$

$$\text{area } FROQ = 196 - 96 = 100$$

From this follows

$$OR^2 = 100 \Leftrightarrow OR = 10$$

and in the same way

$$OQ^2 = 100 \Leftrightarrow OQ = 10$$

Finally we have

$$AE = MR = OM - OR = 14 - 10 = 4$$

and

$$CG = IQ = OI - OQ = 14 - 10 = 4$$

General comment

The method has been used by Cardinael before in §3.1.10 and in §3.2.2. Important are the inscribed quadrilateral, here being a square and the circumscribed quadrilateral also a square. The inscribed and circumscribed quadrilaterals are characteristic for this method which will be further generalized in questions **Q81** to **Q84**. Initially for a rectangle in **Q81**, then for right-angled trapezia in **Q82** and **Q83** and finally for a general convex quadrilateral in **Q84.**

A salient theme of this method is that the bisector of the angle at B is also the diagonal of the inscribed square (quadrilateral) $BALK$ and of the circumscribed square (quadrilateral) $BMOI$.

Hence, by truncating a constant boundary area, symmetrically placed around B, from $BMOI$ there remains a similar square (quadrilateral) $FROQ$.

The similarity of inscribed and circumscribed squares (quadrilaterals) allow, by means of extraction of roots of the areas, to calculate the corresponding sides from which next follows the breadth of the constant boundary area.

In the general case of **Q84**, of an arbitrary given convex quadrilateral with its constant boundary area along two succeeding sides, concerns a generalization of the inscribed and circumscribed square.

Then it is possible to calculate the breadth of the required constant boundary area and even to construct one (**Q95**).
Firstly, a schematic outline of Cardinael's calculation in **Answer** will be discussed, elucidating the general method he will develop three following Questions. Then follows an implementation of the outline in six steps, in order to prepare for the general case in **Q84**.

The original rectangle $BCDA$ is not similar to the rectangle $FEDG$ remaining after truncating the boundary area $BCDA$. By using the inscribed square $BALK$ and the similar circumscribed square $BMOI$, resulting from the point multiplication of $BALK$ from point B and multiplication factor

$$k = \frac{BM}{BA} = \frac{BO}{BL} = \frac{BI}{BK} \text{ and with } AM = IC$$

there appears a more favourable situation. Now the diagonals BL and BO of $BALK$ and $BMOI$ coincide with the bisector of the angle at the middle vertex B of the constant boundary. In that favourable case, truncating the boundary area of $BMOI$ results in a square $FROQ$ which is similar to the circumscribed square $BMOI$ and can be considered the result of a point multiplication of $BMOI$ from point O with multiplication factor

$$l = \frac{OR}{OM} = \frac{OF}{OB} = \frac{OQ}{OI} = \sqrt{\frac{FROQ}{BMOI}}$$

In addition, as a consequence of the requirement $AM = IC$ (see above), the rectangles $ICGQ$ and $AMRE$ are also equal and therefore the boundary areas of $BCDA$ and $BMOI$ are also equal (in area).
This leads to the important property that

$$\begin{aligned} &BMOI - \text{ boundary area of } BMOI = \\ &BMOI - \text{ boundary area of } BCDA = \\ &BMOI - \frac{1}{2}BCDA = FROQ \end{aligned}$$

With area $FROQ$ known it is now possible to calculate the multiplication factor l and therefore also OR and OQ and finally AE and CG:

$$OR = l \cdot OM$$

$$OQ = l \cdot OI$$

and so

$$AE = MR = OM - OR$$

$$CG = IQ = OI - OQ$$

Next follows the implementation of the schematic outline in six succeeding steps.

Step 1

$BC = 16,\ CD = 12,\ DA = 16$ and $AB = 12$

so

$$\text{area } BCDA = 196$$

Step 2

all sides of the inscribed square $BALK$ have length 12

so

$$\text{area } BALK = 144$$

Step 3

From

$$\frac{BM}{BA} = \frac{BO}{BL} = \frac{BI}{BK} \text{ and } IC = AM \text{ and } IC + IK = KC = 4$$

follows

$$AM = IC = 2 \text{ and } IK = 4 - 2 = 2$$

and therefore

$$BM = 14 \text{ and } K = \frac{BM}{BA} = \frac{14}{12} = \frac{7}{6}$$

From this follows

$$OM = k \cdot LA = \frac{7}{6} \cdot 12 = 14,\ OI = k \cdot LK = \frac{7}{6} \cdot 12 = 14$$

and

$$BMOI = k^2 \cdot BALK = \frac{49}{36} \cdot 144 = 196$$

Step 4

When a constant boundary area is truncated from square $BMOI$ there remains the square $FROQ$ which is is similar with $BMOI$. Square $FROQ$ can be considered the result of a point multiplication of $BMOI$ from O with the multiplication factor

$$l = \sqrt{\frac{FROQ}{BMOI}}$$

From

$$\begin{aligned}\text{area } FROQ &= \text{area } BMOI \text{ - boundary area}\\ &= 196 - \tfrac{1}{2} \cdot BCDA\\ &= 196 - \tfrac{1}{2} \cdot 192\\ &= 100\end{aligned}$$

now follows

$$l = \sqrt{\frac{100}{196}} = \sqrt{\frac{25}{29}} = \frac{5}{7}$$

Step 5
Finally leading to the calculation of AE and CG

$$OR = l \cdot OM = l \cdot 14 = 10$$

$$OQ = l \cdot OI = l \cdot 14 = 10$$

and so

$$AE = MR = OM - OR = 14 - 10 = 4$$

$$CG = IQ = OI - OQ = 14 - 10 = 4$$

Step 6

The width of the boundary area therefore equals 4.

Theory used

Answer

Circumscribed square, elementary manipulation with rectangles, squares, similarity and point multiplication

Compare with

Q48 for adding a constant boundary area and calculation with the help of right-unfolded border area and the hyperbolic application of area

Q82, 83, 84, 95 for generalizations

[Petri 1583], ex. 73 for an identical question

[Ceulen 1615], 155

[Sitters 2008], §5.2.2.2:

ex. 6 is identical to **Q48**

ex. 10 and ex. 15 are identical to **Q81**

ex. 23 is a variant of **Q81**

ex. 24 is identical to **Q81**

[Huygens Oeuvres], 23 – 27, ex. 4 is an identical question

Question 82

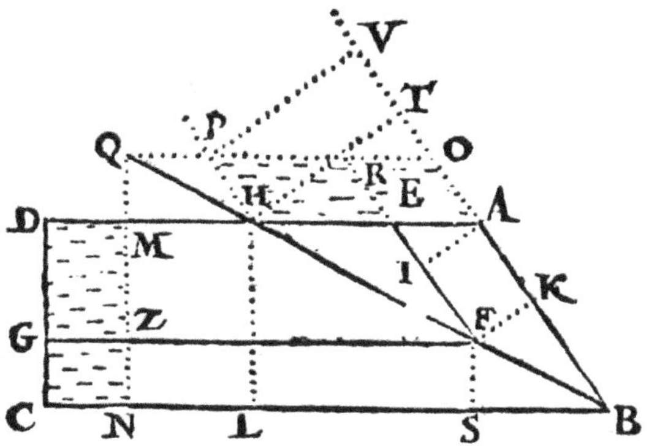

Figure 4.111: Subgroup 11, partitioning a right-angled trapezium into two parts with a constant boundary area along the sides of the acute angle

Comment on Answer

Area $BCDA = \frac{1}{2} \cdot CD \cdot (BC + DA) = 704$. From $AH = AB$ and AD parallel to BC follows that BH is the bisector of the angle at B. We call $BAHL$ the inscribed trapezium. From

$$AD = 38, \ DC = 16 \text{ and } BC = 50$$

follows (Elem.I-47)

$$BA = 20 \ \text{ and for that } AH = 20, \ CL = 18 \text{ and } BL = 32$$

From this follows

$$BAHL = \frac{1}{2} \cdot HL \cdot (AH + BL) = 416$$

N is chosen so that

$$NC : NL = BA : BL$$

so we have

$$NC : NL = 20 : 32$$

From Elem.V-18 and $NC + NL = 18$ follows

$$NC = 6\frac{12}{13} \text{ and } NL = 11\frac{1}{13}$$

Parallelogram $AOPH$ is chosen such that

$$AO = HP \text{ and } HP = MD \text{ so } AO = NC$$

By means of the ratio

$$NC : NL = BA : BL$$

it can be demonstrated that the extensions of NM and OP have their point Q of intersection on the extension of BH (for a general proof see **Q84, Comment on Answer**). For $AOPH$ holds that

$$AOPH = NMDC \tag{4.32}$$

as

$$AO = HP = MD \text{ and } CD = HL = HT$$

For the similarity of $HMQP$ and $BAHL$ follows, with the help of

$$HP = NC = 6\frac{12}{13}, \quad BA = 20 \text{ and } BAHL = 416$$

that for $HMQP$ we arrive at

$$\text{area } HMQP = 49\frac{11}{13}$$

From this follows for the circumscribed trapezium $BOQN$

$$\begin{aligned} BOQN &= BOPHMN + HMQP \\ &= BCDA + HMQP \\ &= 704 + 49\ \tfrac{11}{13} \\ &= 753\ \tfrac{11}{13} \end{aligned}$$

From (4.32) follows

$$\text{boundary area of } BCDA = \text{ boundary area of } BOQN$$

$$\text{so boundary area of } BOQN = \frac{1}{2} \cdot BCDA = 352$$

As BQ is the diagonal of $BOQN$ and also the bisector of the angle at B, the middle vertex of the boundary area, we have

$$BOQN - \text{ boundary area } = FRQZ$$

with the remaining trapezium $FRQZ$ similar to $BOQN$.
For $FRQZ$ we have

$$\text{area } FRQZ = 753\frac{11}{13} - 352 = 401\frac{11}{13}$$

From the similarity of the circumscribed trapezium $BOQN$ to $FRQZ$ follows that their areas are proportional to the squares of corresponding sides, so

$$BOQN : FRQZ = QO^2 : QR^2 = QN^2 : QZ^2$$

and with $BOQN$, $FRQZ$, QN and QO known, QZ and QR can be calculated, and after that AE and CG follow from

$$CG = NZ = QN - QZ$$

$$AE = OR = QO - QR$$

Remark 1. The sides QO and QN of $BOQN$ can be calculated by using the similarity of $BOQN$ and $BAHL$.

Remark 2. Cardinael calculates AE with the help of CG and the similarity of triangles AIE and $AA'B$(AA' is the perpendicular from A on BC).

General comment

The theme of **Q81** is applied again, hereby making use of the specific properties of the trapezium. The given trapezium $BCDA$ with the required boundary area is replaced by the circumscribed trapezium $BOQN$ around the boundary area's bisector BQ of the angle at B, the middle vertex of the boundary area.

By truncating this border area from $BOQN$ there remains a similar trapezium $FRQZ$ and as $BOQN$ and the boundary area ($= \frac{1}{2} \cdot BCDA$) can be calculated, the multiplication factor between $BOQN$ and $FRQZ$ is known and so are the lengths of the sides of $FRQZ$. From this we also have

$$AE = OR = QO - QR, \quad CG = NZ = QN - QZ \text{ and } FK = CG$$

The square $LNOH$ of **Q81** corresponds to trapezium $HPQM$ of **Q82**, and in the same way

$$\begin{array}{rcr} \textbf{Q81}: AM & \Leftrightarrow & \textbf{Q82}: AD \\ IC & \Leftrightarrow & NC \\ IK & \leftrightarrow & NL \\ BALK & \Leftrightarrow & BAHL \\ BMOI & \Leftrightarrow & BOQN \\ FROQ & \Leftrightarrow & FRQZ \end{array}$$

Next, a schematic outline of the above calculation is discussed followed by its implementation in six steps.

The original trapezium $BCDA$ is not similar to the trapezium $FEDG$ remaining after truncating the boundary area $BCDA$.

By using the inscribed trapezium $BAHL$ and the similar circumscribed trapezium $BOQN$, which can be considered a point multiplication of $BAHL$ from point B with the multiplication factor

$$k = \frac{BO}{BA} = \frac{BQ}{BH} = \frac{BN}{BL}$$

under the condition

$$AO = NC$$

has arisen the required configuration in which diagonal BQ coincides with bisector So, truncating the boundary area of $BOQN$ results in a trapezium $FRQZ$ which is similar to the circumscribed trapezium $BOQH$ and can be considered a point multiplication of $BOQN$ from point Q with multiplication factor

$$l = \frac{QR}{QO} = \frac{QF}{QB} = \frac{QZ}{QN} = \sqrt{\frac{FRQZ}{BOQN}}$$

Also, as a consequence of the requirement $AO = NC$ discussed above, the rectangle $NCGZ$ equals the parallelogram $AORE$ and therefore the boundary areas of $BCDA$ and $BOQN$ are equal. This leads to the following important property

$$BOQN - \text{ boundary area of } BMOI$$

$$BOQN - \text{ boundary area of } BCDA$$

$$BOQN - \frac{1}{2} \cdot BCDA$$

$$FRQZ$$

so $FRQZ$ can be calculated.

With area $FRQZ$ known it is now possible to determine the multiplication factor l and therefore also QR and QZ and finally AE and CG

$$QR = l \cdot QO$$

$$QZ = l \cdot QN$$

and so

$$AE = OR = QO - QR$$

$$CG = NZ = QN - QZ$$

Next follows the implementation of the scheme in six consecutive steps.

Step 1

$BC = 50, \quad CD = 16, \quad AD = 38$

from which follows

$BA = 20, \quad HA = 20, \quad BL = 32$

So

$$\text{area } BCDA = \frac{1}{2}(50 + 38) \cdot 16 = 704$$

Step 2

$BA = 20, \quad AH = 20, \quad HL = 16, \quad LB = 52,$

so for the inscribed trapezium we have

$$\text{area } BAHL = \frac{1}{2}(20 + 32) \cdot 16 = 416$$

Step 3

From

$k = \frac{BO}{BA} = \frac{BO}{BH} = \frac{BN}{BL}$ and $AO = NC$

follows

$AO = NC = 6\frac{12}{13}$ and $BO = 26\frac{12}{13}$

so

$k = \frac{BO}{BA} = \frac{26\frac{12}{13}}{20} = \frac{35}{26}.$

From this follows

$OQ = k \cdot HA = 26\frac{12}{13}$, $QN = k \cdot HL = \frac{280}{13} = 21\frac{7}{13}$
and

$$\text{area } BOQN = k^2 \cdot BAHL = (\frac{35}{26})^2 \cdot 416 = \frac{9800}{13} = 753\frac{11}{13}$$

Step 4
When the boundary area is truncated from trapezium $BOQN$ there remains the trapezium $FRQZ$ which is similar to $BOQN$. Trapezium $FRQZ$ can be considered a point multiplication of $BOQN$ from Q with the multiplication factor

$$l = \sqrt{\frac{FRQZ}{BOQN}}$$

From

$$\begin{aligned}\text{area } FRQZ &= \text{ area } BOQN - \text{ boundary area } BOQN\\ &= 753\frac{11}{13} - \frac{1}{2} \cdot BCDA\\ &= 753\frac{11}{13} - 352\\ &= 401\frac{11}{13}\end{aligned}$$

follows

$$l = \sqrt{\frac{401\frac{11}{13}}{753\frac{11}{13}}} = \sqrt{\frac{653}{1225}}$$

Step 5
This leads to the calculation of AE and CG:

$$QZ = l \cdot QN = l \cdot 21\frac{7}{13} = \sqrt{247\frac{49}{169}}$$

$$QR = l \cdot QO = l \cdot 26\frac{12}{13} = \sqrt{386\frac{66}{169}}$$

And so

$$AE = OR = QO - QR = 26\frac{12}{13} - \sqrt{386\frac{66}{169}}$$

$$CG = NZ = QN - QZ = 21\frac{7}{13} - \sqrt{247\frac{49}{169}}$$

Theory used

Answer

Interchanged inner angles equal;
Elem.I-29;
Similarity;
Elem.V-18

Compare with

Q81, 83, 84, 95

Question 83

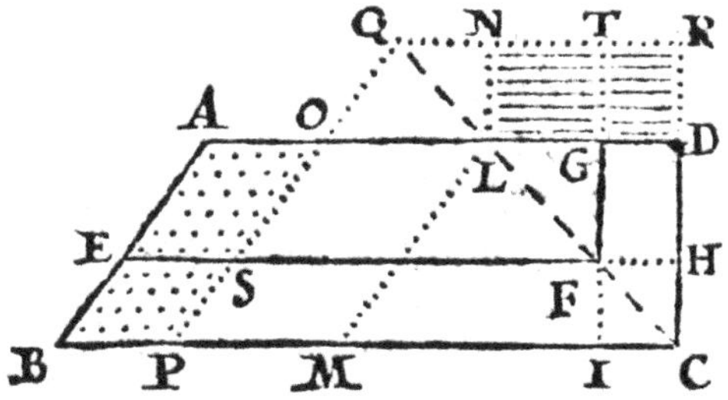

Figure 4.112: Subgroup 11, partitioning a right-angled trapezium into two parts by a constant boundary area along the sides of the right angle

Comment on Answer

area $CBDA = \frac{1}{2} \cdot 16(50 + 38) = 704$

$$AB = \sqrt{DC^2 + (CB - AD)^2} = \sqrt{256 + 144} = 20$$

Choose L on AD so that

$$DL = DC = 16$$

and choose M on BC so that $BM = AL = 22$ and $MC = 28$. Because DA is parallel to CB, CL is the bisector of angle C. $CDLM$ is denoted as the inscribed trapezium of $CBAD$. For $CDLM$ holds

$$CDLM = \frac{1}{2} \cdot CD \cdot (DL + MC) = 352$$

Trapezium $CDLM$ is multiplied from C with a multiplication factor k such that

$$k = \frac{CR}{CD} = \frac{CQ}{CL} = \frac{CP}{CM} \text{ and that also } PB = DR$$

From this choice for the position of P follows

$$\frac{PB}{PM} = \frac{DR}{PM} = \frac{CR - CD}{CP - CM} = \frac{k \cdot CD - CD}{k \cdot CM - CM} = \frac{CD(k-1)}{CM(k-1)} = \frac{CD}{CM} = \frac{4}{7}$$

$CRQP$ is denoted as the circumscribed trapezium of $CBAD$. Following Elem.V-18 and $PB + PM = 22$

$$PB = 8 \text{ and } PM = 14$$

so

$$k = \frac{CR}{CD} = \frac{16+8}{16} = \frac{3}{2}$$

and

$$\text{area } CRQP = (\frac{3}{2})^2 \cdot CDLM = \frac{9}{4} \cdot 352 = 792$$

The constant boundary area of $CRQP$ has the same area as the boundary area of $CBAD$, as

$$PB = DR \text{ and for that } PBES = DRTG$$

so

$$\text{boundary area of } CRQP = \frac{1}{2} \cdot CBAD = 352$$

Moreover $CRQP$ has the required property, unlike from $CBAD$, that by truncating the boundary area, the remaining trapezium $FRQS$ is similar to $CRQP$ and can be considered a point multiplication from point Q with multiplication factor

$$l = \frac{QT}{QR} = \sqrt{\frac{FTQS}{CRQP}}$$

For area $FTQS$ holds

$$FTQS = CRQP - \text{ boundary area} = 792 - 352 = 440$$

so

$$l = \sqrt{\frac{440}{792}} = \sqrt{\frac{5}{9}}$$

Now from

$$QR = k \cdot DL = \frac{3}{2} \cdot 16 = 24 \text{ and } QP = k \cdot LM = \frac{3}{2} \cdot 20 = 30$$

follows

$$QT = l \cdot QR = \sqrt{\frac{5}{9}} \cdot 24 = 8\sqrt{5} \text{ and } QS = l \cdot QP = \sqrt{\frac{5}{9}} \cdot 30 = 10\sqrt{5}$$

so

$$DG = RT = QR - QT = 24 - 8\sqrt{5}$$

and

$$BE = PS = QP - QS = 30 - 10\sqrt{5}$$

General comment

Also here Cardinael relies on the method of **Q81** and **Q82**. The given trapezium $CBAD$ with its required boundary area is replaced through the inscribed trapezium $CDLM$, by a circumscribed trapezium $CRQP$, which has the bisector BQ of the angle at B, being the middle vertex of the boundary area, as diagonal BQ.

By truncating this boundary area there remains a similar trapezium $FTQS$ and as $CRQP$ and his boundary area ($= \frac{1}{2} \cdot CBAD$) can be calculated, the multiplication factor between $CRQP$ and $FTQS$ is known and therefore also the lengths of the sides of $FTQS$. From this follows

$$DG = RT = QR - QT$$

and

$$BE = PS = QP - QS$$

The trapezium $HPQM$ of **Q82** corresponds with the trapezium $LNQO$ of **Q83**, and similarly

$$\begin{array}{rcr} \textbf{Q82}: AD & \Leftrightarrow & \textbf{Q83}: DA \\ NC & \Leftrightarrow & PB \\ NL & \leftrightarrow & PM \\ BAHL & \Leftrightarrow & CDLM \\ BOQN & \Leftrightarrow & CRQP \\ FRQZ & \Leftrightarrow & FTQS \end{array}$$

Also here, first a schematic outline of the above calculation is given, followed by its implementation in six steps.

The original trapezium $CBAD$ and the remaining (not similar) trapezium $FGAE$ are replaced by means of the inscribed trapezium $CDLM$, by the circumscribed trapezium resulting from of a point multiplication of $CDLM$

from point C such that $DR = PB$) $CRQP$ and after truncating the boundary area, there remains a similar trapezium $FTQS$.
Therefore ($\sim$: similarity)

$$CDLM \sim CRQP \text{ (point multiplication with factor} k = \frac{CR}{CD} = \frac{3}{2})$$

$$CRQP \sim FTQS \text{ (point multiplication with factor } l = \sqrt{\frac{5}{9}})$$

from which follows

$$QT = l \cdot LD \text{ and } QS = l \cdot LM \Leftrightarrow$$

$$DG = RT = QR - QT \text{ and } BE = PS = QP - QS$$

and after that the width $FI = FH$ of the boundary area follows directly from

$$FI = FH = DG$$

Summary of the calculation in six steps.

Step 1

$$CB = 50,\ AD = 38,\ DC = 16$$

so

$$\text{area } CBAD = 704 (= \frac{1}{2}(AD + CB) \cdot DC)$$

Step 2

$$CD = 16,\ DL = 16,\ LM = 20,\ MC = 28$$

so

$$\text{area } CDLM = 352 (= \frac{1}{2}(DL + MC) \cdot DC)$$

Step 3
From

$$k = \frac{CR}{CD} = \frac{CQ}{CL} = \frac{CP}{CM} = k \text{ and } DR = PB$$

follows

$$DR = PB = 8 \text{ and } CR = 24$$

so

$$k = \frac{CR}{CD} = \frac{3}{2}$$

From this follows

$$QR = k \cdot DL = \frac{3}{2} \cdot 16 = 24, \quad QP = k \cdot LM = \frac{3}{2} \cdot 20 = 30$$

so

$$\text{area } CRQP = k^2 \cdot CDLM = \frac{9}{4} \cdot 352 = 792$$

Step 4
When the boundary area is truncated from trapezium $CRQP$ there remains the trapezium $FTQS$ which is similar to $CRQP$. Trapezium $FTQS$ can be considered a point multiplication of $CRQP$ from Q with multiplication factor

$$l = \sqrt{\frac{FTQS}{CRQP}}$$

From area $FTQS$ = area $CRQP$ - boundary $CRQP$ = 792 - $\frac{1}{2} \cdot CBAD$ = = 792 - 352 = 440 follows

$$l = \sqrt{\frac{440}{792}} = \sqrt{\frac{5}{9}}$$

Step 5
This leads to the calculation of DG and BE

$$QT = l \cdot QR = \sqrt{\frac{5}{9}} \cdot 24 = 8\sqrt{5}$$

$$QS = l \cdot QP = \sqrt{\frac{5}{9}} \cdot 30 = 10\sqrt{5}$$

so

$$DG = RT = QR - QT = 24 - 8\sqrt{5}$$
$$BE = PS = QP - QS = 30 - 10\sqrt{5}$$

Step 6
From figure 4.112 follows for the width DG of the boundary area

$$DG = 24 - 8\sqrt{5}$$

Theory used

Answer

Similarity;
Elem.V-18

Compare with

Q81, 82, 84, 95

[Sitters 2008], §5.2.2.2 ex. 25

Question 84

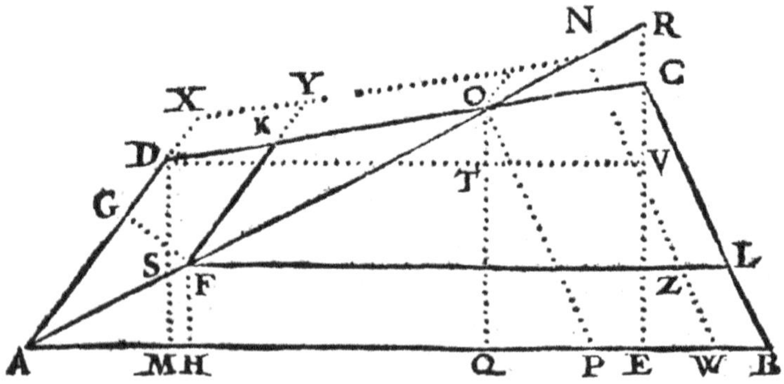

Figure 4.113: Subgroup 11, partitioning any given convex quadrilateral into two parts by constant boundary area along two succeeding sides

Comment on Answer

The bisector of angle A intersects CD in O. Point F is on AO. The perpendicular DM from D on AB intersects AO in S. Point P is on AB such that OP is parallel to BC. $ADOP$ is the so called inscribed quadrilateral.
By partitioning $ABCD$ in three triangles and a rectangle, the area can be calculated. **Q73** is needed to calculate DM. AM is calculated in triangle AMD with AD and MD known: $AM = 12$. So the area of $ABCD$ can be calculated

$$\text{area } ABCD = 1015$$

SD and SM are calculated using that AS is bisector of angle A. Because of the similarity of AMS and AER, line segment ER can be calculated

$$ER = 27 \text{ and then } CR = 5$$

From the similarity of SDO and CRO follows

$$DO = \sqrt{800} \text{ and } CO = \sqrt{200}$$

and for that

$$OT = 4 \text{ and } OQ = 20$$

From the similarity of OQP and CEB follows $PQ = 20$. $TD(= QM)$ is calculated in triangle TDO with OT and DO known

$$TD = QM = 28$$

Then AP can be calculated

$$AP = AM + QM + PQ = 12 + 28 + 10 = 50$$

All sides of $ADOP$ are known now

$$AD = 20, DO = \sqrt{800}, OP = \sqrt{500} \text{ and } PA = 50$$

so the area of $ADOP$ can be calculated

$$ADOP = 700$$

On the extensions of AD and AP are respectively defined the points X and W so that

$$\frac{AD + DX}{AD} = \frac{AP + WP}{AP}, DX = WB \text{ and } WP + WB = PB \tag{4.33}$$

From this definition follows, with a little calculation, the solution to the three equations with three unknowns DX, WB and WP

$$WB = \frac{PB \cdot AD}{AP + AD}, \; DX = \frac{PB \cdot AD}{AP + AD}, \; WP = \frac{PB \cdot AP}{AP + AD}$$

So if k is defined as

$$k = \frac{AD + DX}{AD} = \frac{AX}{AD}$$

and point N on the extension of AO is defined as

$$\frac{AN}{AO} = k$$

then the quadrilateral $AXNW$ is the result of a point multiplication of $ADOP$ from A with the multiplication factor

$$k = \frac{AX}{AD} = \frac{AN}{AO} = \frac{AW}{AP}$$

With $AD = 20,\ AP = 50,\ PB = 15$ we have

$$WB = 4\frac{2}{7},\ DX = 4\frac{2}{7} \text{ and } WP = 10\frac{2}{7}$$

and therefore

$$k = \frac{AX}{AD} = \frac{24\frac{2}{7}}{20} = \frac{17}{14}$$

from which follows

$$AXNW = k^2 \cdot ADOP = (\frac{17}{14})^2 \cdot 700 = 1032\frac{1}{7}$$

Next, it will be demonstrated that the requirement of (4.33) are equivalent with

$$\frac{WB}{WP} = \frac{AD}{AP},\ DX = WB \text{ and } WP + WB = PB \qquad (4.34)$$

So, what needs to be proven is (4.33) $\Leftrightarrow$ (4.34) with abbreviations

$$AX = AD + DX \text{ and } AW = AP + WP$$

(4.33) $\Rightarrow$ (4.34):

$\frac{AX}{AD} = \frac{AW}{AP}$ and $DX = WB$ and $WP + WB = PB \Rightarrow$

$\frac{AD}{AX} = \frac{AP}{AW}$ and $''$ and $''$ $\Rightarrow$

$\frac{AX-AD}{AW-AP} = \frac{AD}{AP}$ and $''$ and $''$ $\Rightarrow$

$\frac{DX}{WP} = \frac{AD}{AP}$ and $''$ and $''$ $\Rightarrow$

$\frac{WB}{WP} = \frac{AD}{AP}$ and $DX = WB$ and $WP + WB = PB$

(4.34) $\Rightarrow$ (4.33):

$\frac{WB}{WP} = \frac{AD}{AP}$ and $DX = WB$ and $WP + WB = PB \Rightarrow$

$\frac{WB}{AD} = \frac{WP}{AP}$ and $''$ and $''$ $\Rightarrow$

$\frac{DX}{AD} = \frac{WP}{AP}$ and $''$ and $''$ $\Rightarrow$

$\frac{DX+AD}{WP+AP} = \frac{AD}{AP}$ and $''$ and $''$ $\Rightarrow$

$\frac{AX}{AW} = \frac{AD}{AP}$ and $''$ and $''$ $\Rightarrow$

$\frac{AX}{AD} = \frac{AW}{AP}$ and $DX = WB$ and $WP + WB = PB$

The original quadrilateral $ABCD$ is not similar to quadrilateral $FKCL$, remaining after truncating the constant border area from $ABCD$. By using the inscribed quadrilateral $ADOP$ and the similar circumscribed quadrilateral $AXNW$, with diagonals AO and AN, starting in the middle boundary vertex A and coinciding with the bisector of the angle at A, the required configuration is reached by truncating the boundary area from $AXNW$ and producing a quadrilateral $FYNZ$ similar to $AXNW$. This quadrilateral $FYNZ$ can be considered a point multiplication from point N with multiplication factor

$$l = \frac{NF}{NA}$$

In addition, as a consequence of the requirement $DX = WB$ (4.33), it holds that

$$DXYK = WBLZ$$

and therefore

$$\text{area of } AXNW = \text{ area of } ABCD = \frac{1}{2} \cdot ABCD = 507\frac{1}{2}$$

which leads to

$$\begin{aligned} \text{area } FYNZ &= \text{area } AXNW \text{ - boundary area of } AXNW \\ &= 1032\,\tfrac{1}{7} - \tfrac{1}{2} \cdot ABCD \\ &= 1032\,\tfrac{1}{7} - 507\tfrac{1}{2} \\ &= 524\tfrac{9}{14} \end{aligned}$$

Therefore $ADOP \sim AXNW$($\sim$: similarity)
(point multiplication from A with factor $k = \frac{AX}{AD} = \sqrt{\frac{AXNW}{ADOP}} = \frac{17}{14}$)
$AXNW \sim FYNZ$
(point multiplication from N with factor $l = \frac{NF}{NA} = \sqrt{\frac{FYNZ}{AXNW}} = \sqrt{\frac{524\frac{9}{14}}{1032\frac{1}{7}}}$)
From this follows

$$NY = l \cdot NX$$

$$NZ = l \cdot NW$$

and also

$$DK = XY = NX - NY$$

$$BL = WZ = NW - NZ$$

and after that, the width of the boundary area follows from the similarity

$$CE : CB = FH : LB \Rightarrow$$

$$\text{width } = FH = \frac{CE \cdot LB}{CB}$$

General comment

Also now, and this time for the most general case, a schematic outline of the derivation of **Answer** will be provided, followed by the implementation in six steps. The seemingly laborious calculation is based on the following idea. If only $ABCD$ and $FKCL$ were similar, in which case the diagonal from the middle vertex of the constant boundary area (here AC) would coincide with the bisector of the angle at that vertex (here AO), then from the known proportion of their areas the ratios of the corresponding sides would follow from the roots of those areas. So, for example

$$\frac{KC}{DC} = \frac{LC}{BC} = \sqrt{\frac{\text{area } FKCL}{\text{area } ABCD}}$$

Furthermore, as the sides of $ABCD$ are known, then KC and LC (and next DK and BL) could be calculated.

Here unfortunately, the situation is not that simple. Reformulating the problem of $ABCD$ and $FKCL$ by the similar quadrilaterals $AXNW$ and $FYNZ$, their difference being a boundary area with the same area and breadth as the area between $ABCD$ and $FKCL$, Cardinael arrives at the ideal configuration of similarity (the quadrilateral $AXNW$ is the generalization of the circumscribed square from §3.1.10 case 18 and also of the circumscribed square $BMOI$ from **Q81**).

He does so by firstly defining $ADOP$ as an inscribed quadrilateral around the bisector AO of angle A with OP parallel to BC. The sides of $ADOP$ and the area can be calculated. The crucial next step is the multiplication of $ADOP$ from point A with a factor k chosen such that $WB = DX$. In that case we have that $WBLZ = DXYK$ so that the boundary area of $ABCD$ has the same outer length, the same breadth and so the same area as the boundary area of $AXNW$. The calculation (as in step 2 below) shows that to this is satisfied when W divides the line segment PB in the ratio $WP : WB = AP : AD$.

With the multiplication factor k known, the sides and the area of $AXNW$ can be calculated. By omitting the boundary area of $AXNW$, there remains the similar quadrilateral $FYNZ$ with known area. As the ratio of the areas of these two quadrilaterals can be calculated and the sides of $AXNW$ are known, the sides of $FYNZ$ can be calculated. Therefore, XY and WZ are known and so DK and LB.

Finally, here is a summary of the above calculation in six steps

Step 1

$AB = 65,\ BC = \sqrt{605},\ CD = \sqrt{1800},\ DA = 20$

so

$$\text{area } ABCD = 1015$$

Step 2

$$AD = 20,\ DO = \sqrt{800},\ OP = \sqrt{500} \text{ and } PA = 50$$

so

$$\text{area } ADOP = 700$$

Step 3

From

$$\frac{AX}{AD} = \frac{AN}{AD} = \frac{AW}{AP} \text{ and } WB = DX \text{ and } WP + WB = PB = 65 - 15 = 50$$

follows

$$DX = 4\frac{2}{7} \text{ and } WP = 15 - 4\frac{2}{7} = 10\frac{5}{7}$$

and therefore

$$AX = 24\frac{2}{7} \text{ and } K = \frac{AX}{AD} = \frac{17}{14} = \sqrt{\frac{AXNW}{ADOP}}$$

From this follows

$$NX = k \cdot OD - \sqrt{1179\frac{29}{49}} \text{ and } NW = k \cdot OP = \sqrt{737\frac{12}{49}}$$

and

$$\text{area } AXNW = k^2 \cdot ADOP = 1032\frac{1}{7}$$

Step 4

$FYNZ$ remains from $AXNW$ when the boundary area of $AXNW$ is truncated, so

area $FYNZ$ = area $AXNW$ - boundary area of $AXNW$
= area $AXNW - \frac{1}{2}\cdot$area $ABCD$

$$= 1032\ \tfrac{1}{7} - 507\tfrac{1}{2} = 524\tfrac{9}{14}$$

from which follows

$$l = \sqrt{\frac{524\frac{9}{14}}{1032\frac{1}{7}}}$$

Step 5

This leads to the calculation of DK and BL

$$NY = l \cdot NX = l \cdot \sqrt{1179\frac{29}{49}} = \sqrt{599\frac{29}{49}}$$

$$NZ = l \cdot NW = l \cdot \sqrt{737\frac{12}{49}} = \sqrt{374\frac{73}{98}}$$

and therefore

$$DK = XY = NX - NY = \sqrt{1179\frac{29}{49}} - \sqrt{599\frac{29}{49}}$$

$$BL = WZ = NW - NZ = \sqrt{737\frac{12}{49}} - \sqrt{374\frac{73}{98}}$$

Step 6

$$FH = FG = 24\frac{2}{7} - \sqrt{299\frac{19}{49}} \text{ (follows from the similarity of triangles)}$$

These steps are also recognizable in the solutions of **Q81, 82** and **83**.

Theory used

Answer

Similarity;

Q73;

Elem.VI-3, V-18

Compare with

Q48, 81, 82, 83, 84, 95;

[Sitters 2008] for a more extensive treatment

Question 85

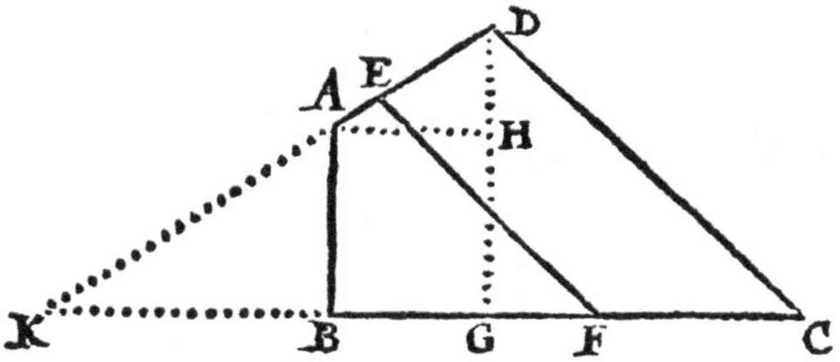

Figure 4.114: Subgroup 11, partitioning a quadrilateral into two equal parts with a line parallel to a side

Comment on Answer

The perpendiculars DG and AH can be found by mean of **Q74**

$$DG = 15, AH = 12 \text{ and } DH = 5$$

From the similarity of DAH and DKG follows

$$KG = 36$$

and with $GC = 20$ follows $KC = 56$.
The area of $ABCD$ can be calculated and equals 300. Half of it is 150, so $EFCD = 150$. The area of triangle KDC equals

$$300 + KAB = 300 + \frac{1}{2} \cdot (56 - 32) \cdot 10 = 420$$

If the area of $EFCD$(=150) is subtracted there results 270 for KEF. From the similarity of KEF and KDC then follows

$$KEF : KDC = 270 : 420 \Leftrightarrow$$

$$KF^2 : 56^2 = 270 : 420 \Leftrightarrow$$

$$KF = \sqrt{2016}$$

so

$$FC = 56 - \sqrt{2016} \text{ and } ED = 39 - \sqrt{977\frac{11}{14}}$$

Theory used

Answer

Q74;

Elem.VI-19: proportion of the areas of similar triangles

Question 86

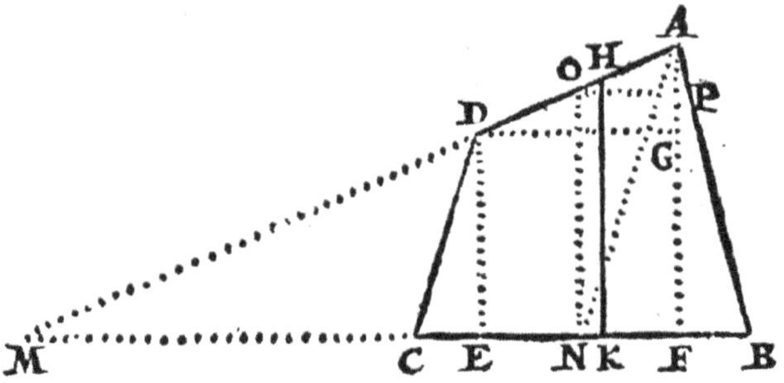

Figure 4.115: Subgroup 11, partitioning a quadrilateral into two equal parts with a line perpendicular to one side of the quadrilateral

Comment on Answer

From the similarity of AGD and DEM follows

$$GA : GD = ED : EM \Leftrightarrow$$

$$10 : 30 = 30 : EM \Leftrightarrow$$

$$EM = 90$$

so for the area of DEM holds

$$DEM = \frac{1}{2} \cdot ED \cdot EM = 15 \cdot 90 = 1350$$

Adding

$$EDHK = 550$$

gives

$$HKM = 1350 + 550 = 1900$$

$EDHK$ can be calculated by means of

$$CDE = 150, AFB = 200, DEFG = 900 \text{ and } AGD = 150$$

Furthermore

$$HK^2 = 1266\frac{2}{3}$$

so

$$HK = \sqrt{1266\frac{2}{3}}$$

The calculation of HD and KC leads to

$$DEM : EM^2 = HKM : KM^2 \Leftrightarrow$$

$$1350 : 8100 = 1900 : KM^2 \Leftrightarrow$$

$$KM = \sqrt{11400}$$

so

$$KC = KM - CM = \sqrt{11400} - 80$$

Elem.I-47 applied in HKM and DEM gives

$$HM = \sqrt{1266\frac{2}{3}} \text{ and } DM = \sqrt{9000}$$

so

$$HD = \sqrt{1266\frac{2}{3}} - \sqrt{9000}$$

Cardinael's calculation here is rather confusing, depending on the identity $HK = \sqrt{1266\frac{2}{3}}$. This may come as a surprise, but Cardinael assumes that the reader understands why, as from the similarity of the triangles HKM and DEM follows that

$$\frac{HK^2}{DE^2} = \frac{HKM}{DEM} \Leftrightarrow$$

$$HK^2 = 900 \cdot \frac{1900}{1350} = 1266\frac{2}{3}$$

Otherwise

From $ABCD = 1400$ follows that $BN = \frac{ABCD}{AF} = \frac{1400}{40} = 35$ so

$$BN = 35 \text{ and } FN = 25$$

From the similarity of AGD and APO follows

$$GD : GA = OP : AP \Leftrightarrow$$

$$30 : 10 = 25 : AP \Leftrightarrow$$

$$AP = 8\frac{1}{3}$$

so

$$PF = 31\frac{2}{3} \text{ and } ON = 31\frac{2}{3}$$

Finally

$$ON \cdot AF = 31\frac{2}{3} \cdot 40 = 1266\frac{2}{3}$$

so

$$HK = \sqrt{1266\frac{2}{3}}$$

because HK is the geometrical mean of AF and ON.

This alternative calculation of HK depends on a point N so that the triangle AFN has the same size as the trapezium $AFKH$. The point N is fixed by the requirement that

$$BN \cdot AF = ABCD$$

from which follows that

$$BN = 35 \text{ and } FN = 25$$

So

$$ABN = ABKH$$

and also

$$AFN = AFKH$$

so that the configuration of **Q80** has been reached from which it can be concluded that HK is the geometrical mean of AF and ON. ON is calculated by means of the similarity of triangles and then HK can be calculated because $HK^2 = AF \cdot ON$. It is remarkable that Cardinael does not refer to **Q80** while the configuration of this question is essential for solving the problem.

Theory used

Answer

Similarity;

Elem.VI-19;

Q80

Question 87

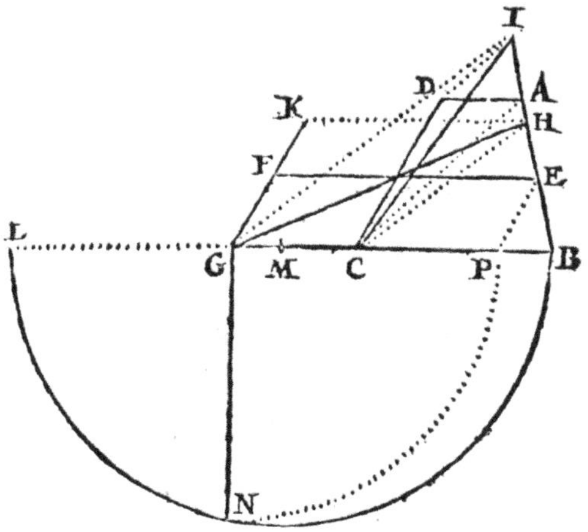

Figure 4.116: Subgroup 10, transforming a trapezium into a trapezium with common angle and equal area

Comment on Answer

$CBAD$ is transformed into a triangle CBI by drawing the line DI parallel to CA (Elem.I-37). Triangle CBI is transformed into a triangle BGH by drawing CH parallel to GI (Elem.VI-2 and VI-15).
Draw a line through G parallel to CD.
Draw HK parallel to BG with K on the previously drawn line through G.

As the required trapezium $GBEF$ has to be equal in area to BGH, the configuration of **Q80** has been reached, with EF as the geometrical mean of

BG and HK.

EF is constructed in the inclosed figure in the usual way according to Elem.VI-13, by means of line segment GL equal to HK, with M center point of BL, with ML radius of the semicircle and perpendicular GN in GB equal to EF. Hence $GBEF$ is found and can be constructed.

Theory used

Answer

Elem.I-37, VI-2 and VI-15, VI-13

Q80

Question 88

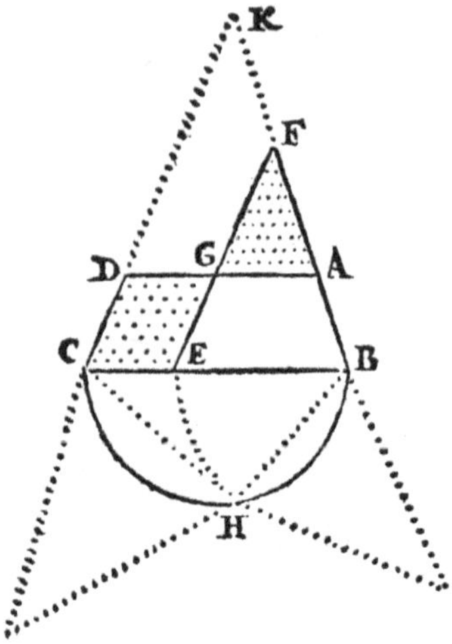

Figure 4.117: Subgroup 10, transforming a trapezium into a triangle with the same base vertex

Comment on Answer

Construct a semicircle with BC as diameter and point H on the arc of a circle such that $CH = DA$. Choose E on BC so that $BE = BH$. Then BHC is a right triangle with the similar triangles at its sides, BKC on BC, DAK on CH and BEF on BH. From Elem.VI-31 follows that

$$BCK = DAK + BEF$$

and from this it can be concluded that

$$AGF = CEGD \text{ and } CBAD = BEF$$

General comment

A fine application of Elem.VI-31.

Theory used

Answer

Elem.VI-31: generalisation of Elem.I-47 (the Pythagorean theorem)

Compare with

Q80

Question 89

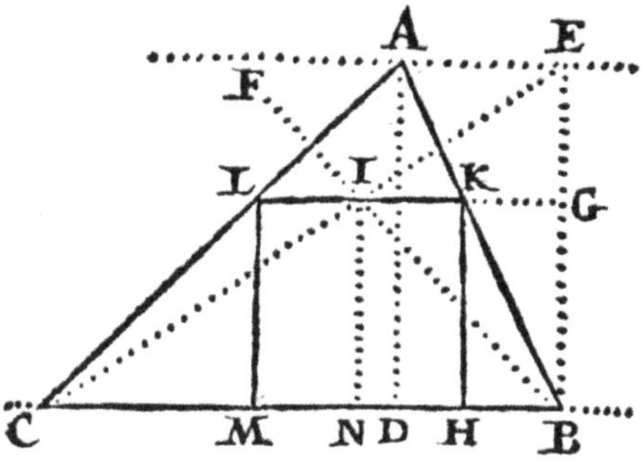

Figure 4.118: Subgroup 10, a square of maximal area inscribed inside a triangle

Comment on Answer

Triangle ABC is transformed into the right triangle EBC with equal area. The required construction of triangle EBC now becomes easy. The bisector of angle EBC intersects EC in I. The perpendiculars from I on BC and BE intersect these sides in N and G. Square $GINB$ is the maximal square in EBC and if EBC is again transformed into the original triangle ABC this square becomes the required maximal square $HKLM$ in ABC.

Given the solution he provides, Cardinael's question is ambiguous. For each side of the triangle there exists exactly one square along the side with both vertices on both the other sides. So, in total there are three of those squares. It cannot be determined in advance which of these three squares is the largest. The length of the side of such a square is equal to

$$\frac{h_a \cdot a}{h_a + a} (= \frac{2a\triangle}{a^2 + 2\triangle})$$

with a the side opposite to vertex A, h_a is the perpendicular from A on a and $\triangle$ the area of triangle ABC. Cardinael constructs (or calculates as in **Otherwise**) only one of these squares without demonstrating that it is the largest of the three possible squares.

Otherwise

The alternative solution uses concrete numbers for the case where $AD = 12$ and $BC = 14$. Calculting the product, i.e. the double surface, and divided by the sum (=26) of $BE(= AD = 12)$ and $BC(= 14)$ this results in $6\frac{6}{13}$ for $GI(= NI)$. And so the sides of the required square $HKLM$ equal $6\frac{6}{13}$.

In this numerical example, ordinary geometrical algebra is applied

$$ABC = EBC \Leftrightarrow$$

$$IBC + IBE = ABC \Leftrightarrow$$

$$IN \cdot BC + IG \cdot BE = 2\triangle \Leftrightarrow \text{(with } IN = IG)$$

$$IN \cdot (BE + BC) = AD \cdot BC \Leftrightarrow$$

$$\text{(with } IN = HK \text{ and } AD = 12 \text{ and } BC = 14 \text{ and } BE = AD)$$

$$HK \cdot 26 = 168 \Leftrightarrow$$

$$HK = KL = \frac{168}{26} = 6\frac{6}{13}$$

Theory used

Answer

Elem.I-37 and I-41

Question 90

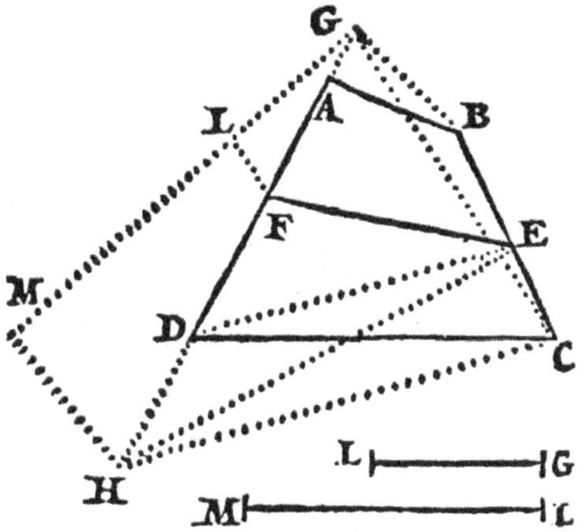

Figure 4.119: Subgroup 11, partitioning a quadrilateral at a point on one side into two parts according to a given ratio

Comment on Answer

BG is drawn parallel to EA and CH parallel to ED. The triangle EAG then is equal to triangle EAB and also EDH is equal to EDC (Elem.I-37). Hence, the whole triangle GHE is equal to the quadrilateral $ABCD$. By means of any line GM drawn to L so that

$$GM = GL + LM$$

GH can now be partitioned at point F such that FL is parallel to HM in the ratio

$$GF : FH = GL : LM \text{ (Elem.VI-2)}$$

Also

$$EGF : EHF = GL : LM \text{ (Elem.VI-1)}$$

And as $GEF = ABEF$ and $EHF = EFDC$, EF is the required line.

General comment

In figure 4.119, EG is mistakenly denoted as CG and EA is missing.

Theory used

Answer

Elem.I-37, VI-2 and VI-1

Question 91

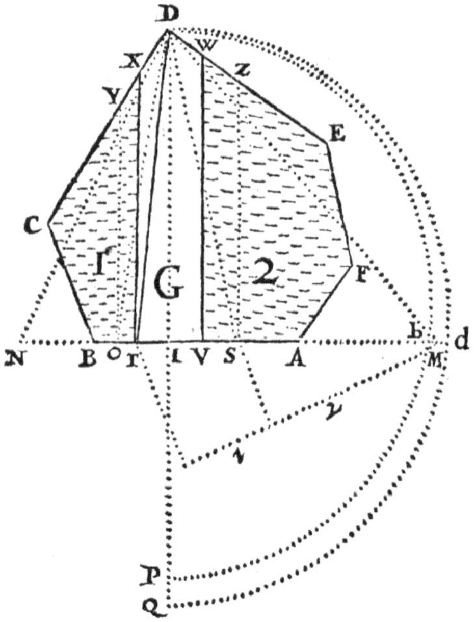

Figure 4.120: Subgroup 11, partitioning a hexagon into two parts with the ratio of one to two by means of a boundary area of a prescribed size

Comment on Answer

The hexagon of figure 4.120 is first transformed into a triangle with equal area having top D, base MN and perpendicular DI. Quadrilateral G in figure 4.121 is also transformed in a right triangle HKL of equal area with $KL = DI$ and right angle K, by means of Elem.I-45. Point T is chosen so that $TN = HK$ and S in such a way on MT that

$$MS : ST = 2 : 1 \text{ (Elem.VI-2)}$$

Now $NO = ST$ from which follows directly that $SO = NT$ and $SO = HK$. Then $MS : NO = 2 : 1$ and $SO = HK$. So

$$DMS : DNO = 2 : 1 \text{ (Elem.VI-1)}$$

and for the intermediating triangle DOS we have $DOS = G$.

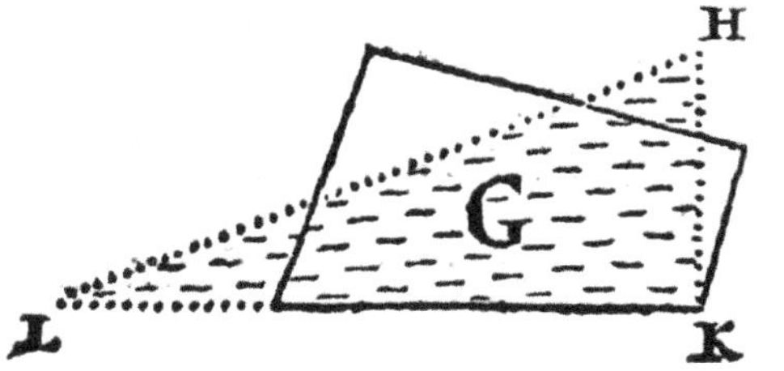

Figure 4.121: Question 91: transforming a quadrangle into a triangle

The problem is to transform triangle DOS to a border area $VWXT$ of the same size. This can be achieved by applying twice **Q80**. DIS is replaced by the trapezium $DIVW$ and DIO is replaced by the trapezium $DITX$. The letter T being used in connection with the perpendicular TX on AB indicates a point different from the letter T initially used to denote line segment NT. The two points accidently coincide in the drawing.

The perpendicular erected in O intersects CD in Y and the perpendicular in S intersects DE in Z. The geometrical means of OY and DI and of SZ and DI are respectively TX and VW.

TX is constructed with the help of the circle with

$$DP = DI + IP = DI + OY$$

as diameter and Ib perpendicular on DP as geometric mean between DI and IP, so for TX is chosen Ib. VW is constructed with the circle that has as diameter

$$DQ = DI + IQ = DI + SZ$$

and has Id perpendicular on DQ as geometric mean between DI and IQ, so for VW is chosen Id. According to **Q80** one arrives at

$$DIO = DITX \text{ and } DIS = DIVW$$

So follows that border area $VWDXT$ is equal to DOS and so to G.
From the construction of M and VW also follows that

$$DMS = DEFAS = WEFAV$$

and also follows from the construction of N and TX that

$$DNO = DCBO = XCBT$$

so that

$$WEFAV : XCBT = DMS : DNO = 2 : 1$$

General comment

Also here Cardinael curiously does refer to **Q80** whcih is essential for the solution and neither in **Q86 Otherwise**.

Theory used

Elem.I-37, I-45, VI-15, VI-21 and VI-13;

Q80

Question 92

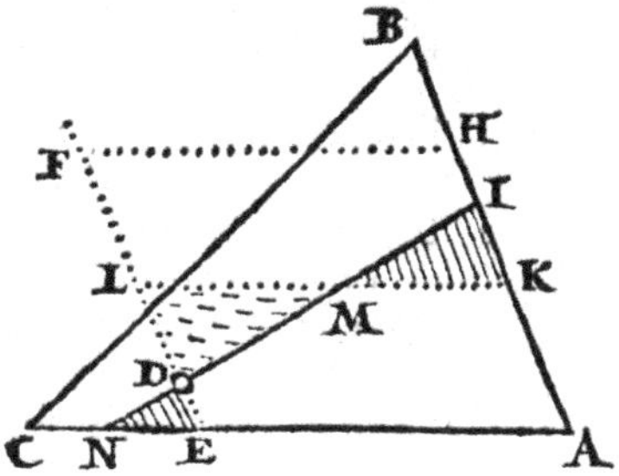

Figure 4.122: Subgroup 11, partitioning a triangle by a line segment through a given point inside the triangle

Comment on Answer

To construct the parallelogram $HAEF$ one can proceed as in figure 4.124. First parallelogram $BACP$ is made equal to twice the triangle ABC

$$BACP = 2 \cdot ABC$$

Then $BACP$ is divided into two equal parts by the line segment TU. The line through point D parallel to AB intersects AC in E and $BP = G$. Quadrilateral $UACT$ is thus equal to ABC

$$UACT = ABC$$

EG and TU intersect in S and the extension of AS intersects CP in W.

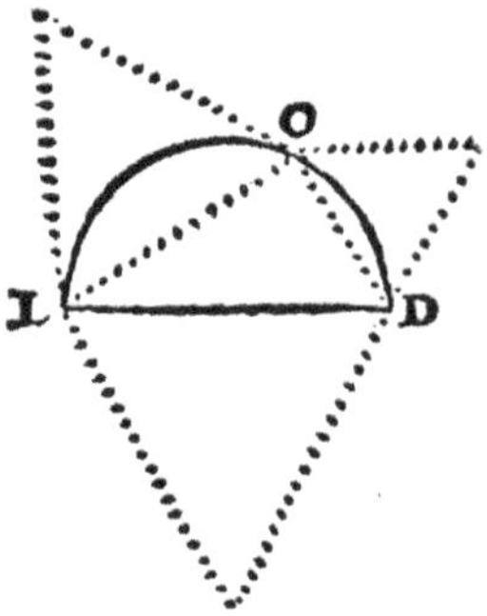

Figure 4.123: Question 92: solution

Now, by drawing WH parallel to CA with F the intersection point of WH and EG, then for parallelogram $HACW$ it holds that (Elem.I-43)

$$SECT = HUSF \Leftrightarrow$$

$$UACT = HAEF \Leftrightarrow$$

$$HAEF = ABC$$

Now, KL is drawn parallel to AE such that $AK = KH$ and $EL = LF$. So KL divides $HAEF$ in two equal parts $KLEA$ and $HFLK$with:

$$KLEA = \frac{1}{2}ABC$$

Let now NI be the line through D that intersects KL in M. Then line NI cuts off triangle MDL from $KLEA$. The problem is now to construct NI in such a way, that holds true

$$MDL = IKM + DEN$$

because in that case we have

$$IAN = KLEA = \frac{1}{2}ABC$$

so NI is the required partitioning . This line NI can be constructed by means of Elem.VI-31. In figure 4.124, DL is the diameter of a circle with semicircle

DOL and point O on this semicircle so that $DO = DE$ (figure 4.122 or 4.124). If point I is chosen on AH such that

$$KI = OL$$

then it follows from the similarity of MDL, DEN and IKM and with the help of Elem.VI-31, that

$$MDL = IKM + DEN$$

and for that NI is the line that was to be constructed.

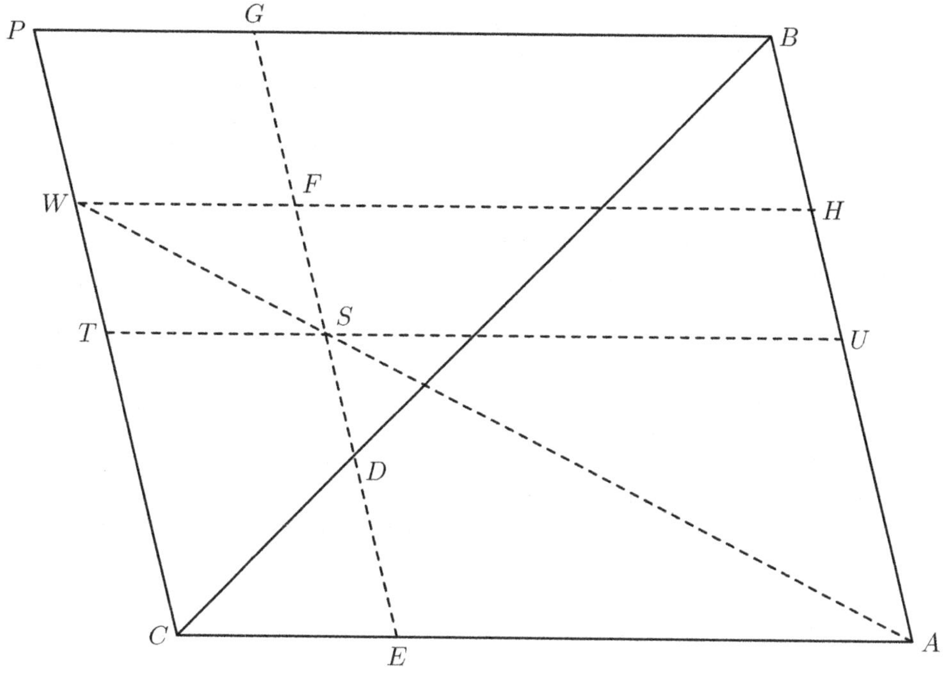

Figure 4.124: Question 92: the construction of a parallelogram

General comment

Cardinael does not mention the number of possible solutions. However, it can be seen that this depends of the location of point D when considering the two extreme cases. If D is a vertex on one of the sides then there is exactly one solution (comparable construction in **Q77**), but if D coincides with the point of gravity of the triangle, then there are three solutions possible, being the three medians.

Van Schooten jr. in his book ([Schooten 1660], 107 – 110) describes the construction of a line that divides a triangle according to a prescribed proportion both with the point inside and outside the triangle. In both cases, he essentially relies on Cardinael's method, though for the this construction he acknowledged Pappus of Alexandria (about 330) who in his *Colectionem Mathematicarum*, (7th book, proposition 164) also uses Elem.VI-31 for a construction analogous to the present case in figure 4.123, that is the construction of OL in order to choose KI in figure 4.122 equal to OL and to conclude that NI is the required line.

In this connection, it would be interesting to know if Cardinael was familiar with Pappus's *Collectionem*, all more so as in **Q0** (Title page, the Pythagorean theorem) he employs transformations of rectangles and parallelograms in a way similar to the Pappus's method in a variation on the Pythagorean theorem ([Heath 1956], I, 366). Nevertheless, Cardinael deserves all the credit for being one of the first to apply Elem.VI-31 to the problem of the partitioning a triangle in this surprising way.

Cardinael's predecessors like Simon Stevin , Christoph Clavius and Ludolf van Ceulen dwelled at length on such problems. Clavius explicitly refers to Stevin's preparatory work ([Gravelaar 1902],146) and Van Ceulen corresponded on it. In this exchange Stevin acknowledged not being able to solve these important problems satisfactory. After his acquaintance in 1585 with the work of Nicolas Tartaglia and his pupil Johannes Baptista Benedictus, he is capable to conclude his general theory on the problems of partitioning polygons which appears in his *Wisconstige gedachtenissen* [Stevin 1605-8]. Cardinael's simple, elegant and purely geometrical solution from 1612 for the case of the triangle is superior to Stevin's rather artificial approach from the point of pure geometry.

Theory used

Answer

Elem.I-43

Elem.VI-31

Compare with

Q53

Q88

[Brasser 1663], 362, problem 32: algebraic solution

[Ceulen 1615], 213

[Schooten 1660], 107, 110

[Stevin 1605-8], 144, propositions 8 and 9

§5.4 Partitioning a triangle

Question 93

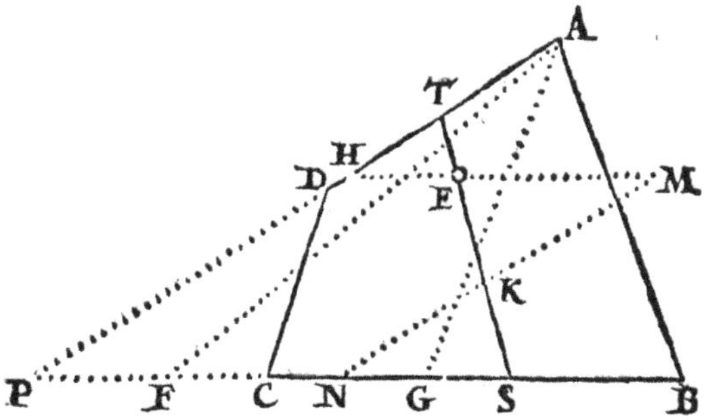

Figure 4.125: Subgroup 11, partitioning a quadrilateral at a point inside the quadrilateral into two equal parts

Comment on Answer

The extension of AD intersects the extension of BC in P. Point F is constructed so that

$$ABF = ABCD$$

so that AC is parallel to DF (Elem.I-37). G is chosen so that $GF = GB$, so AG divides ABF into two equal parts (Elem.I-38). Construct HM parallel to PB and MN parallel to PA so that

$$HMNP = AGP \text{ (as in } \mathbf{Q92}\text{)}$$

The line segment TS intersects MH in E and MN in K so that in the similar triangles HET, MEK and NSK holds true

$$MEK = HET + NSK (\text{as in } \mathbf{Q92})$$

To construct NS (and so to know point S) the semicircle is constructed with ME diameter and within the right triangle with HE one of the right-angled sides. The other right-angled side is NS. From the construction of TS follows

$$TSP = HMNP$$

and so

$$TSP = AGP$$

but

$$TSCD = AGCD$$

and because we have

$$ABCD = ABF$$

also holds true

$$AGCD = AGF = \frac{1}{2} \cdot ABCD$$

from which follows that $TSCD = \frac{1}{2} \cdot ABCD$ so that TS must be the line to be constructed through E.

Theory used

Answer

Elem.I-37 (construction F), I-38 (construction G),

I-42 (construction $HMNP$), VI-31 (construction NS);

Q92

Compare with

Q92

Question 94

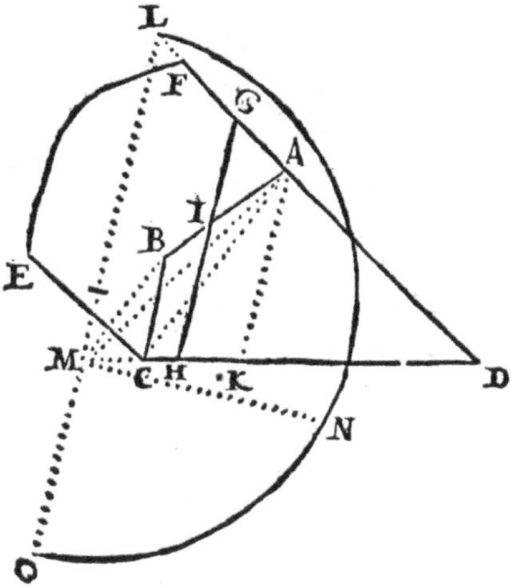

Figure 4.126: Subgroup 11, replace an angled ditch dividing a land into two parts by a rectilinear one so that both parts keep their size

Comment on Answer

Firstly, quadrilateral $ABCD$ is transformed into the triangle AMD with equal area (Elem.I-37). The line segment AK with K on DC is parallel to BC. Drawing from M, the line ML parallel to BC, there arises the trapezium $AKML$ with the inscribed triangle MAK with

$$MAK = ABCK$$

According to the method of **Q80** but now using the geometrical mean MN of ML and AK, which can be constructed according to Elem.VI-13 by means of $MO = AK$, the semicircle with LO as diameter and MN perpendicular on LO, the trapezium $GHML$ can be constructed with

$$GH = MN$$

and

$$GHML = MAL$$

So

$$GHKA = MAK$$

and

$$GHKA = ABCK$$

And so GH is the rectilinear ditch to be constructed.

Theory used

Answer

Elem.I-37, VI-13;

Q80

Question 95

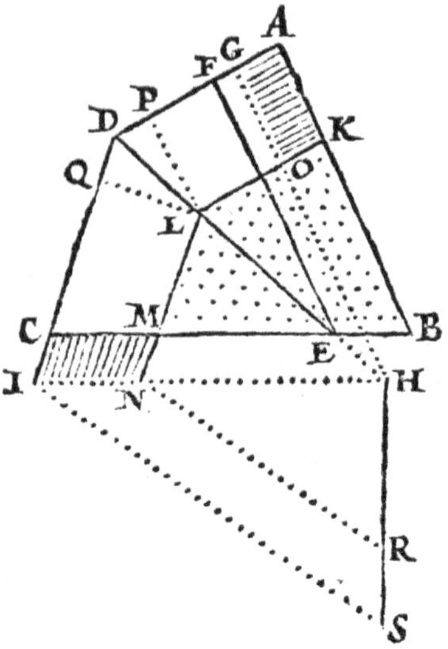

Figure 4.127: Subgroup 11, partitioning any convex quadrangle into two parts by means of a constant border area along two succeeding sides

Comment on Answer

Point G is constructed by means of Elem.VI-10, such that G divides FA in the ratio

$$DC : DF = GA : GF$$

and I is chosen on the extension of DC such that

$$CI = GA$$

We then have also (see **Q84**)

$$\begin{aligned}
&DC : DF = CI : GF \Leftrightarrow \\
&DC : CI = DF : GF \Leftrightarrow \\
&(DC + CI) : (DF + GF) = DC : DF \Leftrightarrow \\
&DI : DG = DC : DF \Leftrightarrow \\
&DI : DC = DG : DF.
\end{aligned}$$

If point H is chosen on the extension of DE such that

$$\frac{DH}{DE} = \frac{DI}{DC} = \frac{DG}{DF} = k$$

then follows from this ([Heath 1956], 2, 231) that $DIHG$ has been constructed from $DCEF$ by means of a point multiplication from D with multiplication factor k, so

$$DCEF \sim DIHG \text{ (multiplication factor } k = \frac{DH}{DE}) \Leftrightarrow$$

$$DH = k \cdot DE,\ DI = k \cdot DC \text{ and } DG = k \cdot DF$$

As $CI = GA$, the resulting constant border area $DIHG$ equals and has the same breadth as the border area $KADCMLK$ of $DCBA$, that has to be constructed. By means of Elem.I-45 the squares I and II can be constructed, such that:

$$I = DABC$$

and

$$II = \frac{1}{2} \cdot DABC \text{ (= area } KADCMLK = \text{ area } OGDINLO)$$

Let HT be the constructed side such that

$$HT^2 = II = \frac{1}{2} \cdot DABC = \text{ constant border area of } DIHG$$

In the same way a line segment HS can be constructed from the known $DIHG$, such that

$$HS^2 = DIHG$$

Now if we subtract the square HT^2 from the square HS^2 then we get the square HR^2 such that

$$HR^2 = HS^2 - HT^2$$

Adding and subtracting the squares always happen in the standard way by using Elem.I-47 in right-angled triangles, arriving at

$$HR^2 = HS^2 - HT^2 \Leftrightarrow$$

$$HR^2 = \text{ quadrilateral } LNHO$$

with

$$DIHG \sim LNHO \text{ (with factor } \frac{HO}{HG} = \frac{HL}{HD} = \frac{HN}{HI} = l)$$

The next step now is to construct the quadrilateral $LNHO$ similar to $DIHG$ and with an area that equals HR^2. The claim is that by drawing RN, NL and LK parallel to respectively SI, ID and DA as in figure 4.127, the required quadrilateral $LNHO$ has been constructed. From figure 4.127 follows

$$\frac{HS}{HR} = \frac{HI}{HN} = \frac{HD}{HL} = \frac{HG}{HD}$$

so $LNHO$ is the result of a point multiplication of $DIHG$ from point H with multiplication factor $l = \frac{HO}{HG}$. From the construction follows that

$$DIHG - LNHO$$

is a constant boundary area of $DIHG$ for which we have

$$\text{area } = HS^2 - HT^2 = DIHG - \frac{1}{2} \cdot DABC$$

But because of $CINM = GAKO$ we can conclude that also for the constant boundary area $KADCMLK$ of $DABC$ we have

$$KADCMLK = \frac{1}{2} \cdot DABC$$

General comment

The construction above is completely analogous to the calculation of **Q84**, up to the point of the construction of quadrilateral $DIHG$

$DCEF$(**Q95**) corresponds to $ADOP$(**Q84**)

$DIHG$(**Q95**) corresponds to $AXNW$(**Q84**)

point N(**Q95**) corresponds to point Y(**Q84**)

The construction of point N by means of squares HS^2 and HT^2 in **Q95** is mirroring the use of the areas $AXNW$ and $FYNZ$ in **Q84**. The salient feature of this construction is the possibility of constructing $DIHG$ with H on the bisector of the angle at D, the middle vertex of the boundary area of $DIHG$, and with G such that $CI = GA$ and so $CINM = GAKO$.

So, just like in **Q84** the original problem of the constant boundary area for $DABC$ is transformed into the much simpler problem for a constant boundary area (with equal area) for $DIHG$.

Further remark that the the partitioning of triangle ACB into two equal parts with a line parallel to AC is an extreme case of **Q95**, with the angle at D equal to 180 degrees. If in this case, DE becomes the perpendicular BD from B on AC, then point L is situated such that $BL : BD = 1 : \sqrt{2}$, because then we have

$$KMB : ACB = 1 : 2$$

Hence, the construction of point L is a standard procedure.

Theory used

Answer

Elem.I-45, VI-10

[Heath 1956], 2, 231

Q84

Compare with

Q84

Question 96

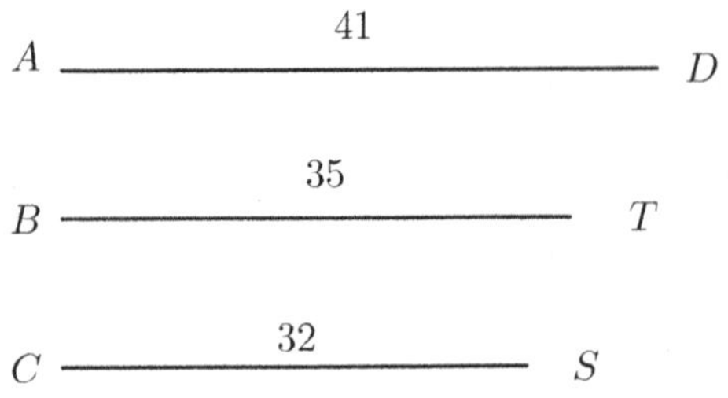

Figure 4.128: Subgroup 12, constructing a triangle with given catheti

Comment on Answer

AD, BT and CS in figure 4.128 are the perpendiculars from A, B and C of triangle ABC (Cardinael's references to AB, BC and CD are not correct and must be read as AD, BT and CS).
So we have

$$AD \cdot BC = BT \cdot AC = CS \cdot AB \tag{4.35}$$

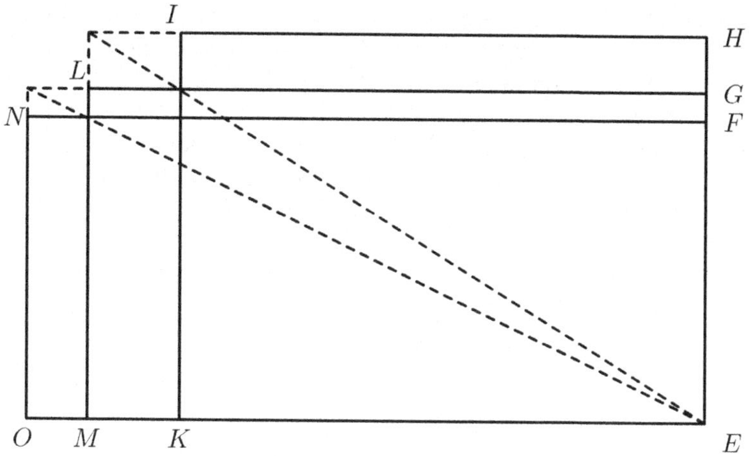

Figure 4.129: Question 96: the construction of three line segments

Figure 4.129 shows the construction of the three line segments, based on Elem.I-43 with $EH = AD, EG = BT$ and $EF = CS$, EK, EM and EO, with EK arbitrarily chosen, satisfying the relation (4.35), so

$$AD \cdot EK = BT \cdot EM = CS \cdot EO \tag{4.36}$$

From (4.35) and (4.36) follows

$$EK : BC = EM : AC = EO : AB \tag{4.37}$$

From another way of performing the above construction , this time of the three line segments EK', EM' and EO', with EK' again arbitrarily chosen, also satisfying the relation (4.35), so

$$AD \cdot EK' = BT \cdot EM' = CS \cdot EO' \tag{4.38}$$

follows, again this time with the help of (4.35) and (4.38)

$$EK' : BC = EM' : AC = EO' : AB \tag{4.39}$$

This implies that the required ABC, still being unknown, is similar with such a triangle PQR, determined by $EK(= PR)$, $EM(= QR)$ and $EO(= QP)$

$$QPR \sim ABC$$

The triangle ABC can therefore be constructed, first by a triangle QPR as in Elem.I-22 and then by the perpendicular QD on PR as in figure 4.38. On QD can be constructed point A such that AD equals the given line segment.
The lines through A parallel to QP and QR intersect PR in the points B and C. From B and C can be constructed the perpendiculars BT and CS on the sides AC and AB. From this construction follows directly that AD, BT and CS equal the given line segments AD, BT and CS in figure 4.128.

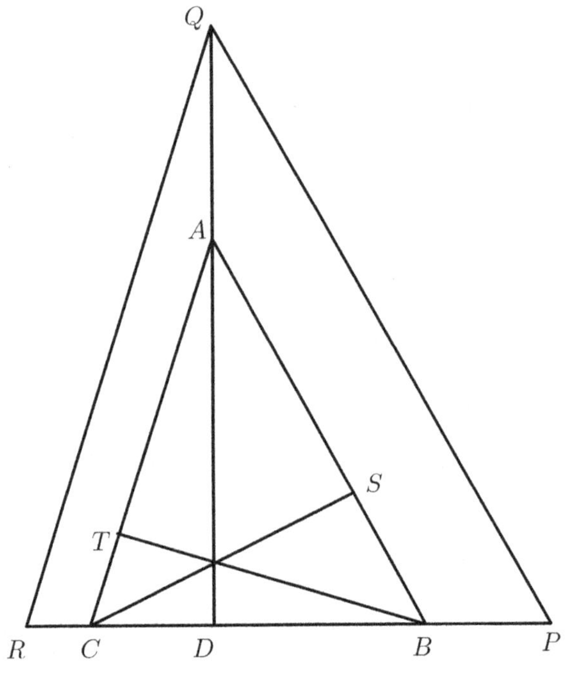

Figure 4.130: Question 96: solution

General comment

One can consider ABC a point multiplication from D of triangle QPR with multiplication factor $\frac{AD}{QD}$. This same type of construction is used in **Q97**. The construction in Cardinael's figure is not correct. The line segments QR and AC are interchanged with the segments QP and AB as in our figure 4.130).

Theory used

Answer

Elem.I-43, I-22, similarity

Compare with

Q97, with also a figure is constructed similar to the required one

Q30, where in analogue question the calculation is required instead of a construction

Question 97

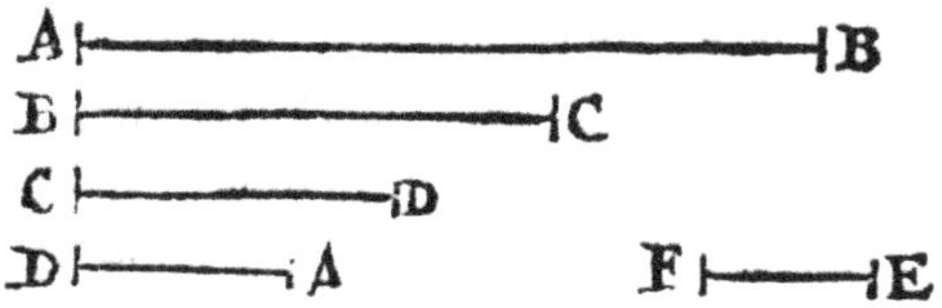

Figure 4.131: Subgroup 12, construction quadrilateral inscribed in a circle

Comment on Answer

The line segment FE could be any segment, though in this question, the square $KLMN$ to be constructed needs to be larger than the square $OPQL$ which equals $BC \cdot DA + AB \cdot CD$).

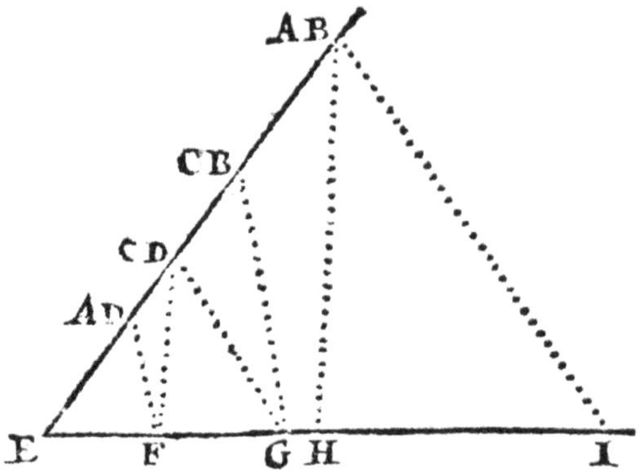

Figure 4.132: Question 97: constructing the diagonals

As shown in figure 4.132, the segments EG, EI and EH are constructed on basis of the ratios

$$\mathbf{AD} : \mathbf{BC} = ED : EC = \mathbf{EF} : \mathbf{EG}$$
$$\mathbf{CD} : \mathbf{AB} = EC : EB = \mathbf{EG} : \mathbf{EI}$$
$$\mathbf{CD} : \mathbf{AB} = ED : EA = \mathbf{EF} : \mathbf{EH}$$

These ratios are derived from figure 4.134 where the required quadrilateral $ABCD$ is multiplied from the intersection E of the diagonals with a certain factor and transformed into the similar quadrangle $HIGF$. Within the quadrilateral $ABCD$, triangle DEC is similar to triangle AEB and triangle BEC similar to triangle AED.

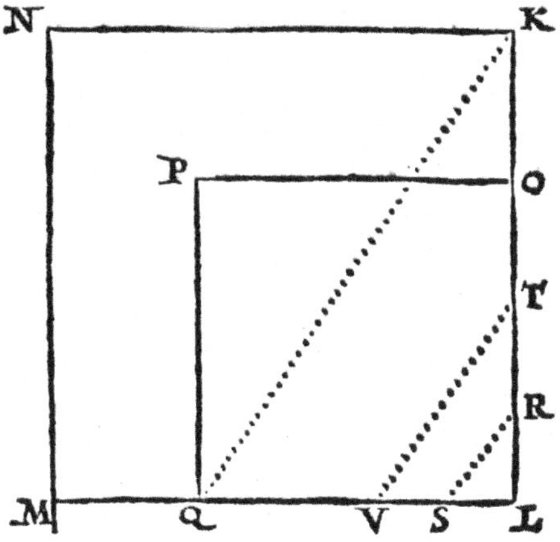

Figure 4.133: Question 97: the construction of the squares

The ratios which are rendered bold are used to construct from the assumed line segment EF the segments EG, EI and EH as in figure 4.132.
This facilitates the construction of diagonals FI and HG, according to

$$FI = EF + EI$$

$$HG = EG + EH$$

Using these diagonals, belonging to quadrangle $HIGF$ similar to $ABCD$, a rectangle can be constructed which can be transformed to the equal square $KLMN$.

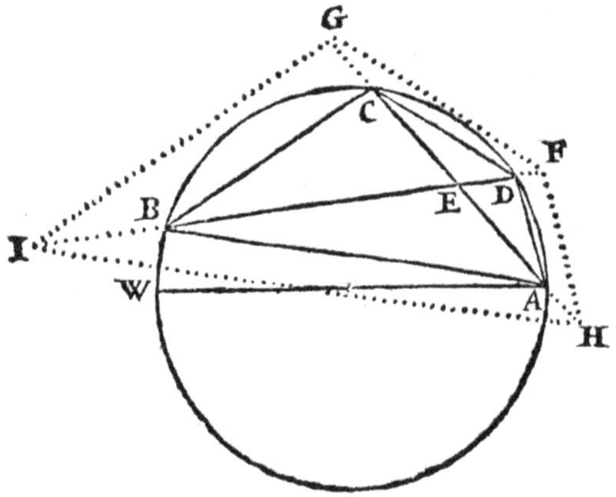

Figure 4.134: Question 97: the quadrangle inscribed in a circle

If the rectangles $AB \cdot CD$ and $BC \cdot DA$ are added together, they form a square $OPQL$ which, according to **Q68** (Ptolemy, Copernicus) has its area equal to the product of the diagonals AC and BD

$$AC \cdot BD = AB \cdot CD + BC \cdot DA$$

From

$$KLMN : OPQL = HIGF : ABCD$$

follows

$$LK : LQ = FE : DE$$

and as in figure 4.133 LR equals FE and RS is parallel to KQ:

$$LK : LQ = LR : LS = FE : LS$$

so

$$DE = LS$$

In the same way, by choosing LT equal to EG and TV parallel with KQ the line segment LV can be constructed

$$CE = LV$$

As all its sides are known, triangle DEC can now be constructed. In the same way also triangles DEA and AEB can be constructed and so quadrilateral $ABCD$ is known.

General comment

The requirement that the sum of three of the given line segments is larger than the fourth segment implies a single solution [Molenbroek 1943, 373, ex. 66]. The construction is based on the following idea. From the required quadrilateral $ABCD$, which can be inscribed in a circle, the sides are known but not the diagonals AC and BD. Call the intersection point of the (unknown) diagonals E and assume that with a point multiplication from E the unknown line segment ED transforms in an arbitrarily chosen line segment EF and $ABCD$ transforms into $HIGF$ as in figure 4.134.

The properties of the cyclic quadrilateral make it possible, as in figure 4.132, to construct the diagonals HG and FI of quadrilateral $HGIF$, similar to $ABCD$, without knowing its sides. Figure 4.132 can be confusing as all segments have to be seen with their starting point at E, which makes that FI can not be seen as the diagonal FI. The diagonal FI arises from the addition of line segments EF and EI.

Because now the product of the diagonals of the two similar cyclic quadrilaterals is known the lengths of the diagonals AC and BD can be constructed. E.g., the product of AC and BD is known because of the expression $AC \cdot BD = AB \cdot CD + BC \cdot DA$) and thus the proportion of the corresponding line segments in the two quadrilaterals is known, as shown in figure 4.133. As the sides of triangles EDC, ECB, EAB and EAD are known, they now can be constructed.

Theory used

Answer

Ptolemy's theorem for quadrilateral inscribable in a circle;

Elem.VI-20, VI-2, III-25;

Similarity

Compare with

Q68;

[Molenbroek 1943], 366 – 373

Question 98

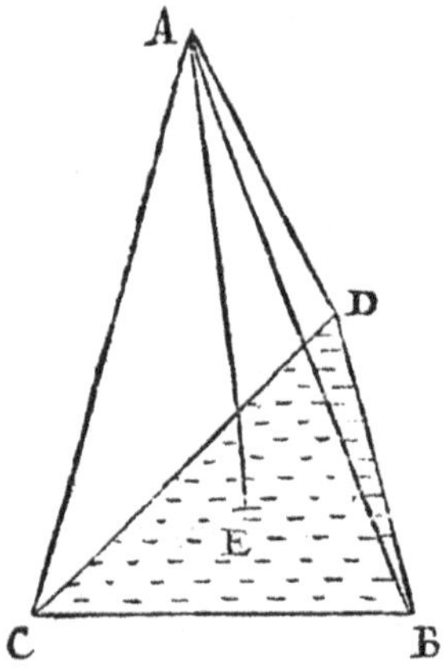

Figure 4.135: Subgroup 13, the volume of a tetrahedron with two congruent lateral faces and three pairs of edges of 13, 14 and 15

Comment on Answer

From the solution method of **Q76** follows that $AE = \sqrt{123\frac{3}{4}}$. Its third being $\sqrt{13\frac{3}{4}}$. The area of the base BCD is equal to 84. So the volume of the pyramid will be

$$\text{volume} = 84 \cdot \sqrt{13\frac{3}{4}} = \sqrt{97020} \approx 311\frac{1}{2}$$

Theory used

Answer

Q76

Q1

Elem.XII-7 (volume pyramid)

Compare with

Q99

Question 99

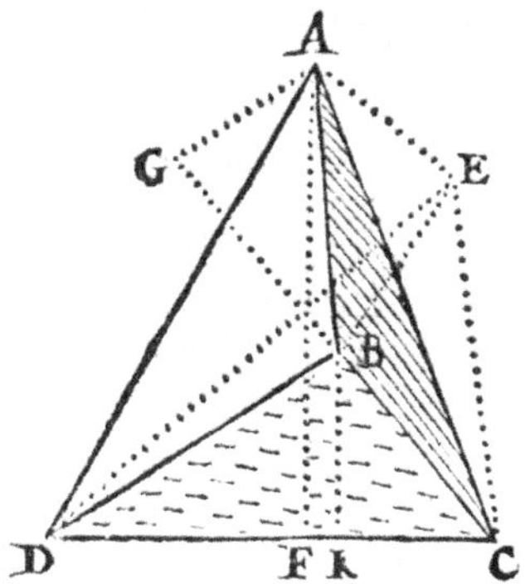

Figure 4.136: Subgroup 13, the volume of any inclining tetrahedron, with the perpendicular from the top outside of the base

Comment on Answer

AF is the perpendicular from A on CD. According to **Q1**, $FD = 5\frac{5}{13}$ and so $CF = 7\frac{8}{13}$. BK is the perpendicular of B on CD and AE is the perpendicular of A on the plane of the base BCD. Also, AG is the perpendicular from A on BC and so, according to **Q2**, $BG = 6\frac{11}{34}$ and $CG = 8\frac{1}{2} + 6\frac{11}{34} = 14\frac{14}{17}$. The line FE is perpendicular to DC as $DC \perp AF$ and $DC \perp AE$ and intersects CG in H. To determine CH, first CK and BK have to be calculated. From **Q1** follows

$$CK = 5\frac{5}{13}$$

and from Elem.I-47 follows

$$BK = 6\frac{15}{26}$$

Now

$$CK : CB = CF : CH \Leftrightarrow$$

$$5\frac{5}{13} : 8\frac{1}{2} = 7\frac{8}{13} : CH \Leftrightarrow$$

$$CH = 12\frac{3}{140}$$

So

$$HG = CG - CH = 14\frac{14}{17} - 12\frac{3}{140}$$

And also, because of the similarity of BKC and EGH, for $CG \perp AG$ and $CG \perp AE$, so $CG \perp EG$

$$KB : CK = HG : EG \Leftrightarrow$$

$$6\frac{5}{26} : 5\frac{5}{13} = 2\frac{1909}{2380} : EG \Leftrightarrow$$

$$EG = 2\frac{5}{17}$$

From

$$BG = 6\frac{11}{13} \text{ and } AB = \sqrt{445\frac{1}{4}}$$

follows for triangle ABG with a right angle G that $AG = \sqrt{405\frac{76}{289}}$. In the same way, from $EG = 2\frac{5}{17}$ and $AG = \sqrt{405\frac{76}{289}}$ follows for triangle AEG with a right angle E, that

$$AE = \sqrt{AG^2 - EG^2} = \sqrt{405\frac{76}{289} - (2\frac{5}{17})^2} = 20$$

According to **Q98**, the volume of the pyramid is thus equal to the area of BCD times $\frac{1}{3}$ part of the height so

$$BCD = \frac{1}{3} \cdot \frac{1}{2} \cdot 13 \cdot 6\frac{15}{26} \cdot 20 = 285$$

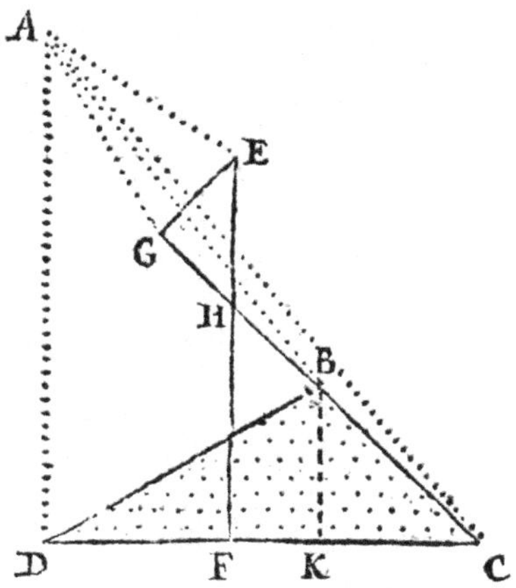

Figure 4.137: Question 99: solution

General comment

As **Q76** and **Q98** are the three dimensional generalisations of **Q1** therefore **Q99** is the spacial generalisation of **Q2**. The allusion to a 3-dimensional space in figures 4.136 and 4.137 is rather misleading. BCD is the base of a tetrahedron and on that base are point G, the perpendicular projection of the top A on BC, and point E, the projection of A on the base.

In figure 4.137 the dashed lines provide a birds view of the base with the lines inside a plane. The dashed lines schematically represent the spaciously connecting lines with A.

Theory used

Answer

Q1, 2, 98

Compare with

Q76 and **98**

Question 100

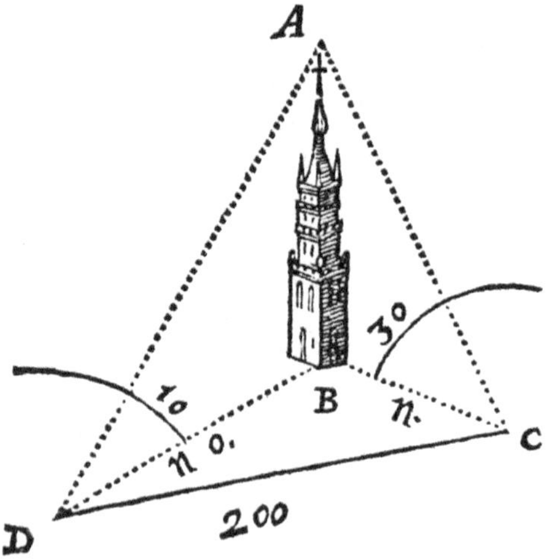

Figure 4.138: Subgroup 13, determining the height of a tower and its distance to two points

Comment on Answer

At the beginning of the seventeenth century, the common conception of trigonometric functions (tan, sin or cosine) was that of the ratio of the length of a right-angled triangle with the other side of that triangle being 100.000 in the case of tan, or of the side of a right angle with a hypotenuse of 100.000 in the case of sin or cosine. So the assumption $AB = 100.000$ leads to

$$AB = 100.000 \Rightarrow \text{ in figure 4.138}$$

$$BC = \text{ tan } 60° = 173205 \text{ and } BD = \text{ tan } 80° = 567128$$

In the same way, the assumption $BC = 100.000$ leads to

$$BC = 100.000 \Rightarrow \text{ in figure 4.139}$$

$$EC = EB = \text{ sin } 45° = 70711$$

From this follows

$$AB = 100.000 \Rightarrow EC = EB = \frac{173205}{100.000} \cdot 70711 = 122475$$

and so

$$AB = 100.000 \Rightarrow ED = BD - EB = 567128 - 122475 = 44653$$

In the right triangle DEC holds following Elem.I-47

$$DC = \sqrt{ED^2 + EC^2} = \sqrt{444653^2 + 122475^2} = \sqrt{212716416 \cdot 10^{11}}$$

$$DC \approx 461213$$

So from $AB = 100.000$ follows $DC = 461213$, but since $DC = 200$ a multiplication factor of $\frac{200}{461213}$ follows, so

$$AB = \frac{200}{461213} \cdot 100.000 \approx 43\frac{3}{8}$$

And so for the real value of BD and BC

$$BD = \frac{200}{461213} \cdot 567128 \approx 246 \text{ and } BC = \frac{200}{461213} \cdot 173205 \approx 75$$

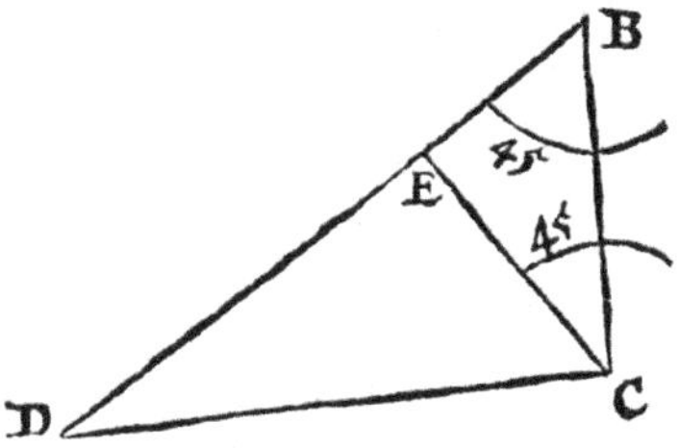

Figure 4.139: Question 100: solution

Otherwise

While not being explicitly mentioned by Cardinael, it is here assumed that

$$AD = 100.000$$

so

$$AB = \sin 10° = 17365$$

For triangle ABC holds

$$\sin 30° = 50.000$$

implying

$$AB = 50.000 \text{ if } AC = 100.000$$

so the multiplication factor is $\frac{17365}{50.000}$, from which follows

$$BC = \sin 60^\circ \cdot \frac{17365}{50.000} = \frac{17365}{50.000} \cdot 86603 = 30.077$$

For triangle ABD holds

$$BD = \sin 80^\circ = 98481$$

For the right angled triangle BEC of figure 4.139 holds

$$\sin \angle EBC = \sin 45^\circ = 70711$$

which means that

$$EC = 70711 \text{ if } BC = 100.000$$

So, the multiplication factor equals $\frac{30077}{100.000}$, which leads to

$$EC = \frac{30077}{100.000} \cdot 70711 = 21267 = EB$$

From this results

$$ED = BD - EB = 98481 - 21267 = 77214$$

so, following Elem.I-47 one arrives at

$$CD = 80080$$

which means that the assumption $AD = 100.000$ has introduced a multiplication factor $\frac{80080}{200}$, so the real values of AB, BD and BC are

$$AB = \frac{200}{80080} \cdot 17365 \approx 43\frac{3}{8}$$

$$BD = \frac{200}{80080} \cdot 98481 \approx 246$$

$$BC = \frac{200}{80080} \cdot 30077 \approx 75$$

General comment

The trigonometric calculation in Cardinael's era up to the nineteenth century were far less efficient than the current ones. Basically, all trigonometric ratios were given (as in Cardinael's manuscript [hs-Nann 1647]-1 f. 186vv), based on circles (mostly) with a diameter of 200.000.

The sines, cosines, tangents, etc., are the sides of a right angle in a right triangle, with hypotenuse 100.000 in the case of sin or cosine and with a side of the right angle 100.000 in the case of tan. Therefore calculations with proportions were always needed to get the real values of these trigonometric functions.

Theory used

Answer and Otherwise

Similarity, goniometric tables

Compare with

[Leeuwen 1663] 52, Question 69

Arithmetical Question

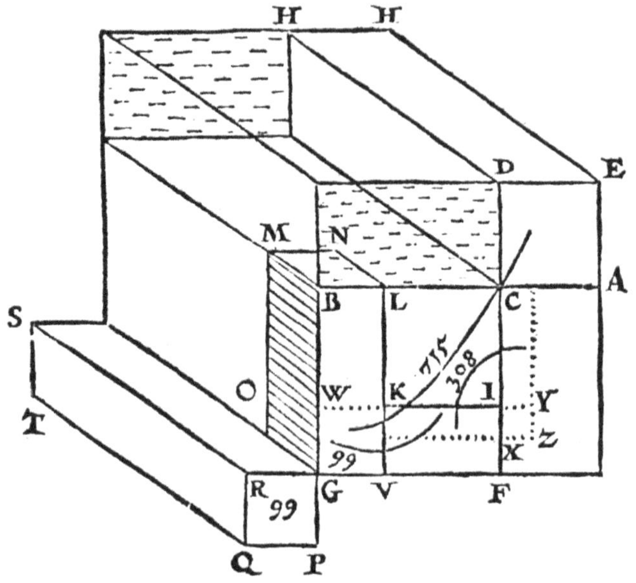

Figure 4.140: Solving an arithmetical question in a geometrical way

Comment on Answer

Using a 3-dimensional version of geometrical algebra for solving this problem, at some point Cardinael arrives at a configuration of solids all having the same unknown length (= depth)

$$AB = EH = AC + CB$$

with their right angled faces in the plane ABG.

By constructing on these areas of these right-angled faces the volumes of the corresponding solids, it becomes possible to replace the cutting, pasting and shifting operations on these solids by the corresponding operations on rectangles. The starting point of the calculation is the geometrical translation of the two equations for cuboids with length $EH = AB$ and with right-angled faces. The data can be represented as a System I with two equations and two unknowns, in which AC is the unknown number of type A, being the price of a piece of A, CB the unknown number of type B, being the price of a piece of B, $AC+CB$ the price of a piece of C and $(CB-AC)$ the price of a piece of D

the costs of $C: (AC^2 + CB^2) \cdot (AC + CB) = 715$
the costs of $D: (CB^2 - AC^2) \cdot (CB - AC) = 99.$

Basing on figure 4.140, Cardinael shows that these equations correspond to the geometrical interpretation of

$$\textbf{System I} \left\{ \begin{array}{l} (AC^2 + CB^2) \cdot (AC + CB) = 715 \\ (CB^2 - AC^2) \cdot (CB - AC) = 99 \end{array} \right.$$

The left-hand side of the first equation corresponds to the volume of the sum of the cuboids with depth $AC + CB$ and with respective faces the squares $ACDE$ and $BCFG$. Given the data, this volume equals 715. The left-hand side of the second equation corresponds to the volume of the solid that has as face the gnomon $IFGBLKI$ and depth BM with

$$BM = BL = BC - AC$$

Given is that the volume of this solid equals 99.

The solid with the gnomon as a face and depth $BM = BL$, can be replaced by the cuboid of the same volume, with square $GPQR$ as a face and with depth $RS = AB = EH = AC + CB$.

$$GPQR = LBMN = KVGW$$

This is possible because $RS = BG + FV$ and

$$\begin{aligned} RS \cdot GPQR7 \quad &= (BG + VF) \cdot GPQR \\ &= (BG + VF) \cdot LBMN \\ &= IFGBLKI \cdot BM \end{aligned}$$

System I is equivalent to the System II as follows

$$\textbf{System II} \left\{ \begin{array}{l} AC^2 \cdot EH + BC^2 \cdot EH = 715 \\ GR^2 \cdot RS = 99 \end{array} \right.$$

Because all solids (cuboids) have the same depth $AB = AC + CB$, they can be further omitted and to the right-angled faces of the beams an area shall be attached equal to the volume of the corresponding cuboid.

Notation These enlarged rectangles will further be denoted by an overline.

So, the calculation with cuboids can be replaced by a calculation with rectangles with their areas multiplied by a factor AB. System II thus is replaced by

$$\textbf{System III}\begin{cases} \overline{ACDE} + \overline{BCFG} = 715 \\ \overline{GPQR} = 99 \end{cases}$$

and from this system AC and CB have to be solved. Replacing the second equation by the difference of the first and the second equation, System III transforms into

$$\textbf{System IV}\begin{cases} \overline{ACDE} + \overline{BCFG} = 715 \\ \overline{ACDE} + \overline{BCFG} - \overline{GPQR} = 616 \end{cases}$$

Replacing $\overline{GPQR}$ by $\overline{KVGW}$ produces

$$\textbf{system V}\begin{cases} \overline{ACDE} + \overline{BCFG} = 715 \\ \overline{ACDE} + \overline{BCFG} - \overline{KVGW} = 616 \end{cases}$$

Splitting $\overline{BCFG} - \overline{KVGW}$ into three parts gives

$$\textbf{System VI}\begin{cases} \overline{ACDE} + \overline{BCFG} = 715 \\ \overline{ACDE} + \overline{CIKL} + \overline{IKVF} + \overline{LKWB} = 616 \end{cases}$$

As $\overline{CIKL} = \overline{ACDE}$ and $\overline{IKVF} = \overline{LKWB}$ follows

$$\textbf{System VII}\begin{cases} \overline{ACDE} + \overline{BCFG} = 715 \\ \overline{CIKL} + \overline{IKVF} = 308 \end{cases}$$

Figure 4.140 shows that $\overline{CIKL} + \overline{IKVF} + \frac{1}{4} \cdot \overline{KVGW}$ can be written as $\overline{CX}^2$ with X as the center point of $\overline{IF}$, so

$$\overline{IX} = \frac{1}{2}\overline{KV} = \frac{1}{2}\sqrt{99} = \sqrt{24\frac{3}{4}}$$

and so

$$\frac{1}{4} \cdot \overline{KVGW} = \overline{IYZK} = \overline{IX}^2 = (\frac{1}{2}\overline{KV})^2 = \frac{1}{4} \cdot \overline{KV}^2 = \frac{1}{4} \cdot 99 = 24\frac{3}{4}$$

Resulting in

$$\overline{CK}^2 = 308 + 24\frac{3}{4} = 33234 \text{ and for that } \overline{CX} = \sqrt{332\frac{3}{4}}$$

System VII thus transforms into

$$\textbf{System VIII}\left\{\begin{array}{l}\overline{ACDE}+\overline{BCFG}=715\\ \overline{CX}^2=308+24\tfrac{3}{4}=332\tfrac{3}{4}\end{array}\right.$$

From $\overline{CX}^2=\sqrt{332\frac{3}{4}}$ follows

$$\overline{AC}=\overline{CI}=\overline{CX}-\overline{IX}=\sqrt{332\frac{3}{4}}-\sqrt{24\frac{3}{4}}=\sqrt{176}$$

and also

$$\overline{CB}=\overline{AC}+\overline{BL}=\sqrt{176}+\sqrt{99}=\sqrt{539}$$

from which follows

$$\overline{AB}=\overline{AC}+\overline{CB}=\sqrt{176}+\sqrt{539}=\sqrt{1331}$$

From

$$\overline{AB}=\sqrt{AB}\cdot AB$$

we have

$$\overline{AB}^2=AB^3 \text{ so that } AB^3=1331 \text{ and so } AB=11$$

Then also AC and CB can be calculated

$$AC=\frac{1}{\sqrt{AB}}\cdot\overline{AC}=\frac{\sqrt{176}}{\sqrt{11}}=4$$

and

$$CB=\frac{1}{\sqrt{AB}}\cdot\overline{CB}=\frac{\sqrt{539}}{\sqrt{11}}=7$$

The numbers of acquired pieces of cloth of C and D therefore are

$$C=49+16=65 \text{ and } D=49-16=33$$

An algebraic solution as we would currently do can be as follows
Assuming the merchant purchases a pieces of kind A, b pieces of kind B, c pieces of kind C and d pieces of kind D this leads to the system

$$(1)\left\{\begin{array}{l}(a^2+b^2)\cdot(a+b)=715\\ (b^2-a^2)\cdot(b-a)=99\end{array}\right\}\Leftrightarrow$$

The first equation gives the costs for C, the second equation the costs for D. In the next solution the transitions show expected similarities with Cardinael's geometrical approach of solving the problem

$$(2)\left\{\begin{array}{l}(a^2+b^2)\cdot(a+b)=715\\(a+b)\cdot(b-a)^2=99\end{array}\right\}\Leftrightarrow$$

$$(3)\left\{\begin{array}{l}((a+b)^2)-2ab)\cdot(a+b)=715\\(a^2+b^2)\cdot(a+b)-(a+b)\cdot(b-a)^2=616\end{array}\right\}\Leftrightarrow$$

$$(4)\left\{\begin{array}{l}(a+b)^3=2ab\cdot(a+b)+715\\(a+b)\cdot 2ab=616\end{array}\right\}\Leftrightarrow$$

$$(5)\left\{\begin{array}{l}(a+b)^3=616+715=1331\\(a+b)\cdot ab=308\end{array}\right\}\Leftrightarrow$$

$$(6)\left\{\begin{array}{l}a+b=11\\ab=28\end{array}\right\}\Leftrightarrow$$

$$(7)\left\{\begin{array}{c}a=4\\b=7\end{array}\right\}\text{and so } C=65 \text{ and } D=33$$

Theory used

Answer

geometrical algebra: a spacial variation and ordinary geometrical algebra;
multiplication factor: areas representing a volume

Compare with

[Stampioen 1634]

Jan Nanningsz. 1647 manuscript, II, 288

[De Graaf 1672], 142 quest. CIX;

[De Graaf 1706], 252

Pell 1611-1685 , British Library MS Add. 4398, f. 180^r

[Pell 2005], 132 – 133

[Vorsterman 1868] 65, 66, quest. 54

Appendix A

Cross-References

A.1 Propositions from Euclid's *Elements* used by Cardinael

Alternative solutions, in the translation described as Otherwise 1 and 2, are denoted by one or two accents, as for example in **Q27'** or **Q4"**.

Euclid Elements Book I

Prop.	Description	[Heath]	Used
I-22	constructing a triangle given three line segments	292	**Q96**
I-29	equality of interchanged interior angles	311	**Q82**
I-32	sum of the angles of a triangle	316	**Q69**
I-37	triangles with same base and altitude	332	**Q77**
			Q78
			Q87
			Q90
			Q91
			Q93
I-38	triangles on equal bases and with equal altitude	333	**Q78**
			Q93
I-39	equal triangles on the same base	336	**Q11**
			Q80
I-42	constructing a parallelogram with given angle	339	**Q93**
	and area equal to a given triangle (Golden Section)		§3.2.2
I-43	equal complements of parallelograms	340	**Q27'**
	around the diameter		**Q44**
			Q66
			Q92
			Q96
I-45	constructing a solution for the equation $ax = bc$		**Q95**
			Q96
I-47	The Pythagorean theorem	349	passim

Euclid Elements Book II

Prop.	Description	[Heath]	Used
II-1	$a \cdot (b + c) = ab + ac$	372	**Q68**
II-3	$(a + b) \cdot ab + a^2$	372	**Q56**
II-4	$(a + b)^2 = a^2 + b^2 + 2ab$	372	**Q56**
			Q57
II-5	elliptic application	382	**Q5'**
			Q32
			Q44
II-6	hyperbolic application	385	**Q4"**
			Q14'
			Q35
			Q40
			Q41
			Q42
			Q43
			Q48
II-5,6	$(a + b)\dot{(}a - b) = a^2 - b^2$	383	**Q1**
			Q2
			Q56
			Q56'
II-9	$(a + b)^2 + (a - b)^2 = 2(a^2 + b^2)$	373,392	**Q57**
II-11	division in mean and extreme ratio	403	**Q25**
			Q35
			Q69
II-12	the Rule of cosine for the obtuse triangle	403-4	**Q2'**
II-13	the Rule of cosine for the acute triangle	406	**Q1'**
II-14	constructing a square equal to a given	409	§3.1.2c
	triangle, i.e. the construction of the		**Q4**
	geometrical mean without using ratios		**Q7**

Euclid Elements Book III

Prop.	Description	[Heath]	Used
III-21	inscribed angles on the same arc are equal	49	**Q63**
			Q65
			Q68
III-25	constructing a circle given a segment of a circle	54	**Q97**
III-31	a right inscribed angle resting on a semicircle and	61	**Q4**
	vice versa, around an inscribed angle		**Q4”**
	on a semi circle rest a right angle		**Q5**
			Q7
			Q9
			Q46
			Q63
			Q64
III-37	if a line through P touches a circle in Q		§2.3.3
	and another line through P intersects the circle		
	in A and B then $PQ^2 = PA \cdot PB$		

Euclid Elements Book IV

Prop.	Description	[Heath]	Used
IV-5	every triangle has a circumscribed circle		**Q4**
			Q4”
			Q5
			Q27
			Q46
			Q65
IV-10	partitioning in extreme and mean reason	96	**Q35**
	constructing a pentagon and a decagon		**Q69**

Euclid Elements Book V

Prop.	Description	[Heath]	Used
V-9,10	definitions on double and triplicate ratio	114	**Q38**
	see also [Dijksterhuis 1930], 83 – 4		**Q39**
V-17	rule for ratio $a : b = c : d \Leftrightarrow$	169	**Q8**
	$(b-a) : (d-c) = a : c = b : d$ and		**Q78'**
V-18	for ratio $a : b = c : d \Leftrightarrow (a+b) : b = (c+d) : d$		**Q79'**
			Q82
			Q83
			Q84
			Q95

Euclid Elements Book VI

The general propositions on the similarity of triangles [Heath 1956], 187 – 276 are used in

Q3, 4', 4"', 5', 8, 9, 10, 12, 13, 17, 18, 21, 26, 30, 31, 38, 38', 46, 53, 54, 58, 59, 63, 65, 66, 66', 66", 68, 73, 74, 74',76, 77', 78, 79, 80, 86, 88, 92, 93.

and on the similarity of polygons [Heath 1956], 231 – 5 in

Q82, 83, 84

Prop.	Description	[Heath]	Used
VI-1	ratios of surfaces of triangles having	190	**Q77**
	the same altitude as their bases		**Q78**
			Q78'
			Q89
			Q90
			Q91
			Q95
VI-2	a line parallel to the base of a triangle	194	**Q11**
			Q21'
			Q54
			Q82
			Q90
			Q97
VI-3	bisecting an angle of a triangle	195	**Q13**
			Q84
VI-8	perpendicular from a right angle	209	**Q4'**
			Q5'
			Q6
			Q25'
			Q38
			Q39
	Conversion of Elem.VI-8		**Q46**
VI-9	cutting a given part off a line segment	211	**Q77**
VI-10	partitioning a line to a given ratio	213	**Q95**
VI-11	constructing the third proportional	214	**Q5**
	$a : b = b : x$, together with		**Q14**
VI-12	constructing the fourth proportional	215	**Q27'**
	or the rule of three: $a : b = c : x$, and		**Q44'**
VI-13	construction the mean proportional	216	**Q45**
	$a : x = x : b$		**Q55**
			Q80
			Q87
			Q91
			Q97

Prop.	Description	[Heath]	Used
VI-15	equal triangles with one angle equal	219	**Q78'**
			Q79'
			Q82
			Q91
VI-16	principal rule of proportions in the form:	221	**Q3**
	$a : b = c : x \Leftrightarrow ax = bc \Leftrightarrow x = \frac{bc}{a}$ in		**Q6**
	Q8, 16, 17, 18', 21, 24, 25, 26, 30, 34, 39		
	47, 49, 53, 54, 58, 59, 60, 62, 65, 66', 66"		
	73, 74, 74', 76, 77', 78, 82, 83, 84		
	85, 86, 99, 100		
	or in the form $a : b = c : d \Leftrightarrow bc = ad$		**Q27'**
			Q38'
			Q66"
			Q68
VI-17	calculating the mean proportion	228	**Q4'**
	$a : b = b : c \Leftrightarrow b^2 = ac$		**Q4"'**
	Q5', 9, 27', 38, 38', 39, 45, 55, 63, 80, 86'		
VI-18	similar polygons	229-32	**Q82**
			Q83
			Q84
VI-19	ratios of surfaces of similar triangles	232	**Q85**
VI-20	ratios of similar figures equal	235	**Q47**
	the squares of corresponding line segments		**Q53**
			Q59
	figure = skin of the circle		**Q60**
			Q82
			Q83
			Q84
			Q97
VI-30	partitioning in extreme and mean ratio	267	**Q25**
			Q35
			Q64
			Q69
VI-31	generalization of the Pythagorean theorem	268	**Q18**
			Q53
			Q88
			Q92
			Q93

Euclid Elements Book VII

Elem.XII, formula for the volume of a tetrahedron: $I = \frac{1}{3}Gh$ ([Heath 1956], p.394):
Q98

Rules not directly derived from Euclids' *Elements*

Description	[Heath]	Used
$r = \frac{O}{s}$ with r the radius of the inscribed circle of a triangle O the surface and s half the circumference used by Cardinael without explanation		**Q15** **Q65**
$R = \frac{bc}{2h}$ for the radius of the circumscribed circle of a triangle with h the altitude, consequence of Elem.VI-16	224	**Q65**
Proposition on the circumference of a circle (Archimedes) circumference $= 2\pi r$ with $\pi = \frac{22}{7}$		**Q54**
Proposition on the area of a circle (Archimedes) area $= \pi \cdot r^2$		**Q55** **Q60**
Circle of Apollonius , geometric locus of the points A so that $AB = k \cdot AC$ with B en C fixed given points ([Molenbroek 1943],199, [Pijls 2004^b], 86)		**Q32**
Ptolemy's theorem on a quadrilateral inscribed in a circle, simple version with a diagonal coinciding with the diameter		**Q66"**
Ptolemy's theorem on the quadrilateral inscribed in a circle		**Q68**
Proposition: $BD^2 + ED^2 =$ constant if D moves over a circle with radius DC and C the centre of BE proven in Q57		**Q70**
The volume of a tetrahedron, Elem.XII-7, $I = \frac{1}{3}Gh$	394	**Q98**

A.2 Correspondence between Cardinael's and Rudd's problems

Thomas Rudd used 79 problems from Cardinael for his English edition. The first table shows the correspondence from Cardinael to Rudd, the second from Rudd to Cardinael.

Cardinael	0	10	20	30	40	50	60	70	80	90
1	1	11	16		90	97		19	88	
2	3	48			91	39		20	93	75
3	64	13	17		92	42	60	21	94	77
4	2	38	18	40	86	56	32	22	95	78
5	4	7	50	41	85	58	61	24	73	80
6	65	8	28	81	31	59	62		74	
7	5	49	35	82	84		70			
8	33	9		29	87	44	63	71	79	
9	34	6		30	96	46		72		
10	10	14	43	89		57	36	37	76	

Rudd	0	10	20	30	40	50	60	70	80	90
1	1	11	73	46	35	25	65	78	36	42
2	4		74	64	53		66	79	37	43
3	2	13		8	30		68	85		82
4	5	20	75	9	58		3	86	47	83
5	7			27			6	92	45	84
6	19	21		70	59	54		90	44	49
7	15	23		80		60		93	48	51
8	16	24	26	14	12	55		94	81	
9	18	71	38	52	17	56		88	40	
10	10	72	39	34	25	63	67	95	41	

Appendix B

Boundary extension

There are basically two types of boundary area problems of which we provide a short survey.

B.1 Method 1. Pasting boundary areas in extension of each other

Q48:
Here is considered the problem of adding a boundary area to a rectangle and the solution (**method 1**) consists of pasting one part of the area on the extension of the other. On the resulting configuration 'application of area' is applied to determine the unknown width.

B.2 Method 2. Geometrical algebra, circumferential square or quadrilateral and similarity

Q81:
Here is considered truncating a boundary area from a rectangle. The method (**method 2**) consists of elementary geometrical algebra transforming the rectangle into a square with the same circumference (the circumscribed square) around the bisector from the inner vertex of the boundary area.

Q82-83:
Also in this problem a boundary area is truncated. Both exercises consider a generalization of method 2 for right-angled trapezia. It is also a preparation for the general case of **Q84**.

Q84:
In this general case a boundary area is truncated from (any) convex quadrilateral. The generalization of method 2 also consists of elementary geometrical algebra with a special role for the bisector from the middle vertex of the boundary area which also involves a generalization of the circumferential square into a circumferential quadrilateral, but now in combination with several similarity requirements.

Q95:
This is the same case as in **Q84**, but now a construction of the boundary area is required.

Appendix C

Partitioning a triangle

Euclid, already $\pm$ 300 BC, considered in detail the partitioning of figures. This is shown by R.C. Archibald's reconstruction of Euclid's lost treatise titled *On Division of Figures* [Archibald 1915]. The reconstruction is based on comments by Proclus (410-485), and on Woepcke's translation of an Arabic manuscript in the Bibliothèque National in Paris (952.2, supp. Arab.) [Woepcke 1851]. The manuscript explicitly mentions Euclid as the author and it corresponds to Proclus' description. Unfortunately the proofs of only four of the in total thirty-six propositions are discussed because, as Woepcke writes, the Arabic translator judged the other proofs too elementary. Archibald mentions that for the reconstruction , he also made use the *Practica Geometriae* of Leonardo Pisano (also called Fibonacci, c.1170-1250), written in 1220. Prince Baldessaro Boncompagni published an edition of this work in 1862 based on a single manuscript, conserved at the Vatican library. More recently, a critical edition has been published by Barnabas Hughes [Hughes 2008]. Especially part IV with seventy five propositions is of great importance in connection with Euclid's book on the division of figures.

Also Luca Pacioli (about 1445- about 1514) devoted a section of his *Summa de Arithmetica et Geometria* (Venice 1494) to the division of figures, acknowledging to base himself on Leonardo's work.

According to Archibald, twenty two propositions in Woepcke are identical to propositions in Leonardo, eight other propositions in Woepcke correspond more to Fibonacci's style and the remaining six in Woepcke are by Leonardo considered as well known. The proofs missing in Woepcke's edition of the Arabic translation appear in Leonardo's work and are, according to Archibald, all in the Euclidean style and use in a typical way the theory from the *Elements*. Hughes provides a cross table collating identical propositions in four extant sources ([Hughes 2008], 185) including a translation from Plato of Tivoli. He concludes that much of Fibonacci's text is based on the Arabic translation of Euclid's *Division of figures*, with 28 propositions that were adapted from Euclid's text and 32 were added.

Bibliography

[Aa 1845] Aa, A. J. van der (1845) *Geschiedkundige beschrijving van de stad Breda en hare omstreken.* Gorinchem: Noorduyn en Zoon.

[Archibald 1915] Archibald, Raymond Clare, (1915) *Euclid's book on divisions of figures: ... with a restoration based on Woepcke's text and on the Practica geometriae of Leonardo Pisano.* Cambridge [England]: University Press.

[Biagioli 1989] Biagioli, Mario (1989) The Social Status of Italian Mathematicians, 1450-1600, *History of Science*, 27, 1 – 75.

[Bierens de Haan 1878-87] Bierens de Haan, David (1878-1887) *Bouwstoffen voor de geschiedenis der wis- en natuurkundige wetenschappen in Nederland* (2 vols.), Amsterdam.

[Bierens de Haan 1960] Bierens de Haan, David (1960) *Bibliographie Néerlandaise historique-scientifique: des ouvrages importants dont les auteurs sont nés aux 16e, 17e et 18e siècle, sur les sciences mathématiques et physiques, avec leurs applications.* Nieuwkoop: De Graaf.

[Boncompagni 1862] Boncompagni, Baldassarre (1862) *Scritti di Leonardo Pisano.* Vol. II. La Practica Geometriae. Rome: Tipografia delle scienze matematiche e fisiche.

[Bos 2001] Bos, Henk (2001) *Redefining geometrical exactness: Descartes' transformation of the early modern concept of construction.* New York: Springer.

[Brasser 1663] Brasser, Jacob R., Nicolaus Petri, and Gerrit Evertsz Backer (1663) *Regula Cos, of Algebra, zijnde de alder-konstrijcksten regel om het onbekende bekent te maken.* Amsterdam: Gerrit van Goedesbergh.

[Busard 1968] Busard, Hubertus L. L.(1868) L'algèbre au Moyen Âge: Le 'Liber mensurationum' d'Abū Bekr, *Journal des Savants*, (AvrilJuin 1968), 65 – 125.

[Ceulen 1615] Ceulen, Ludolf van (1615) *De arithmetische en geometrische fondamenten.* Tot Leyden: Ioost van Colster, ende Iacob Marcus.

[Curtius 1604] Curtius, Sebastian (1604) *Ein newes wolgegründtes Rechenbuch: nach rechter Art der Practic, mit manchem schonen geschwinden Vortheil, neben gründlicher Aussfuhrung der gantzen und gebrochenen: sampt den Proportionen Speciebus ... Dessgleichen zum Beschluss eine Aufflösung mancherley künstlicher cossischer unnd polygonalischer Exempla, etlicher furnemer und kunstreicher Rechenmeistern, so zu End irer Rechenbucher one Facit gesetzt seyn: dergleichen vorhin in Druck nie aussgangen, nun aber der lieben Jugend zu gedeylicher Wolfart mit Fleiss zusammen colligirt und jetzt erstmals in Druck gegeben.* Nurnberg: In Verlegung Conrad Baurn.

[Curtius 1610] Curtius, Sebastian (1610) Compendium Arithmeticae, *Das ist: Ein Neues, Kurtzes und Wolgegründtes SchulRechenbüchlein, von allerley Hauß- und Kauffmanns rechnungen ...: Meinen lieben Discipulis.* Nurnberg: Fuhrmann.

[Curtius 1615] Sebastian Curtius, Jan. Pieterszoon Dou (1616) *Tractat vom machen und Gebrauch eines Neugeordneten mathematischen Instruments: inn welchem underschiedliche Kunstliche stuck die Geometriae betreffende verfasset und begriffen seind.* Gedruckt zu Ambsterdam: Bey Wilhelm Janss.

[Curtius 1615] Curtius, Sebastian, Sebastian Curtius, and Jan Pieterszoon Dou (1616) *Practica des Landmessens: darinn gelehrt wird, wie man alle recht-, und krumseitige Land, Wälder, Baumgärten ... sowoh mit dem Quadranten, als ohne demselben messen soll.* Amsterdam: Jansson.

[Dillen 1929] Dillen, J.G. van (1929-1974) *Bronnen tot de geschiedenis van het bedrijfsleven en het gildewezen van Amsterdam.* 's-Gravenhage: Nijhoff.

[Dijksterhuis 1929-30] Dijksterhuis, Edward Jan (1929-30) *De elementen van Euclides.* (2 vols.) Groningen: P. Noordhoff.

[Ernst 2002] Ernst, Bruno (2002) *De interessantste bewijzen voor de stelling van Pythagoras.* Utrecht: Epsilon Uitgaven.

[Friberg 2002] Friberg, Jöran (2009) A Geometric Algorithm with Solutions to Quadratic Equations in a Sumerian Juridical Document from Ur III Umma, Cuneiform Digital Library Journal, 3, 1 – 27.

[Friberg 2007] Friberg, Jöran (2007) Amazing traces of a Babylonian origin in Greek mathematics. Singapore: World Scientific Press.

[Leeuwen 1663] Gietermaker, Claes Hendricksz. (1663) *Den Amsterdamschen belachelijcken geometrischen bril-maker Cornelis van Leeuwen.: Voorgestelt, in een vertooningh der opgeraepte vuyle leugenen, ende opgeblasen quacksalvers-redenen ontrent sijnen bril, voor de Amsterdamsche belachelijcke geometristen.* t'Amstelredam: gedruckt voor den autheur.

[De Graaf 1672] Graaf, Abraham de (1672) *De beginselen van de algebra of stelkonst, volgens de manier van Renatus Des Cartes verklaart met uytgelezene voorbeelden; zoo wel in de meetkonst, als in de rekenkonst.* [Amsterdam]: [Rieuwertsz, Jan].

[De Graaf 1706] Graaf, Abraham de (1706) *Analysis of stelkunstige ontknoping in de meetkunstige werkstukken: vindende van hen de grootste en kleenste ; de raaklynen op de kromme; de plaatzen; de ontbinding der bepaalde werkstukken door de plaatsen; en de quadratura van eenige kromlinische grootheden.* Amsterdam: Joannes Loots.

[De Jong 1996] Jong, R.M.A. De (1996) Genealogie van de familie De Gavere. genealogysk jierboek. Ljouwert: Fryske Akademie.

[Freudenthal 1977] Freudenthal, Hans (1977) What is algebra and what has been its history?,*Archive for History of Exact Sciences*,16, 189 – 200.

[Gravelaar 1902] Gravelaar, N. L. W. A. (190) Stevin's *Problemata Geometrica, Nieuw Archief voor Wiskunde* (2), 5, 106 – 191.

[Hauff 1803] Hauff, Johann K. (1803) *Lehrbegriff der reinen Mathematik*: Bd. 1. Elementargeometrie, Frankfurt am Main: Guilhauman.

[Heath 1956] Heath, Thomas Little (1956) *The thirteen books of Euclid's Elements.* New York: Dover Publications.

[Heeffer 2006] Heeffer, Albrecht (2006) *From Precepts to Equations. The Conceptual Development of Symbolic Algebra in the Sixteenth Century.* PhD dissertation, Ghent, Belgium: Ghent University.

[Heeffer 2009] Heeffer, Albrecht (2009) On the nature and origin of algebraic symbolism. in B. Van Kerkhove (ed.) *New Perspectives on Mathematical Practices. Essays in Philosophy and History of Mathematics*, Singapore: World Scientific Publishing, 1-27.

[Heeffer 2010] Heeffer, Albrecht (2010) The Symbolic Model for Algebra: Functions and Mechanisms, in Magnani, Lorenzo; Carnielli, Walter; Pizzi, Claudio (eds.) *Model-Based Reasoning in Science and Technology, Abduction, Logic, and Computational Discovery, Studies in Computational Intelligence*, vol. 314, Heidelberg: Springer, 519-532.

[Hoogendoorn 2006-7] Hoogendoorn, Klaas (2006-2007) De werken van de meetkundige rekenmeester Sybrandt Hansz. Cardinael (1578-1647) Proeve van een bibliografie Klaas Hoogendoorn, De boekenwereld. 23, 276_290.

[Høyrup 2002] Høyrup, Jens (2002) *Lengths, Widths, Surfaces: A Portrait of Old Babylonian Algebra and Its Kin.* (Studies and Sources in the History of Mathematics and Physical Sciences). New York: Springer.

[Høyrup 2016] Høyrup, Jens (2016) What is 'geometric algebra', and what has it been in historiography? Contribution to the session Histoire de l'historiographie de l'algèbre, Séminaire Histoire et Philosophie des Mathématiques, CNRS, SPHERE, & projet ERC SAW, Université Paris Diderot, 11 janvier 2016. Preprint, 15.1.2016.

[Hughes 2008] Hughes, Barnabas (2008) *Fibonacci's De practica geometrie.* New York: Springer.

[Huygens Oeuvres] Huygens, Christiaan (1888-1950) *Oeuvres complètes de Christiaan Huygens publiées.* Koninklijke Hollandsche maatschappij der wetenschappen (eds.) (23 vols.) La Haye: Martinus Nijhoff.

[Jansen 1960] Jansen, L. (1960) De Derde Vergroting van Amsterdam. In *52nd Yearbook of the Amstelodamum Historical Society.* Amsterdam: J.H. de Bussy. 1959, pp.42-89.

[Karpinski and Kokomoor 1928] Karpinski, L. and Kokomoor, F. (1928). The Teaching of Elementary Geometry in the Seventeenth Century. *Isis: A Journal of the History of Science*, 10, 21 – 32.

[Kinckhuysen 1661] Kinckhuysen, Gerard (1661) *Algebra, ofte Stel-konst, beschreven tot dienst van de leerlinghen.* Haerlem: Passchier van Wesbusch.

[Kline 1972] Kline, Morris (1972) *Mathematical thought from ancient to modern times.* (3 vols.) Oxford: Oxford University Press.

[Ogilvy 1969] Ogilvy, Charles Stanley (1969) Excursions in geometry. Oxford: Oxford University Press.

[Schaefer 2001] Schaefer, Christoph (2001) 'Krygsvernuftelingen': Militäringenieure und Fortifikation in den Vereinigten Niederlanden. PhD dissertation. Giessen.

[Van Maanen 1987] Maanen, Jan van (1987) *Facets of seventeenth-century mathematics in The Netherlands*, PhD dissertation, Utrecht: Utrecht University.

[Van Maanen 2003] Maanen, Jan van (2003) Cardinael in de geschiedenis van de wiskunde, *Nieuw archief voor wiskunde*, series 5, (4) 1, 51 – 55.

[Mahoney 1973] Mahoney, Michael S. (1973) *The mathematical career of Pierre de Fermat (1601-1665).* Princeton, N.J.: Princeton University Press.

[Molenbroek 1934] Molenbroek, Pieter (1934) *Leerboek der stereometrie.* Groningen: Noordhoff.

[Molenbroek 1943] Molenbroek, Pieter, and Wijdenes, P. (1943) *Leerboek der vlakke meetkunde.* Groningen: Noordhoff.

[Neugebauer and Sachs 1945] Neugebauer, Otto en Abraham Sachs (1945) *Mathematical Cuneiform Texts.* American Oriental Series, vol. 29. New Haven: American Oriental Society.

[Pell 2005] Pell, John, Charles Cavendish, Noel Malcolm, and Jacqueline A. Stedall (eds.) (2005) *John Pell (1611-1685) and his correspondence with Sir Charles Cavendish: the mental world of an early mathematician.* Oxford: Oxford University Press.

[Petri 1583] Petri, Nicolaus (1583) *Practicque, om te leeren rekenen, cijpheren ende boeckhouwen, met die regel coss ende geometrie seer profijtelijcken voor alle coopluyden.* Amsterdam: Cornelis Claesz.

[Pijls 2004] Pijls, Wim (2004) Gelijkvormigheid, jonge inzichten bij een oud concept, *Euclides*, 80, part I in (2), pp. 48-51, part II in (3), pp. 86-89.

[Rodet 1878] Rodet, Léon) (1878) L'algèbre d'Al-Khārizmi et les méthodes indienne et grecque, Journal asiatique, 11, 5-98.

[Schooten 1660] Schooten, Frans van (1660) *Mathematische oeffeningen, begrepen in vijf boecken: ... Waer by gevougt is een tractaet, handelende van reeckening in speelen van geluck door Christianus Hugenius.* T'Amsterdam: By Gerrit van Goedesbergh.

[Sitters 2003] Sitters, Mathijs, H. (2003) Sybrandt Hansz. Cardinael (1578-1647) Meester in de meetkunde, Niew Archief voor wiskunde, series 5, (4) 4, 309 – 316.

[Sitters 2007] Sitters, Matthijs, H. (2007) *Sybrandt Hansz. Cardinael 1578-1647: Rekenmeester en wiskundige: zijn leven en zijn werk.* Groningen: PhD dissertation.

[Sitters 2008] Sitters, Matthijs, H. (2008) *Sybrandt Hansz. Cardinael 1578-1647: Rekenmeester en wiskundige: zijn leven en zijn werk.* Hilversum: Verloren.

[Stampioen 1634] Stampioen, Johan (1634) *Resolutie ende ontbindinghe der twee vraegh-stucken, in den jare 1632.* Rotterdam: op de weduwe van Matthiis Bastiaensz.

[Stampioen 1639] Stampioen, Johan (1639) *Algebra, ofte Nieuwe stel-regel, waer door alles ghevonden wordt inde wiskonst, wat vindtbaer is.* 's Graven-Hage: ghedruckt ten huyse vanden autheur.

[Stevin 1605-8] Stevin, Simon. (1605-1608) *Wisconstige gedachtenissen, Inhoudende t'ghene daer hem in gheoeffent heeft den ... vorst ende heere,* Mavrits prince van Oraengien. Tot Leyden: Inde Druckerye van Ian Bouvvensz.

[Stewart 1746] Stewart, Matthew (1746) *Some general theorems of considerable use in the higher parts of mathematics.* Edinburgh: Printed by W. Sands, A. Murray, and J. Cochran, sold by said W. Sands ... and by J. and P. Knapton.

[Versluys 1914] Versluys, J. (1914) *Zes en negentig bewijzen voor het theorema van Pythagoras: Verzameld en gerangschikt.* Amsterdam: Versluys.

[Vorsterman 1868] Vorsterman van Oyen, George Auguste (1868) *Honderd vier en veertig vraagstukken van Nederlandsche wiskundigen der XVIIe eeuw, voorafgegaan door eene bijdrage tot de geschiedenis van de beoefening der wiskunde in Nederland, van Stevin tot Huyghens.* Schoonhoven: Van Nooten.

[Alberti 2010] Williams, Kim, Lionel March, and Stephen R. Wassell (eds.) (2010) *The mathematical works of Leon Battista Alberti.* [Basel]: Birkhauser.

[Unguru 1975] Unguru, Sabetai (1975) On the need to rewrite the history of Greek mathematics,*Archive for History of Exact Sciences*,15, 67 – 114.

[Unguru 1979] Unguru, Sabetai (1979) History of Ancient Mathematics: Some reflections on the state of the art,*Isis*,70, 555565.

[van der Waerden 1976] Waerden, B. L. van der (1976) Defense of a 'shocking' point of view,*Archive for History of Exact Sciences*,15, 1976, 199–210.

[Weil 1978] Weil, André (1978) Who betrayed Euclid?,*Archive for History of Exact Sciences*,19, 91 – 93.

[Woepcke 1851] Woepcke, Franz (1851) Notice sur les traductions arabes de deux ouvrages perdus d'Euclide, *Journal Asiatique*, IV (18): 217-47.

[Yates 1972] Yates, Frances A. 1972. *The Rosicrucian enlightenment.* London: Routledge and Kegan Paul.

Index

AH-system, **193**, 194–196, 229, 231, 243, 244, 246, 278, 318
Apollonius, 24, 208, 290, 294, 474
application of area, 166, 168–171, **171**, **173**, 177, **182**, 183, 190, 194, 196, 199, 206, 208, 215, 228, 231, 244–247, 249, 266, 278, 290, 307, 309, 311, 313, 314, 318, 327, 329, 404, 477
Arabic, 22, 25, 479, 480
Archimedes, 94, 96, 101, 337, 338, 341, 352, 354, 474

bisector, 205, 206, 240, 298, 299, 350, 351, 373, 400–402, 405, 407, 408, 411, 413, 416, 419, 420, 430, 445, 477, 478
Bombelli, 207
book-keeping, 6, 23
boundary extension, 204, 205, **205**, 206, 211, 215, 337, 404, 405, 407, 408, 411–413, 415, 416, 419, 420, 433, 445, **477**

Cardano, 207
circumferential, 210, **477**, 478
circumscribed, 204–206, 215, 333, 355, 356, 360–362, 365, 368, 370, 376, 400–402, 404, 406–408, 412, 413, 419, 420, 470, 474
construction, 166–168, 170–172, 178, 179, 181, 182, 185, 194, 196, 204–208, 211, 216, 263, 271, 281, 293, 298, 320, 344, 367, 373, 391, 393, 394, 398, 430, 434, 437, 438, 440, 444, 445, 447–450, 452, 469, 472, 478, 479
convex, 204, 206, 211, 215, 394, 401, 416, 442, 478

decagon, 210, 298, 373, 376
Descartes, 6, 214

elliptic application, 166, 169, 177, 178, **182**, 183, **189**, 190, 193, 197, 200–204, 206–209, 227, 231, 232, 245, 247, 248, 288, 290, 311, 313, 314, 316, 318, 469
equivalence, 193, 195, 197, 346, 347
exceeding, 169–171, 178–180, 199–201, 203, 208, 227, 245, 263, 265, 306, 309, 323, 329, 331

falling short, 169, 171, 177, 197, 201, 202, 208, 227, 231, 245, 288, 311, 313

geometrical configuration, 170, 177, 178, 203, 204, 207, 306, 309, 311, 324, 408, 419
geometrical algebra, 2, 16, **23**, 24–26, 166, **166**, 169, 171, 182, 211, 216, 220, 222, 225, 228, 231,

247, 257, 258, 260, 261, 266, 267, 273, 275, 278, 294, 296, 300, 302, 307, 309, 311, 314, 318, 320, 324, 327, 329, 333, 335, 337–340, 342–344, 376, 430, 461, 465, 477, 478
geometrical identity, 171, 220, 282
geometrical mean, 166–170, 207, 210, 235, 236, 304, 305, 319, 320, 397–399, 426, 428, 433, 441, 469
golden section, 177, **179–181**, 210, 213, 262, 266, 298, 299, 376, 468

hexagon, 211, 432, 433
hyperbolic application, 169, 177–180, 185, 186, 189, 191, 192, **197**, **199**, **200**, 201, 203, 206, 208, 209, 245, 247, 248, 263, 265, 298, 306, 307, 309, 311, 323, 324, 329, 331, 332, 376, 404, 469

inscribed, 97, 98, 100–104, 106, 107, 110, 111, 114, 141, 152, 154, 155, 204–206, 210, 248, 249, 343, 348, 350, 351, 353–356, 360, 361, 365, 369, 371, 372, 400–403, 405, 408, 409, 411, 413, 416, 419, 420, 429, 441, 449, 451, 452, 470, 474

Kinckhuysen, 5, 6

Law of Cosine, **173**, 221, 224, 225, 286, 469
linear equation, 170, 172, **172**, 327

mean and extreme ratio, **179**, 214, 298, 359, 373, 469
metrical, 166

Old-Babylonian, 2, 17, 18, 20–22, 25, 26

parabolic application, 165, **169**, 171, 172, **172**, 208
partitioning, 14, 15, 215, 390–392, 394, 395, 397, 400, 405, 411, 416, 423, 424, 431, 432, 435, 436, 438, 439, 442, 445, 470, 472, 473, 479
pentagon, 210, 470
perimeter, 184, 197, 199
Petri, 207, 222, 232, 266, 285, 290, 299, 338, 377, 404
point multiplication, **166**
polygon, 14, 166, 168, 210, 211, 215, 335, 438, 471, 473
Ptolemy's theorem, 210, 365–367, 370–372, 451, 452, 474
Pythagoras, 222, 225, 228, 231, 233, 241, 244, 247, 250, 254, 255, 257–261, 267, 273, 275, 290, 318, 321, 343, 367, 369

quadratic equation, 15, 17, 18, 169, 170, 186–189, **189**, 190, 192, 206–208, 324, 329
quadrilateral, 211, 213, 215, 216, 365, 369, 371, 372, 393, 397, 401, 416, 417, 419–421, 423, 424, 431, 433, 435, 439, 441, 444, 445, 449, 450, 452, 474, 477, 478

regula falsi, 165, 228, 250, 253, 281, 285, 335, 356, 372
Rule of Coss, 23, 165, 266, 285
ruler and compass, 23, 293

similarity, 165, 167, 206, 210, 225–228, 230–232, 234–237, 240, 241,

251, 253, 261, 264, 266, 267, 271, 273, 279, 281, 298, 299, 303, 304, 306, 324, 334, 335, 338, 349–352, 357–359, 361, 363, 366, 367, 369, 371, 372, 381–384, 387, 390–393, 395, 396, 401, 404, 406, 407, 411, 414–417, 419, 420, 422–427, 437, 448, 452, 455, 460, **471**, 477, 478
Stampioen, 6, 465
Stevin, 6, 23, 207, 438, 439
Stewart's theorem, 209, 213, **213**, 214, 294, 345
surveying, 6, 8, 11, 14, 22, 23, 207, 215, 260
system I, **182**, 183, 184, 187, 190, 193, 195–197, 199, 201–204, 210, 231, 243, 246, 249, 277, 278, 288, 314, 315, 317, 318, 331, 462
system II, 17, **185**, 186, 191, 193, 197, 199, 200, 462
system III, 187, **187**, 463
system IV, 188, **188**, 192, 463
system V, 463
system VI, 463
system VII, 463
system VIII, 464

tetrahedron, 211, 386, 389, 453, 454, 456, 474
Thales, 195, 196, 228, 231, 233, 244, 247, 273, 318, 321, 361, 364, 367
transversal, 209, 213, 234, 291, 295
trapezium, 211, 252, 337, 392, 395, 397–399, 405–415, 426–428, 433, 441
type I, 184, 189–191
type II, 170, 186, 189, 191–193
type III, 189
type IV, 189, 191–193

Made in the USA
Monee, IL
05 November 2026

67816264R00282